T0033773

Nonlinear System Identification – Input-Output Modeling Approach

MATHEMATICAL MODELLING:
Theory and Applications

VOLUME 7/2

This series is aimed at publishing work dealing with the definition, development and application of fundamental theory and methodology, computational and algorithmic implementations and comprehensive empirical studies in mathematical modelling. Work on new mathematics inspired by the construction of mathematical models, combining theory and experiment and furthering the understanding of the systems being modelled are particularly welcomed.

Manuscripts to be considered for publication lie within the following, non-exhaustive list of areas: mathematical modelling in engineering, industrial mathematics, control theory, operations research, decision theory, economic modelling, mathematical programming, mathematical system theory, geophysical sciences, climate modelling, environmental processes, mathematical modelling in psychology, political science, sociology and behavioural sciences, mathematical biology, mathematical ecology, image processing, computer vision, artificial intelligence, fuzzy systems, and approximate reasoning, genetic algorithms, neural networks, expert systems, pattern recognition, clustering, chaos and fractals.

Original monographs, comprehensive surveys as well as edited collections will be considered for publication.

Editor:
R. Lowen (*Antwerp, Belgium*)

Editorial Board:
G.J. Klir (*New York, USA*)
J.-L. Lions (*Paris, France*)
P.G. Mezey (*Saskatchewan, Canada*)
F. Pfeiffer (*München, Germany*)
H.-J. Zimmerman (*Aachen, Germany*)

The titles published in this series are listed at the end of this volume.

Nonlinear System Identification –
Input-Output Modeling Approach

Volume 2: Nonlinear System Structure Identification

by

Robert Haber
Laboratory for Process Control,
Department of Plant and Process Engineering,
University of Applied Sciences Cologne (Fachhochschule Köln),
Köln, Germany

and

László Keviczky
Computer and Automation Institute,
Hungarian Academy of Sciences,
Budapest, Hungary

KLUWER ACADEMIC PUBLISHERS
DORDRECHT / BOSTON / LONDON

A C.I.P. Catalogue record for this book is available from the Library of Congress.

ISBN 0-7923-5857-0
ISBN 0-7923-5858-9 (Set)

Published by Kluwer Academic Publishers,
P.O. Box 17, 3300 AA Dordrecht, The Netherlands.

Sold and distributed in North, Central and South America
by Kluwer Academic Publishers,
101 Philip Drive, Norwell, MA 02061, U.S.A.

In all other countries, sold and distributed
by Kluwer Academic Publishers,
P.O. Box 322, 3300 AH Dordrecht, The Netherlands.

Printed on acid-free paper

All Rights Reserved
© 1999 Kluwer Academic Publishers
No part of the material protected by this copyright notice may be reproduced or
utilized in any form or by any means, electronic or mechanical,
including photocopying, recording or by any information storage and
retrieval system, without written permission from the copyright owner.

CONTENTS

VOLUME 2
NONLINEAR SYSTEM STRUCTURE IDENTIFICATION

4. Nonlinearity Test Methods

4.1 INTRODUCTION

The primary task of the identification is to find the best fitting structure. The essential task is to decide from input–output measurements whether a process is linear or nonlinear. A method called the nonlinearity test is required for this purpose and introduced here. The nonlinearity test should not require a long computation time and has to be independent of the concrete nonlinear structure and the order of the linear dynamic part of the process; i.e., a mainly nonparametric method is needed.

The following methods are presented:
1. time domain and steady state tests;
2. a test based on the normalized variance of the noise-free output signal around the output signal of the fitted best linear model;
3. average value test of the output signal;
4. Gaussian distribution test of the output signal;
5. a frequency method;
6. a linear spectral density method;
7. a linear correlation method;
8. second-order (nonlinear) cross-correlation method;
9. second-order (output) auto-correlation method;
10. second-order (output) spectral density method;
11. a dispersion method;
12. a method based on the comparison of the variances of the noise-free model output signals of the best linear model identified and of an alternative nonlinear model;
13. a method based on the identification of orthogonal subsystems;
14. a method based on the identification of the first-order Uryson model;
15. a method based on parameter estimation and F-test of the simple linear and nonlinear structures;
16. a method based on residual analysis of the best fitting linear model.

The time domain methods and those methods that are based on parameter estimation are investigated by sampled data signals. The correlation and dispersion methods have been spread originally in the literature as continuous time methods. Because of digital computers applications a discrete time approach will be presented here. The frequency and spectral density methods are presented as continuous time procedures.

The systems may contain not only linear and nonlinear parts but also a constant term. To eliminate the effect of the constant term the correlation and dispersion methods are performed not on the measured input $u(k)$ and output $y(k)$ sequences but on their normalized values

$$u'(k) = \frac{u(k) - \mathrm{E}\{u(k)\}}{\sigma\{u(k)\}} = \frac{u(k) - \bar{u}}{\sigma_u} \tag{4.1.1}$$

$$y'(k) = \frac{y(k) - \mathrm{E}\{y(k)\}}{\sigma\{y(k)\}} = \frac{y(k) - \bar{y}}{\sigma_y} \tag{4.1.2}$$

In (4.1.1) and (4.1.2) $\mathrm{E}\{...\}$ and $\overline{(...)}$ mean the expected and the mean value, respectively, and $\sigma\{...\}$ is the standard deviation.

Some of the methods work only for systems that are of neither differentiating nor integrating type. These systems are called proportional with linear systems.

Definition 4.1.1   A system is called non-integrating system and non-differentiating system if the final value of its step response for a non-zero input is finite and differs from zero.   ∎

For any system and its components of different degrees corresponding to the Definition 4.1.1 the Volterra kernels have to fulfill the following assumption.

Assumption 4.1.1   If the sums of the Volterra kernels for all existing degrees $(i = 0, ..., n)$ are finite, i.e.,

$$\sum_{j_1=0}^{\infty} \cdots \sum_{j_i=0}^{\infty} h_i(j_1, ..., j_i) = C_i \qquad 0 < |C_i| < \infty \qquad (4.1.3)$$

and

$$\sum_{i=0}^{n} U^i C_i \neq 0 \qquad\qquad\qquad (4.1.4)$$

where U is the value of the input signal, then the system itself and its components of different degrees are neither of differentiating nor of integrating type.   ∎

Lemma   4.1.1

If a system satisfies Assumption 4.1.1 then the system itself and its components of different degrees are of neither differentiating nor integrating type as defined in Definition 4.1.1.

Proof.   Assume the system is described by its Volterra kernels. The output signal of the individual kernels for a step input of level $U \neq 0$ is

$$y_i = \sum_{j_1=0}^{\infty} \cdots \sum_{j_i=0}^{\infty} h_i(j_1, ..., j_i) u(k - j_1) ... u(k - j_i)$$

$$= U^i \sum_{j_1=0}^{\infty} \cdots \sum_{j_i=0}^{\infty} h_i(j_1, ..., j_i) = U^i C_i \qquad (4.1.5)$$

The sum of the output signals of the parallel channels of different degrees is not zero as assumed in (4.1.4).   ∎

Korenberg and Hunter (1990) pointed out that both the second-order (nonlinear) cross- and (output) auto-correlation nonlinearity test methods fail at the simple Wiener model if the sum of the weighting function series of the linear dynamic part is zero. Assuming a polynomial nonlinear static term, the sums of the Volterra kernels of all degrees are also zero and Assumption 4.1.1 is not satisfied. The final value of the step response is zero and the system does not fulfill Definition 4.1.1.

Any nonlinearity test results in a nonlinearity measure/index. The relation between the nonlinearity measure and the linear/nonlinear feature of a process is summarized below:

- All nonlinearity measures are defined in such a way that the measure is zero for linear systems only.
- From the fact that the nonlinearity measure differs from zero it follows that the process investigated is nonlinear.
- The fact that the nonlinearity measure is zero is not sufficient condition for being the process linear except the case when the process was excited by all possible test signals concerning shape and magnitude.
- Consequently, a test with finite number of excitations can provide information concerning the nonlinearity of the system, but nothing can be said concerning the linearity.
- The nonlinearity test can be performed only in a given input signal domain, for other regions different test results can be calculated.

Nonlinearity test methods were surveyed in Haber (1985) and in the report Haber (1989) in detail. The latter is the basis of Chapter 4.

4.2 TIME DOMAIN AND STEADY STATE TESTS

Assumption 4.2.1   Excite the system with a test signal $u_1(t)$ and then repeat the experiment by a new one being constant (γ) times bigger than the first

$$u_2(t) = \gamma u_1(t) \tag{4.2.1}$$

Proposition 4.2.1   Perform the experiment by satisfying Assumption 4.2.1. Measure the constant term y_0 by applying a zero input signal and subtract this value from the measured output signals $y_1(t)$ and $y_2(t)$, respectively. Then sample the output signals so that the series $\{y_1(k)\}$ and $\{y_2(k)\}$ are available. (Only those records have to be considered where the output signals decremented by the constant term differ from zero.) Relate the output signals decremented by the constant term to each other

$$\rho(k) = \frac{y_2(k) - y_0}{y_1(k) - y_0} \tag{4.2.2}$$

If $\rho(k)$ is not constant or not equal to γ then the system is nonlinear.

Proof.   The proposed nonlinearity index (4.2.2) is based on the definition of the nonlinear systems.

Corollary 4.2.1   An analytical nonlinearity index can be calculated as follows

$$v = \max_k \left\{ \left\| \frac{\rho(k) - \gamma}{\gamma} \right\| \right\} \tag{4.2.3}$$

If v differs from zero then the system is nonlinear.   ∎

If the system cannot be driven by a zero input signal then y_0 can be calculated, e.g., by Gardiner's method (Gardiner, 1966, 1973).

If the test signal is a constant then the method is suitable for detecting only the linear or nonlinear stationary character of a system.

Assumption 4.2.2   Besides Assumption 4.2.1 let the test signals be step functions and excite the process so long that the transients are settled.

Corollary 4.2.2   Calculate the index (4.2.2) only for the last sampling $k = N$. The steady state behavior of the process is nonlinear if the nonlinearity index

$$v = \left| \frac{\rho(N) - \gamma}{\gamma} \right| \qquad N \to \infty \qquad\qquad (4.2.4)$$

differs from zero.

Proof.   The components of different degrees of the process must not be of differentiating or integrating type to obtain finite end values of the step responses of the components of different degrees. According to Assumption 4.1.1 the system has a finite, non-zero end value to a non-zero excitation. (4.2.4) follows then from (4.2.3).   ∎

It means that, e.g., a *quasi-linear* dynamic process with constant gain but a working point dependent time constant would be detected as linear.

If the system is disturbed by noise the average values of several records have to be used instead of $y_1(k)$ and $y_2(k)$.

Example 4.2.1   *Nonlinearity test of the simple Wiener and Hammerstein models based on time domain and steady state tests (Haber, 1985)*

The time domain test was applied for detecting the nonlinear character of the simple Hammerstein (Figure 4.2.1a) and Wiener (Figure 4.2.1b) cascade models. Both models have a first-order linear dynamic part with the transfer function

$$G(s) = \frac{1}{1 + 10s}$$

whose step response is seen in Figure 4.2.1e, and a static quadratic polynomial (Figure 4.2.1d)

$$Y = 2 + U + 0.5U^2$$

(Capitals U and Y denote steady state values.) The test signals were steps with amplitudes 1 and 2

$$u(t) = 1(t) \qquad\qquad y(t) = y_1(t)$$
$$u(t) = 2 \cdot 1(t) \qquad\quad y(t) = y_2(t)$$

The step responses $y_1(t)$ and $y_2(t)$ of the nonlinear Hammerstein and Wiener models having first-order time lags are plotted in Figures 4.2.2a and 4.2.2b, respectively. The output signals were recorded up to 50 [s], and $N = 9$ (not equidistant) samples were taken. The constant term $(y_0 = 2)$ can be well seen as the common initial value of the step responses. The sampled output and the ratio values according to (4.2.2) are listed in Table 4.2.1.

TABLE 4.2.1 Step response values and the ratios of their values reduced by the constant term

Time	Simple Hammerstein model			Simple Wiener model		
[s]	$u(t): 0 \to 1$	$u(t): 0 \to 2$		$u(t): 0 \to 1$	$u(t): 0 \to 2$	
t	$y_1(t)$	$y_2(t)$	$\rho(t)$	$y_1(t)$	$y_2(t)$	$\rho(t)$
0	2.000	2.000	-	2.000	2.000	-
2.5	2.332	2.885	2.666	2.246	2.540	2.195
5	2.590	3.574	2.666	2.471	3.097	2.329
10	2.948	4.528	2.666	2.832	4.063	2.470
15	3.165	5.108	2.666	3.079	4.716	2.559
20	3.297	5.459	2.666	3.283	5.225	2.605
25	3.377	5.672	2.666	3.339	5.521	2.630
30	3.425	5.801	2.666	3.402	5.717	2.651
40	3.473	5.927	2.666	3.464	5.891	2.658
50	3.490	5.973	2.666	3.486	5.960	2.665

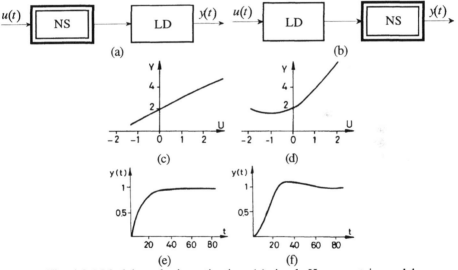

Fig. 4.2.1 Models under investigation: (a) simple Hammerstein model; (b) simple Wiener model; (c) linear static characteristic curve; (d) quadratic static characteristic curve; (e) step response of the first-order lag term; (f) step response of the second-order lag term

The nonlinearity indices were calculated according to (4.2.3) and the following results were obtained:
- $v = 0.333$ for the Hammerstein model;
- $v = 0.333$ for the Wiener model.

The steady state test was evaluated from the final values of the records and the following measures were obtained:

- $v = 0.333$ for the Hammerstein model;
- $v = 0.333$ for the Wiener model.

All four indices show the nonlinear character of the processes. ∎

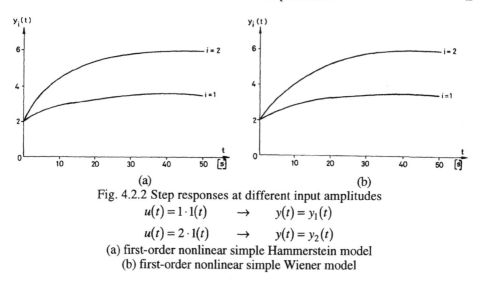

(a)   (b)
Fig. 4.2.2 Step responses at different input amplitudes

$$u(t) = 1 \cdot 1(t) \qquad \rightarrow \qquad y(t) = y_1(t)$$

$$u(t) = 2 \cdot 1(t) \qquad \rightarrow \qquad y(t) = y_2(t)$$

(a) first-order nonlinear simple Hammerstein model
(b) first-order nonlinear simple Wiener model

4.3 TEST BASED ON THE NORMALIZED VARIANCE OF THE NOISE-FREE OUTPUT SIGNAL AROUND THE OUTPUT SIGNAL OF THE BEST FITTING LINEAR MODEL

It follows from the definition of the linear systems that the normalized variance of the noise-free output signal around the output signal of the best fitting linear model is zero for linear processes and it differs from zero for nonlinear systems. The following theorem can be stated.

Theorem 4.3.1   Denote the estimated noise-free output signal by $\hat{w}(k)$ and the noise-free output signal of the best fitting linear model by $\hat{w}_{\mathrm{lin}}(k)$. The nonlinearity index

$$v = \frac{E\left\{\left[\hat{w}(k) - \hat{w}_{\mathrm{lin}}(k)\right]^2\right\}}{\mathrm{var}\left\{\hat{w}(k)\right\}} \qquad (4.3.1)$$

is zero if the process is linear and differs from zero if the system is nonlinear.

Corollary 4.3.1   It can be seen from (4.3.1) that the nonlinearity measure is zero if and only if

$$\hat{w}(k) = \hat{w}_{\mathrm{lin}}(k) \qquad (4.3.2)$$

i.e., the best linear model output coincides with the noise-free output signal of the process. Then the mean values and the standard deviations of the two signals are equal to each other

$$E\{\hat{w}(k)\} = E\{\hat{w}_{\text{lin}}(k)\} \qquad (4.3.3)$$
$$\sigma_{\hat{w}} = \sigma_{\hat{w}_{\text{lin}}} \qquad (4.3.4)$$

■

Usually it has advantages if the normalized values are used in the criteria instead of the not normalized values. Lemma 4.3.1 will say how we can substitute the original values by its normalized ones in (4.3.1).

Lemma 4.3.1
The nonlinearity index of (4.3.1) is equivalent to

$$v = E\left\{ \left[\hat{w}'(k) - \frac{\sigma_{\hat{w}_{\text{lin}}}}{\sigma_{\hat{w}}} \hat{w}'_{\text{lin}}(k) \right]^2 \right\} \qquad (4.3.5)$$

Proof. Substitute the output signal with its normalized values

$$\hat{w}(k) = \sigma_{\hat{w}} \hat{w}'(k) + E\{\hat{w}'(k)\}$$

and

$$\hat{w}_{\text{lin}}(k) = \sigma_{\hat{w}'_{\text{lin}}} \hat{w}'_{\text{lin}}(k) + E\{\hat{w}_{\text{lin}}(k)\}$$

into (4.3.1). In addition to that, the mean value of the best linear model has to be equal to the mean value of the noise-free process output (4.3.3). Then we obtain (4.3.2).   ■
Then nonlinearity index (4.3.5) is zero if and only if

$$\hat{w}'(k) = \hat{w}'_{\text{lin}}(k) \qquad (4.3.6)$$

Then

$$\sigma_{\hat{w}_{\text{lin}}} = \sigma_{\hat{w}} \qquad (4.3.7)$$

is also valid.
The identification of the best fitting linear model can be interpreted also as the minimization of (4.3.1) or (4.3.5) by the parameters of the best linear process model. On the other hand the nonlinearity index can be defined as the minimum of (4.3.1) or (4.3.5) in the parameter space. These thoughts lead to the following theorem which is closely related to Theorem 4.3.1

Theorem 4.3.2   Denote the vector of the estimated parameters of the best fitting linear model by $\hat{\theta}_{\text{lin}}$. The nonlinearity index

$$v = \min_{\hat{\theta}_{\text{lin}}} \frac{E\left\{ \left[\hat{w}(k) - \hat{w}_{\text{lin}}\left(k, \hat{\theta}_{\text{lin}}\right) \right]^2 \right\}}{\text{var}\{\hat{w}(k)\}} \qquad (4.3.8)$$

is equal to that of (4.3.1). Consequently it is zero if the process is linear, and it differs from zero otherwise.

Proof. Theorem 4.3.2 is a consequence of Theorem 4.3.1, Corollary 4.3.1 and Lemma 4.3.1.

Lemma 4.3.2
Denote the vector of the estimated parameters of the best fitting linear model between the normalized input and output signals by $\hat{\theta}'_{\text{lin}}$. The nonlinearity index

$$\nu = \min_{\hat{\theta}'_{\text{lin}}} \mathrm{E}\left\{ \left[\hat{w}'(k) - \frac{\sigma_{\hat{w}_{\text{lin}}}}{\sigma_{\hat{w}}} \hat{w}'_{\text{lin}}\left(k, \hat{\theta}'_{\text{lin}}\right) \right]^2 \right\} \tag{4.3.9}$$

is equal to that of (4.3.5). It is zero if the process is linear and it differs from zero otherwise.

Proof.   Lemma 4.3.2 is a synonym of Theorem 4.3.2 and is deduced from Lemma 4.3.1 and Equation (4.3.3).   ∎
     The parameters of the best linear model can be estimated by computing the cross-correlation function. Consequently a nonlinearity measure based on the cross-correlation function can be formulated instead of the parameters of the fitted linear model. Now a random test signal has to be applied.

Assumption 4.3.1   The test signal is white noise with zero mean and unit standard
                            deviation.

Theorem 4.3.3   If the test signal satisfies Assumption 4.3.1 then the nonlinearity index (4.3.9) is equivalent to

$$\nu = \mathrm{E}\left\{ \left[\hat{w}'(k) - \sum_{\kappa=0}^{m} r_{u'y'}(\kappa)u'(k-\kappa) \right]^2 \right\} \tag{4.3.10}$$

where $r_{u'y'}(\kappa)$ is the cross-correlation function of the normalized input and output signals

$$r_{u'y'}(\kappa) = \mathrm{E}\left\{ u'(k-\kappa)y'(\kappa) \right\}$$

and m is the memory of the weighting function of the process model. The nonlinearity index is zero for linear systems and differs from zero for nonlinear systems.

Proof.   Describe the linear model by its weighting function $\hat{h}(k)$ model with memory m. Then (4.3.9) becomes

$$\nu = \min_{\hat{h}(\kappa), \kappa=0,\dots,m} \mathrm{E}\left\{ \left[\hat{w}'(k) - \sum_{\kappa=0}^{m} \hat{h}(\kappa)u'(k-\kappa) \right]^2 \right\} \tag{4.3.11}$$

Differentiate the functional (4.3.11) by $\hat{h}(k)$ and equate the partial derivatives to zero

$$\frac{\partial}{\partial \hat{h}(\kappa)} \mathrm{E}\{\dots\} = -2\mathrm{E}\left\{ u'(k-\kappa)\left[\hat{w}'(k) - \sum_{\kappa=0}^{m} \hat{h}(\kappa)u'(k-\kappa) \right] \right\} = 0$$

The estimated parameters are the terms of the cross-correlation function

$$\hat{h}(\kappa) = E\{u'(k - \kappa)\hat{w}'(k)\} = r_{u'y'}(\kappa) \qquad (4.3.12)$$

∎

The noise-free output signal can be also computed as the expected mean value of the measured output signal over the input sequence

$$\hat{w}(k) = E\{y(k)|u(k), \ldots, u(k - m)\} \qquad (4.3.13)$$

Rajbman and Chadeev (1980) presented Theorem 4.3.3 and gave Theorems 4.3.1 and 4.3.2 in a modified form. They used the variance of the measured output signal instead of the variance of the computed noise-free output signal. As the variance of the output signal occurs only as a scaling factor, both definitions can be used for a nonlinearity test.

The following example helps to understand the different but equivalent forms of the nonlinearity index.

Example 4.3.1   *Calculation of the nonlinearity measure of a static quadratic process excited by a white noise signal with non-zero mean value (Haber, 1989)*

The test signal $u(k)$ has a constant $u_0 = \bar{u}$ and a stochastic $\Delta u(k)$ component. $\Delta u(k)$ has zero mean value and standard deviation σ_u, i.e.,

$$u(k) = u_0 + \Delta u(k) \qquad (4.3.14)$$

The equation of the process is

$$y(k) = cu^2(k) + e(k)$$

where $e(k)$ is a random noise of zero mean. The computed noise-free output signal is

$$\hat{w}(k) = cu^2(k) \qquad (4.3.15)$$

and the linear model to be fitted is

$$\hat{w}_{\text{lin}}(k) = \hat{c}_0 + \hat{c}_1 u(k) \qquad (4.3.16)$$

According to (4.3.8) of Theorem 4.3.2 the following minimization has to be performed

$$E\{[\hat{w}(k) - \hat{c}_0 - \hat{c}_1 u(k)]^2\} \underset{\hat{c}_0, \hat{c}_1}{\Rightarrow} \min \qquad (4.3.17)$$

By taking the derivative of the left side of (4.3.17) equal to zero we obtain

$$E\{\hat{w}(k)\} - \hat{c}_0 - \hat{c}_1 E\{u(k)\} = 0 \qquad (4.3.18)$$

$$E\{u(k)\hat{w}(k)\} - \hat{c}_0 E\{u(k)\} - \hat{c}_1 E\{u^2(k)\} = 0 \qquad (4.3.19)$$

From (4.3.18) and (4.3.19) the estimated parameters are as follows

$$\hat{c}_1 = \frac{E\{u(k)\hat{w}(k)\} - E\{u(k)\}E\{\hat{w}(k)\}}{E\{u^2(k)\} - E\{u(k)\}E\{u(k)\}}$$

(4.3.20)

$$\hat{c}_0 = E\{\hat{w}(k)\} - \hat{c}_1 E\{u(k)\}$$

(4.3.21)

The expected values of the different components are, in turn,

$$E\{u(k)\} = u_0$$

(4.3.22)

$$E\{u^2(k)\} = E\{[u_0 + \Delta u(k)]^2\} = u_0^2 + \sigma_u^2$$

(4.3.23)

$$E\{\hat{w}(k)\} = E\{c[u_0 + \Delta u(k)]^2\} = c[u_0^2 + \sigma_u^2]$$

(4.3.24)

$$E\{u(k)\hat{w}(k)\} = E\{[u_0 + \Delta u(k)]c[u_0 + \Delta u(k)]^2\} = c[u_0^3 + 3u_0\sigma_u^2]$$

(4.3.25)

On the basis of Equations (4.3.22) to (4.3.25) the estimated parameters are as follows

$$\hat{c}_1 = \frac{c[u_0^3 + 3u_0\sigma_u^2] - cu_0[u_0^2 + \sigma_u^2]}{u_0^2 + \sigma_u^2 - u_0^2} = 2cu_0$$

(4.3.26)

and

$$\hat{c}_0 = c[u_0^2 + \sigma_u^2 - 2cu_0^2] = c[\sigma_u^2 - u_0^2]$$

(4.3.27)

Further on the variance of the noise-free output signal is

$$\text{var}\{\hat{w}^2(k)\} = E\{[c[u_0 + \Delta u(k)]^2 - c[u_0^2 + \sigma_u^2]]^2\} = 2c^2\sigma_u^2[2u_0^2 + \sigma_u^2]$$

(4.3.28)

and the nonlinearity index (4.3.8) becomes

$$\nu = \frac{E\{[c[u_0 + \Delta u(k)]^2 - c[\sigma_u^2 - u_0^2] - 2cu_0[u_0 + \Delta u(k)]]^2\}}{2c^2\sigma_u^2[2u_0^2 + \sigma_u^2]} = \frac{\sigma_u^2}{2u_0^2 + \sigma_u^2}$$

(4.3.29)

The process is detected linear $(\nu = 0)$ if:
- the working point of the excitation is far from the origin of the parabolic characteristics $(u_0 \to \infty)$; or
- the perturbation is very small $(\sigma_u \to 0)$ and the working point is not at the origin $(u_0 \neq 0)$.

The maximum nonlinearity measure is detected if the test signal has a zero mean $(u_0 = 0)$.

Further on we show that the application of (4.3.10) of Theorem 4.3.3 leads to the same nonlinearity index.

The normalized input signal is

$$u'(k) = \frac{u_0 + \Delta u(k) - u_0}{\sigma_u} = \frac{\Delta u(k)}{\sigma_u} \tag{4.3.30}$$

and by using (4.3.24) and (4.3.28) the normalized computed output signal is

$$\hat{w}'(k) = \frac{c[u_0 + \Delta u(k)]^2 - c[u_0^2 + \sigma_u^2]}{\sqrt{2}c\sigma_u\sqrt{2u_0^2 + \sigma_u^2}} = \frac{2u_0\Delta u(k) + \Delta u^2(k) - \sigma_u^2}{\sqrt{2}\sigma_u\sqrt{2u_0^2 + \sigma_u^2}} \tag{4.3.31}$$

The parameter of the linearized model can be obtained by computing the cross-correlation

$$\hat{c}_1' = r_{u'\hat{w}'}(0) = \frac{E\{\Delta u(k)[2u_0\Delta u(k) + \Delta u^2(k) - \sigma_u^2]\}}{\sqrt{2}\sigma_u\sqrt{2u_0^2 + \sigma_u^2}} = \frac{\sqrt{2}u_0}{\sqrt{2u_0^2 + \sigma_u^2}} \tag{4.3.32}$$

The nonlinearity measure of (4.3.10) is

$$v' = E\left\{\left[\frac{2u_0\Delta u(k) + \Delta u^2(k) - \sigma_u^2}{\sqrt{2}\sigma_u\sqrt{2u_0^2 + \sigma_u^2}} - \frac{\sqrt{2}u_0}{\sqrt{2u_0^2 + \sigma_u^2}}\frac{\Delta u(k)}{\sigma_u}\right]^2\right\} = \frac{\sigma_u^2}{2u_0^2 + \sigma_u^2} \tag{4.3.33}$$

which is equal to (4.3.8), as expected.

Finally, we show that the application of (4.3.9) of Lemma 4.3.2 would lead to the same result.

A comparison between (4.3.9) and (4.3.11) shows that the nonlinearity index of (4.3.9) is equal to that of (4.3.11) if

$$\frac{\sigma_{\hat{w}_{lin}}}{\sigma_{\hat{w}}}\hat{w}'_{lin}(k) = \hat{c}_1'u'(k) = r_{u'y'}(0)u'(k) \tag{4.3.34}$$

It is easy to prove that the normalized output signal of a static linear system is equal to the normalized input signal. From (4.3.16) we obtain

$$E\{\hat{w}_{lin}(k)\} = \hat{c}_0 + \hat{c}_1u_0$$

$$\sigma_{\hat{w}_{lin}}^2 = E\left\{[\hat{w}_{lin} - E\{\hat{w}_{lin}\}]^2\right\} = E\left\{[\hat{c}_0 + \hat{c}_1u(k) - \hat{c}_0 - \hat{c}_1u_0]^2\right\} = \hat{c}_1^2\sigma_u^2 \tag{4.3.35}$$

$$\hat{w}'_{lin}(k) = \frac{\hat{c}_0 + \hat{c}_1u(k) - [\hat{c}_0 + \hat{c}_1u_0]}{\hat{c}_1\sigma_u} = \frac{\Delta u(k)}{\sigma_u} = u'(k) \tag{4.3.36}$$

Substituting (4.3.36) into (4.3.34), it remains to prove that

$$\frac{\sigma_{\hat{w}_{lin}}}{\sigma_{\hat{w}}} = \hat{c}_1' \tag{4.3.37}$$

By means of (4.3.35), (4.3.28) and (4.3.26) we can see the identity with (4.3.32)

$$\frac{\sigma_{\hat{w}_{lin}}}{\sigma_{\hat{w}}} = \frac{2cu_0\sigma_u}{\sqrt{2}\sigma_u c\sqrt{2u_0^2+\sigma_u^2}} = \frac{\sqrt{2}u_0}{\sqrt{2u_0^2+\sigma_u^2}} = \hat{c}_1' \qquad\blacksquare$$

Remark:
Allgöwer (1995a, 1995b) defined two nonlinearity measures which compare the difference between the output of the nonlinear system and its best linear approximation as

$$v_1 = \inf_{\hat{w}_{lin}}\sup_u \frac{\|\hat{w}-\hat{w}_{lin}\|}{\|u\|} \qquad (4.3.38)$$

$$v_2 = \inf_{\hat{w}_{lin}}\sup_u \frac{\|\hat{w}-\hat{w}_{lin}\|}{\|\hat{w}\|} \qquad (4.3.39)$$

In both cases the nonlinearity measures have to be calculated for all possible input signals in the domain investigated and the best approximating linear system has to be sought. The norm can be e.g., the L_2-norm.

As is seen, the nonlinearity measure (4.3.1) and (4.3.39) are strongly related. Allgöwer showed that the measure (4.3.39) is between 0 and 1, i.e. (4.3.39) is not only a qualitative but also a quantitative measure.

4.4 AVERAGE VALUE TEST OF THE OUTPUT SIGNAL

Assumption 4.4.1   Assume that the process is excited first by a constant non-zero level u_0, and second by a stationary stochastic signal $\Delta u(k)$ of zero mean added to the same level:

$$u(k) = u_0 + \Delta u(k) \qquad (4.4.1)$$

where

$$u_0 = E\{u(k)\}$$

and

$$E\{\Delta u(k)\} = 0 \qquad\blacksquare$$

Billings and Voon (1983) showed that the system has a nonlinear steady state characteristic if the system's response to a constant (non-zero) input differs from the expected value of the output signal to a stochastic input signal with the same mean level as before.

Theorem 4.4.1   (Billings and Voon, 1983) Excite a process satisfying Assumption 4.1.1 by a constant signal u_0 first and then by a stochastic signal $u_0 + \Delta u(k)$ corresponding to Assumption 4.4.1. The process has a nonlinear steady state characteristic if and only if

$$E\{y(u_0 + \Delta u(k))\} \neq E\{y(u_0)\}$$

and

$$u_0 \neq 0$$

Proof. It is assumed that the system itself and the components of different degrees are not of integrating or differentiating type. Describe the system by its n-degree discrete time Volterra kernels $h_n(\kappa_1, ..., \kappa_n)$. The noise-free output signal $w(k)$ can be disturbed by noise $e(k)$ with zero mean. Denote the noise series by $e_1(k)$ during the constant excitation and by $e_2(k)$ when the stochastic test signal is used. The difference between the expected values of the output signals in the two experiments is

$$\Delta Y = E\{y(u_0 + \Delta u(k))\} - E\{y(u_0)\}$$

$$= E\{w(u_0 + \Delta u(k)) + e_2(k)\} - E\{w(u_0) + e_1(k)\} = E\{w(u_0 + \Delta u(k)) - w(u_0)\}$$

Consequently, it has to be proved that

$$\Delta Y = E\{y(u_0 + \Delta u(k))\} - E\{y(u_0)\}$$

$$= \sum_{n=0}^{\infty} \sum_{\kappa_1=0}^{\infty} ... \sum_{\kappa_n=0}^{\infty} h_n(\kappa_1, ..., \kappa_n) E\left\{ \prod_{i=1}^{n} [u_0 + \Delta u(k - \kappa_i)] - \prod_{i=1}^{n} u_0 \right\} \neq 0 \tag{4.4.2}$$

if $u_0 \neq 0$

Consider (4.4.2) for little n values starting by $n = 0$.

1. *Constant channel*

$$\Delta Y = 0$$

since the output value does not depend on the input signal.

2. *Linear channel*
   Since

$$E\{[u_0 + \Delta u(k - \kappa_1)] - u_0\} = E\{\Delta u(k - \kappa_1)\} = 0$$

therefore
$$\Delta Y = 0$$

3. *Quadratic channel*

$$E\{[u_0 + \Delta u(k - \kappa_1)][u_0 + \Delta u(k - \kappa_2)] - u_0^2\}$$

$$= E\{u_0 \Delta u(k - \kappa_1) + u_0 \Delta u(k - \kappa_2) + \Delta u(k - \kappa_1) + \Delta u(k - \kappa_2)\}$$

$$= E\{\Delta u(k - \kappa_1) \Delta u(k - \kappa_2)\} \neq 0$$

because the auto-correlation function does not vanish for every $(\kappa_1 - \kappa_2)$ and there will be a bias in the steady state output values.
Consequently

$$\Delta Y \neq 0$$

4. *Cubic channel*

$$
\begin{aligned}
\mathrm{E}\left\{\prod_{i=1}^{3}\left[u_0 + \Delta u(k - \kappa_1)\right] - u_0^3\right\} &= \mathrm{E}\left\{u_0^2 \Delta u(k - \kappa_1) + u_0^2 \Delta u(k - \kappa_2) + u_0^2 \Delta u(k - \kappa_3)\right. \\
&\quad + u_0 \Delta u(k - \kappa_1)\Delta u(k - \kappa_2) + u_0 \Delta u(k - \kappa_1)\Delta u(k - \kappa_3) \\
&\quad + u_0 \Delta u(k - \kappa_2)\Delta u(k - \kappa_3) + \Delta u(k - \kappa_1)\Delta u(k - \kappa_2)\Delta u(k - \kappa_3)\Big\} \\
&= u_0 \mathrm{E}\left\{\Delta u(k - \kappa_1)\Delta u(k - \kappa_2)\right\} + u_0 \mathrm{E}\left\{\Delta u(k - \kappa_2)\Delta u(k - \kappa_3)\right\} \\
&\quad + u_0 \mathrm{E}\left\{\Delta u(k - \kappa_1)\Delta u(k - \kappa_3)\right\} \neq 0
\end{aligned}
$$

because only the mean values of the odd degree products vanish. Therefore, there will be a bias in the steady state output values if and only if the mean value of the input signal differs from zero,

$$\Delta Y \neq 0 \quad \text{if} \quad u_0 \neq 0.$$

5. *Channels of higher even degree*
   The bias in the steady state output values differs from zero because the occurring product terms of the odd order auto-correlation functions of the stochastic part of the test signal multiplied by the even degree powers of the constant level of the test signal are not zero.

6. *Channels of higher odd degree*
   The bias in the steady state output values contains also products of odd order auto-correlation functions of the stochastic part of the test signal multiplied by the odd degree powers of the constant level of the test signal. These terms differ from zero if and only if the constant level of the test signal is not zero.   ■

Corollary 4.4.1   If the mean value of the input signal is zero

$$u_0 = 0$$

then only nonlinear systems consisting of even degree subsystems can be detected nonlinear by the average value test of the output signal.

Proof.   This fact was shown for the first- and third-degree channels in the proof of Theorem 4.4.1 in detail. It was shown for higher odd degree channels at the end of the proof.   ■

Lemma 4.4.1
(Haber, 1989) An analytical nonlinearity index is introduced which detects the linear or nonlinear steady state relationship of a process satisfying Assumption 4.1.1

$$v = 2 \frac{E\{y(u_0 + \Delta u(k))\} - E\{y(u_0)\}}{\text{var}\{u(k)\}} \qquad (4.4.3)$$

where

$$\text{var}\{u(k)\} = \sigma_u^2$$

The normalization by the variance of the input excitation in (4.4.3) leads to the result that the nonlinearity index of a static constant, linear, quadratic or cubic system is equal to the second derivative of the nonlinear static function.

Proof. A static cubic degree polynomial is described by

$$y(k) = c_0 + c_1 u(k) + c_2 u^2(k) + c_3 u^3(k)$$

The nonlinearity index can be calculated as follows:

$$v = \frac{2}{\text{var}\{u(k)\}} E\Big\{ c_0 + c_1[u_0 + \Delta u(k)] + c_2[u_0 + \Delta u(k)]^2$$

$$+ c_3[u_0 + \Delta u(k)]^3 \Big\} - E\Big\{ c_0 + c_1 u_0 + c_2 u_0^2 + c_3 u_0^3 \Big\}$$

$$= \frac{2}{\text{var}\{u(k)\}} \Big[c_1 E\{\Delta u(k)\} + c_2 E\{2u_0 \Delta u(k) + \Delta u^2(k)\}$$

$$+ c_3 E\Big\{ 3u_0^2 \Delta u(k) + 3u_0 \Delta u^2(k) + \Delta u^3(k) \Big\} \Big]$$

$$= \frac{2}{\text{var}\{u(k)\}} \Big[c_2 E\{\Delta u^2(k)\} + 3c_3 u_0 E\{\Delta u^2(k)\} \Big] = 2c_2 + 6c_3 u_0$$

The second derivative of the cubic polynomial at the constant input signal u_0 is

$$\frac{d^2 y(k)}{du^2(k)}\bigg|_{u(k) = u_0} = 2c_2 + 6c_3 u_0$$

which completes the proof.   ∎

It can be seen from the above example that if the process contained only odd degree terms and the mean value of the test signal is zero then the test could not detect the nonlinear character, i.e.,

$$v = 0 \qquad \text{if} \qquad c_2 = 0 \qquad \text{and} \qquad u_0 = 0$$

As is seen already in the proof of Theorem 4.4.1, the additive output noises are filtered by the averaging process.

Example 4.4.1 *Average value test of the output signal of various linear, simple Hammerstein and simple Wiener models (Haber and Zierfuss, 1987)*

The average value test of the output signal was simulated at the noise-free simple

Hammerstein (Figure 4.2.1a) and at the simple Wiener (Figure 4.2.1b) models. The cascade models contain different linear dynamic (LD) and nonlinear static (NS) parts. Two types of static characters were considered: a linear static character (Figure 4.2.1c)

$$Y = 2 + U$$

and a quadratic nonlinear static character (Figure 4.2.1d)

$$Y = 2 + U + 0.5U^2$$

The transfer function of the first-order time lag was

$$G(s) = \frac{1}{1 + 10s}$$

and that of the second-order oscillating system was

$$G(s) = \frac{1}{1 + 2 \cdot 0.5 \cdot 10s + 10^2 s^2}$$

Their step responses are drawn in Figures 4.2.1e and 4.2.1f, respectively.

The input and output signals were simulated by sampling time 2 [s] at the models having a first-order time lag and by 4 [s] at those having a second-order time lag. The pulse transfer functions of the first- and second-order linear terms used for the simulation were

$$G\!\left(q^{-1}\right) = \frac{0.1813q^{-1}}{1 - 0.8187q^{-1}}$$

and

$$G\!\left(q^{-1}\right) = \frac{0.0694q^{-1} + 0.06072q^{-2}}{1 - 1.5402q^{-1} + 0.6703q^{-2}}$$

respectively.

The systems were first excited by a constant test signal of the value 1 and then by a stochastic test signal with the same mean value as the constant level. The stochastic test signal was a PRTS of maximum length 26 with amplitude ±2 around the above mean value 1. The minimum switching time of the PRTS signal was 5 times the sampling time.

The nonlinearity tests were applied at the following simple Hammerstein and Wiener models:

(a)   linear static characteristic and no dynamic part;
(b)   linear static characteristic and first-order linear dynamic part;
(c)   linear static characteristic and second-order linear dynamic part;
(d)   quadratic static characteristic and no dynamic part;
(e)   quadratic static characteristic and first-order linear dynamic part;
(f)   quadratic static characteristic and second-order linear dynamic part.

The simulated input and output signals, both for constant level and stochastic test signals are plotted on Figure 4.4.1 for the simple Hammerstein model and on Figure 4.4.2 for the simple Wiener model.

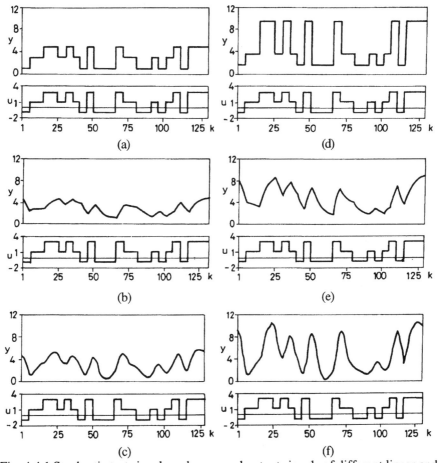

Fig. 4.4.1 Stochastic test signals and measured output signals of different linear and quadratic simple Hammerstein models (according to Figure 4.2.1):
(a) linear static characteristic and no dynamic part; (b) linear static characteristic and first-order linear dynamic part; (c) linear static characteristic and second-order linear dynamic part; (d) quadratic static characteristic and no dynamic part; (e) quadratic static characteristic and first-order linear dynamic part; (f) quadratic static characteristic and second-order linear dynamic part

The nonlinearity indices calculated according to (4.4.3) are summarized in Table 4.4.1. The test detects correctly the linear or the nonlinear feature of the processes. It is worth while mentioning that the nonlinearity index is equal to one not only in the static case – as stated already at presenting the method – but also at the simple Hammerstein model. The reason for this fact is that the static gain K of the linear dynamic term in the simple Hammerstein models under investigation was always $K = 1$. The relation between the mean values of the inner signal $v(k)$ before the linear term and the output signal is

$$E\{y(k)\} = K\, E\{v(k)\}$$

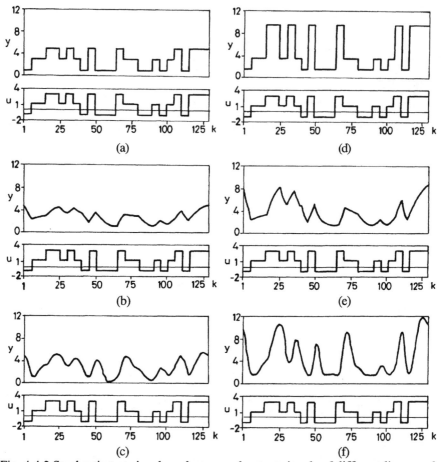

Fig. 4.4.2 Stochastic test signals and measured output signals of different linear and quadratic simple Wiener models (according to Figure 4.2.1):
(a) linear static characteristic and no dynamic part; (b) linear static characteristic and first-order linear dynamic part; (c) linear static characteristic and second-order linear dynamic part; (d) quadratic static characteristic and no dynamic part; (e) quadratic static characteristic and first-order linear dynamic part; (f) quadratic static characteristic and second-order linear dynamic part

and therefore the nonlinearity measure of the simple Hammerstein model v_{sH} is K times that of the static nonlinear term v_{sn} in the simple Hammerstein model

$$
\begin{aligned}
v_{sH} &= 2\frac{E\{y(u_0 + \Delta u(k))\} - E\{y(u_0)\}}{\text{var}\{\Delta u(k)\}} \\
&= 2K\frac{E\{v(u_0 + \Delta u(k))\} - E\{v(u_0)\}}{\text{var}\{\Delta u(k)\}} = K\,v_{sn}
\end{aligned}
$$

TABLE 4.4.1 Nonlinearity indices according to the average value of the output signal

Model	Static charac-teristic	Order of the linear dynamic part	Figure	Output at constant test signal	Mean value of the output to the stochastic test signal	Nonline-arity index
simple Hammer-stein	linear	0	4.4.1a	3.0	3.0	0.0
		1	4.4.1b	3.0	3.0	0.0
		2	4.4.1c	3.0	3.0	0.0
	quadratic	0	4.4.1d	3.05	4.885	1.0
		1	4.4.1e	3.05	4.885	1.0
		2	4.4.1f	3.05	4.885	1.0
simple Wiener	linear	0	4.4.2a	3.0	3.0	0.0
		1	4.4.2b	3.0	3.0	0.0
		2	4.4.2c	3.0	3.0	0.0
	quadratic	0	4.4.2d	3.5	4.885	1.0
		1	4.4.2e	3.5	4.015	0.372
		2	4.4.2f	3.5	4.597	0.792

4.5 GAUSSIAN DISTRIBUTION TEST OF THE OUTPUT SIGNAL

The following lemma is well known and it can be used for a normality test if the process is excited by a test signal with Gaussian distribution.

Lemma 4.5.1
If the normalized input signal of a system is Gaussian then:
- the normalized output signal has a Gaussian distribution if the process is linear;
- the normalized output signal does not have a Gaussian distribution if the process is nonlinear.   ∎

This fact is described by several publications, (e.g., Schetzen, 1980; Bendat and Piersol, 1971; Peebles, 1980; Kaminskas and Rimidis, 1982; Bendat, 1990).

Proof.   Describe the linear system by its weighting function series model

$$y(k) = h_0 + \sum_{\kappa=0}^{\infty} h_1(\kappa)u(k-\kappa)$$

Express $u(k)$ from

$$u'(k) = \frac{u(k) - u_0}{\sigma_u}$$

and put it into the expression of the output signal. We obtain

$$y(k) = h_0 + u_0 \sum_{\kappa=0}^{\infty} h_1(\kappa) + \sigma_u \sum_{\kappa=0}^{\infty} h_1(\kappa)u'(k-\kappa)$$

The mean value of the output signal is

$$\bar{y} = h_0 + u_0 \sum_{\kappa=0}^{\infty} h_1(\kappa)$$

It has to be proved that the normalized output signal

$$y'(k) = \frac{y(k) - \bar{y}}{\sigma_y} = \frac{\sigma_u}{\sigma_y} \sum_{\kappa=0}^{\infty} h_1(\kappa) u'(k - \kappa)$$

has a normal distribution if the input signal is a Gaussian one.

First it can be seen that an individual term $h_1(\kappa)u'(k - \kappa)$ is Gaussian. This is satisfied because the input sequence is Gaussian and multiplication by a constant factor does not alter the distribution. Now the normalized output signal is the sum of Gaussian signals, and therefore it also has a Gaussian distribution independently from the fact of whether or not the terms are independent from each other.

The fact that a nonlinear (e.g., quadratic) system can produce an output signal with non-Gaussian distribution, can be shown by a simple example (e.g., a static square term). ■

It has to be underlined that Lemma 4.5.1 is valid both for white noise and color input signals, i.e., it is independent of the form of the auto-correlation function of the input signal.

There are several, similar procedures known for the normality test (e.g., Bendat and Piersol, 1971; Draper and Smith, 1966; Kennedy and Neville, 1976):
- plot of the empirical probability distribution density or cumulative distribution function;
- plot of the empirical inverse Gaussian cumulative distribution function;
- chi-square test;
- calculation of the mean deviation;
- calculation of the skewness and the kurtosis.

It is practical to perform the test on the normalized output signal $y'(k)$.

1. Plot of the empirical probability density or distribution function

The amplitude range has to be divided into N_q equal intervals. The recommended N_q values are given as a function of the number of the data N in Table 4.5.1. Because of the possible outliers the number of intervals where most of the data fall can be much less than the recommended value N_q. Therefore it is recommended to divide the interval [-3, 3] into $(N_q - 2)$ intervals and consider the tails as the first and the N_q-th interval, respectively.

The following empirical functions can be calculated and plotted against the interval:
- *absolute frequency:*
  the number of observations falling into the intervals;
- *relative frequency:*
  the absolute frequency divided by the number of observations;
- *empirical distribution density:*
  the relative frequency divided by the interval width;
- *cumulative absolute frequency:*

the absolute frequency cumulated from the first interval;
- *cumulative relative frequency:*
the relative frequency cumulated from the first interval;
- *empirical cumulative distribution:*
the empirical density cumulated from the first interval.

TABLE 4.5.1 Minimum number of intervals N_q for the calculation of
the amplitude histogram as a function of the number of data N (Bendat
and Piersol, 1971)

N	200	400	600	800	1000	1500	2000
N_q	16	20	24	27	30	35	39

The calculated empirical functions have to be compared with the probability function of the normal distribution. The probability density function of a Gaussian, normalized random variable x is

$$p(x) = \frac{1}{\sqrt{2\pi}} e^{-x^2}$$

and the probability cumulative distribution function is

$$P(x) = \frac{1}{\sqrt{2\pi}} \int_{-\infty}^{x} e^{-x_1^2} dx_1$$

2. *Plot of the inverse Gaussian cumulative distribution function of the empirical cumulative distribution function*
Instead of using the empirical cumulative distribution function itself, plot its transformation through the inverse Gaussian distribution function (with the same mean value and standard deviation as the investigated signal). If the residuals have a normal distribution then the plotted points lie on a straight line. The usage of the inverse cumulative distribution function instead of the original one is more suitable because the inverse curve should be compared with a straight line and not with an S shape curve. (For manual evaluations papers on the Gaussian probability are available in which the ordinate is scaled so that the inverse transformation is performed automatically.)

3. *Chi-square test of normality*
An analytical measure for the normality test is the chi-square test (e.g., Bendat and Piersol, 1971). The statistics

$$\Xi^2 = \sum_{i=1}^{K} \frac{\left(f_i - f_i^o\right)^2}{f_i^o} \tag{4.5.1}$$

where f_i is the empirical and f_i^o is the theoretically calculated expected frequency of the normalized Gaussian signal, has a chi-square distribution and its value should be less than the tabulated value $\Xi^2\left(N_q - 3, \alpha\right)$ of a chi-square table at the entries: freedom

$(N_q - 3)$ and probability level $100(1 - \alpha)$ percent. (The freedom is less by three than the number of intervals because the mean value and the standard deviation of the output signal are calculated from the observations and the sum of the frequencies is a known value, it is equal to the number of data.)

Corollary 4.5.1   If the normalized input signal is Gaussian the nonlinearity degree can be defined as

$$v \equiv v_{\Xi^2} = \frac{\Xi^2 - \Xi^2(N_q - 3, \alpha)}{\Xi^2(N_q - 3, \alpha)} \tag{4.5.2}$$

The system is linear if v_{Ξ^2} is less than zero and if v_{Ξ^2} is greater than zero then the system is nonlinear.   ∎

4. Calculation of the mean deviation
It can be shown (e.g., Kennedy and Neville, 1976), that if the output signal has a Gaussian distribution then the ratio of the standard deviation and the mean of the absolute deviations is

$$\rho = \frac{\sigma_y}{\frac{1}{N}\sum_{k=1}^{N}|y(k) - E\{y(k)\}|} = \sqrt{\pi/2} = 1.253 \tag{4.5.3}$$

Corollary 4.5.2   The analytical index

$$v \equiv v_\rho = \left|\frac{\rho - 1.253}{1.253}\right| \tag{4.5.4}$$

is a measure of the nonlinearity. The input–output relation is nonlinear if v_ρ significantly differs from zero.   ∎

5. Calculation of the skewness and the kurtosis
An alternative method for the test of normality – and therefore for the test of the nonlinearity – is presented by Kaminskas and Rimidis (1982). It is known from the theory of statistics that both the skewness

$$\beta_1 = E\{y'^3(k)\} \tag{4.5.5}$$

and the kurtosis

$$\beta_2 = E\{y'^4(k)\} - 3 \tag{4.5.6}$$

are zero if the signal $y(k)$ under investigation has a normal distribution (e.g., Cox and Hinkley, 1974). Kaminskas and Rimidis (1982) perform the nonlinearity test by checking whether the above features are zero or not. To decide whether they are zero or not we need the standard deviations of them. Denote the standard deviation of the skewness by σ_{β_1} and that of the kurtosis by σ_{β_2}. They can be calculated by the following formulas:

- *for independent data*
  (i.e., if the auto-correlation function of the output signal is a Dirac pulse) (Cox and Hinkley, 1974):

$$\sigma_{\beta_1}^2 = \frac{6}{N}$$

$$\sigma_{\beta_2}^2 = \frac{24}{N}$$

- *for correlated data*
  (i.e., if the auto-correlation function $r_{y'y'}(\kappa)$ has non-zero values for $\kappa \neq 0$):
  - if the mean value and the standard deviations of the output signal were calculated from the data (Gasser, 1975):

$$\sigma_{\beta_1}^2 = \frac{6}{N} \sum_{\kappa=\kappa_{min}}^{\kappa_{max}} r_{y'y'}^3(\kappa) \tag{4.5.7}$$

$$\sigma_{\beta_2}^2 = \frac{24}{N} \sum_{\kappa=\kappa_{min}}^{\kappa_{max}} r_{y'y'}^4(\kappa) \tag{4.5.8}$$

  - if the mean value and standard deviations of the output signal are known Kaminskas and Rimidis (1982) recommended another form.

The output signal has a normal distribution with 95% probability if

$$-2\sigma_{\beta_1} < \beta_1 < 2\sigma_{\beta_1}$$

and

$$-2\sigma_{\beta_2} < \beta_2 < 2\sigma_{\beta_2}$$

come true. The above mentioned can be summarized in the following corollaries.

Corollary 4.5.3   The nonlinearity index can be defined as

$$\nu \equiv \nu_{\beta_1} = \frac{|\beta_1|}{2\sigma_{\beta_1}} - 1 \tag{4.5.9}$$

The process is linear if the index ν_{β_1} is less than zero and it is nonlinear if the index is greater than zero.   ■

Corollary 4.5.4   The nonlinearity index can be defined as

$$\nu \equiv \nu_{\beta_2} = \frac{|\beta_2|}{2\sigma_{\beta_2}} - 1 \tag{4.5.10}$$

The process is linear if the index ν_{β_2} is less than zero and it is nonlinear if the index is greater than zero.   ■

If the additive output noise also has a Gaussian distribution – as usual – then a linear process with normal output signal distribution will be detected as linear since the sum of two Gaussian signals is also normally distributed. If the additive output noise has a distribution other than Gaussian then the system will not be detected as linear even if the process is a linear one. It can be assumed, however, that the mutual effect of several disturbances has a Gaussian distribution.

TABLE 4.5.2 Nonlinearity indices according to the Gaussian distribution test of the output signal at the linear, simple Hammerstein and Wiener models. Tabulated are the skewnesses β_1, kurtosises β_2, their standard deviations, the mean deviations ρ and the chi-square statistic values Ξ^2 of the test of normality. The nonlinearity index v_ρ is calculated based on the mean deviation value and v_{Ξ^2} is based on the chi-square test.

Model	Static charac- teristic	Order of the linear dynamic part	Skewness			Curtosis			Mean deviation		Chi-square	
			value	standard dev.	nonlinear. index	value	standard dev.	nonlinear. index	value	nonlinear. index	value	nonlinear. index
		n	β_1	σ_{β_1}	v_{β_1}	β_2	σ_{β_2}	v_{β_2}	ρ	v_ρ	Ξ^2	v_{Ξ^2}
simple Hammer- stein	linear	0	-0.049	0.035	-0.30	-0.175	0.069	0.27	1.24	0.010	28.6	-0.487
		1	-0.057	0.050	-0.43	0.325	0.091	0.79	1.25	0.0024	38.3	-0.313
		2	-0.031	0.054	-0.715	0.148	0.102	-0.28	1.26	0.0056	40.5	-0.274
	quadratic	0	1.920	0.035	26.43	4.560	0.069	32.05	1.34	0.0694	7550.0	134.402
		1	0.658	0.050	5.58	1.291	0.092	6.02	1.28	0.0215	645.0	10.567
		2	0.752	0.054	5.97	1.080	0.103	4.22	1.29	0.0295	661.0	10.854
simple Wiener	linear	0	-0.049	0.035	-0.30	-0.175	0.069	0.27	1.24	0.010	28.6	-0.487
		1	0.034	0.050	-0.66	-0.124	0.091	-0.32	1.25	0.0024	45.5	-0.184
		2	0.010	0.054	-0.91	-0.020	0.102	-0.90	1.26	0.0056	37.2	-0.333
	quadratic	0	1.920	0.035	26.43	4.560	0.069	32.05	1.34	0.0694	7550.0	134.400
		1	0.864	0.049	7.82	0.758	0.090	3.21	1.26	0.0056	881.0	14.800
		2	0.805	0.053	6.60	0.678	0.100	2.39	1.29	0.0295	1730.0	30.026

Example 4.5.1   *Gaussian distribution test of the output signal of various linear, simple Hammerstein and simple Wiener models to a Gaussian test signal (Haber, 1989)*

The same processes as in Example 4.4.1, that means
- the simple Hammerstein and
- the simple Wiener models

were simulated
- with linear static parts and
- with quadratic static parts.

The dynamic parts were in turn
- a unit term,
- a first-order term, and
- a second-order oscillating term.

The processes were excited by a white noise random signal with Gaussian distribution, zero mean value and unit standard deviation. $N = 5000$ data were simulated. Table 4.5.2 summarizes:
- the skewnesses, their standard deviations and the nonlinearity index calculated based on them;
- the kurtosises, their standard deviations and the nonlinearity index calculated based on them;
- the chi-square statistic values of the fit of the normality and the nonlinearity

index calculated based on them;
- the mean deviations and the nonlinearity index calculated based on them,

for all simulated cases. The skewnesses and the kurtosises were calculated according to (4.5.5) and (4.5.6). Their standard deviations were computed by (4.5.7) and (4.5.8). Latter formulas use the values of the auto-correlation function of the output signals. They are drawn in Figures 4.5.1 and 4.5.2 for the simple Hammerstein and simple Wiener models, respectively.

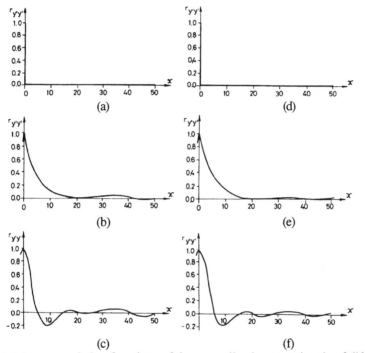

Fig. 4.5.1 Auto-correlation functions of the normalized output signals of different simple Hammerstein models (according to Figure 4.2.1) excited by a white Gaussian test signal of zero mean value: (a) linear static characteristic and no dynamic part; (b) linear static characteristic and first-order linear dynamic part; (c) linear static characteristic and second-order linear dynamic part; (d) quadratic static characteristic and no dynamic part; (e) quadratic static characteristic and first-order linear dynamic part; (f) quadratic static characteristic and second-order linear dynamic part

The degree of freedom was 41, the critical chi-square value is 55.76 at the significance level 95%. The nonlinearity indices were calculated according to (4.5.2), (4.5.4), (4.5.9) and (4.5.10).
The nonlinearity indices computed based on the skewness, the kurtosis and the chi-squares detected the linear or the nonlinear features of the processes always correct. The differences between the nonlinear measures based on the mean deviations were not relevant in the different cases, that means the computation of the mean deviation does not lead to a powerful nonlinearity test.   ∎

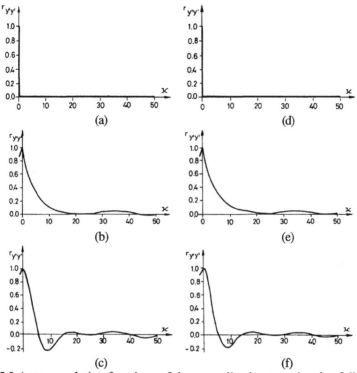

Fig. 4.5.2 Auto-correlation functions of the normalized output signals of different simple Wiener models (according to Figure 4.2.1) excited by a white Gaussian test signal of zero mean value: (a) linear static characteristic and no dynamic part; (b) linear static characteristic and first-order linear dynamic part; (c) linear static characteristic and second-order linear dynamic part; (d) quadratic static characteristic and no dynamic part; (e) quadratic static characteristic and first-order linear dynamic part; (f) quadratic static characteristic and second-order linear dynamic part

4.6 FREQUENCY METHOD

Assumption 4.6.1   Excite the process by a sinusoidal test signal

$$u(t) = U\sin(\omega t)$$ ∎

Then the resulting stationary output signal can be represented in Fourier components

$$y(t) = y_0 + \sum_{i=1}^{n} Y_n \sin(n\omega t + \varphi_n)$$ (4.6.1)

Theoretically n is infinitely large but in practical cases n is a large finite value.

The system is linear if no subharmonics are in the output signal. The following lemma is a consequence of the definition of the linear systems.

Lemma 4.6.1

If Assumption 4.6.1 is satisfied then take the relation between the effective values of the subharmonics and that of all harmonics of the noise-free output signal

$$\nu = \frac{\sum\limits_{i=2}^{n} Y_i^2/2}{\sum\limits_{i=1}^{n} Y_i^2/2} = 1 - \frac{Y_1^2/2}{\sum\limits_{i=1}^{n} Y_i^2/2} = 1 - \frac{Y_1^2}{\sum\limits_{i=1}^{n} Y_i^2} \tag{4.6.2}$$

as an analytical measure of the nonlinearity. The nonlinearity index (4.6.2) is equal to zero for linear systems and differs from zero for nonlinear systems.   ■

Allgöwer (1995b) derived a nonlinearity measure from (4.3.39), which is the square root of (4.6.2). The only difference is that Allgöwer's measure considers the constant (DC) term in the output signal, as well.

The computation of (4.6.2) requires the harmonic analysis of the periodic output signal. It is easier to calculate the effective value of the whole signal than the sum of the effective values of the subharmonics. Therefore a nonlinearity index is defined without requiring a harmonic analysis.

Lemma 4.6.2
If Assumption 4.6.1 is satisfied then the nonlinearity index of (4.6.2) can be rewritten as

$$\nu = \frac{y_{\text{eff}} - y_0^2 - Y_1^2/2 - \text{var}\{e(t)\}}{y_{\text{eff}} - y_0^2 - \text{var}\{e(t)\}} = 1 - \frac{Y_1^2/2}{y_{\text{eff}} - y_0^2 - \text{var}\{e(t)\}} \tag{4.6.3}$$

where

$$y_{\text{eff}} = \frac{1}{T} \int_{t=0}^{T} y^2(t)dt \tag{4.6.4}$$

and $T = 2\pi/\omega$ is the time period.

Proof.   Assume that there is an additive noise at the output

$$y(t) = w(t) + e(t)$$

and $w(t)$ can be expressed in harmonic series. Consequently

$$y_{\text{eff}} = y_0^2 + \frac{Y_1^2}{2} + \sum_{i=2}^{n} \frac{Y_i^2}{2} + \text{var}\{e(t)\} \tag{4.6.5}$$

Having replaced the effective values of the subharmonics in (4.6.3) by the correct formula (4.6.5) we get Lemma 4.6.2.   ■

Accordingly the nonlinearity index of (4.6.4) is valid both in noise-free and noisy cases.

If the additive noise is filtered from the output signal or its influence can be neglected, then the nonlinearity measure (4.6.3) reduces to

$$\nu = 1 - \frac{Y_1^2/2}{y_{\text{eff}} - y_0^2} \tag{4.6.6}$$

In the engineering practice the distortion factor v_d is used as an analytical measure for characterizing the linearity of a system.

Definition 4.6.1   The distortion factor is defined (e.g., Saboke, 1985) by

$$v_d = 1 - \sqrt{\frac{Y_1^2}{\sum_{i=1}^{\infty} Y_i^2}} \qquad (4.6.7)$$

Since (4.6.7) is similar to (4.6.2) the distortion factor can be used as a nonlinearity measure, indeed.

Corollary 4.6.1   Under Assumption 4.6.1 the distortion factor is equal to zero for linear systems and differs from zero for nonlinear systems.

Proof.   The validity of Corollary 4.6.1 is a trivial consequence of the similarity between the definitions (4.6.7) and (4.6.2). The analytical relation between the distortion factor (4.6.7) and the nonlinearity index (4.6.1) is

$$v_d = 1 - \sqrt{1 - v}$$

or

$$v = 1 - (1 - v_d)^2$$

If $v = 1$ then $v_d = 1$ and if $v = 0$ then $v_d = 0$. Both nonlinearity indices are zero if and only if the process is linear.   ∎

The nonlinearity measure (4.6.2) is valid also in the noisy case. If the nonlinearity index is calculated according to the simplified expression (4.6.6), then the noisy measurements can be filtered by the averaging method suggested at the time domain test.

4.7 LINEAR SPECTRAL DENSITY METHOD

It is known from the theory of stochastic systems that for continuous time noise-free linear systems excited by a stochastic signal there is a relation between the continuous auto-spectral density function of the input signal $S_{uu}(j\omega)$, that of the output signal $S_{yy}(j\omega)$, and the cross-spectral density function $S_{uy}(j\omega)$.

Assumption 4.7.1   The process under investigation is excited by a stochastic signal with time invariant statistical features.   ∎

Lemma 4.7.1
Under Assumption 4.7.1 the following relation is valid for a linear system without constant term (e.g., Bendat and Piersol, 1980):

$$\frac{\left| S_{uy}(j\omega) \right|^2}{S_{uu}(j\omega) S_{yy}(j\omega)} = 1 \qquad (4.7.1)$$

Proof. Express the terms in (4.7.1) by means of the auto-spectral density function of the input signal

$$S_{uy}(j\omega) = S_{uu}(j\omega)G(j\omega) \tag{4.7.2}$$

$$S_{yy}(j\omega) = S_{uu}(j\omega)G(j\omega)G(-j\omega) = S_{uu}(j\omega)|G(j\omega)|^2 \tag{4.7.3}$$

Here $G(j\omega)$ is the frequency function of the plant. Having substituted the spectral density functions $S_{uy}(j\omega)$ and $S_{yy}(j\omega)$ in (4.7.1) by the expressions (4.7.2) and (4.7.3), respectively, we get (4.7.1):

$$\frac{|S_{uy}(j\omega)|^2}{S_{uu}(j\omega)S_{yy}(j\omega)} = \frac{S_{uu}(j\omega)|G(j\omega)|^2}{S_{uu}(j\omega)|G(j\omega)|^2 S_{uu}(j\omega)} = 1 \qquad\blacksquare$$

The left side of (4.7.1) can be used for an analytical measure of the linear character of a process.

Theorem 4.7.1 (e.g., Bendat and Piersol, 1980) The nonlinearity index

$$\nu = 1 - \min_{\omega}\left\{ \frac{|S_{u'y'}(j\omega)|^2}{S_{u'u'}(j\omega)S_{y'y'}(j\omega)} \right\} \tag{4.7.4}$$

is zero if the process under investigation is linear and the measurements are noise-free. If the index (4.7.4) differs from zero then either the process is nonlinear or the measurements are noisy.

Proof. The theorem is a consequence of Lemma 4.7.1. $\qquad\blacksquare$

Unfortunately, the spectral density function of the output signal can be effected by the noise even if the noise is independent from the input signal.

The frequency function

$$\nu_c = \frac{|S_{u'y'}(j\omega)|^2}{S_{u'u'}(j\omega)S_{y'y'}(j\omega)}$$

is called the coherence function (e.g., Bendat and Piersol, 1971; Saboke, 1985). The coherence function shows the linearity $(\nu_c = 1)$ of the system as a function of the frequency.

4.8 LINEAR CORRELATION METHOD

Lemma 4.8.1
(Bendat and Piersol, 1980) Under Assumption 4.7.1 the following relation is valid for a linear system (without constant term)

$$\frac{\left[\int\limits_{\tau=0}^{\infty} r_{uy}(\tau)d\tau\right]^2}{\left[\int\limits_{\tau=0}^{\infty} r_{uu}(\tau)d\tau\right]\left[\int\limits_{\tau=0}^{\infty} r_{yy}(\tau)d\tau\right]} = 1 \qquad (4.8.1)$$

Proof. Use in (4.7.1) the corresponding continuous correlation functions $r(\tau)$ instead of the spectral density functions according to the Wiener–Hinchin equations and consider (4.7.1) at the zero frequency. ∎

Szücs *et al.*, (1975) used the square root of the left side of (4.8.1) for detecting the linear feature of a system.

A similar expression to (4.8.1) can be derived for discrete time linear dynamic systems, as well.

Theorem 4.8.1   If the test signal satisfies Assumption 4.7.1 then the analytical index

$$\nu = 1 - \max_{\omega}\left\{\frac{\left[\sum\limits_{\kappa=\kappa_{min}}^{\kappa_{max}} r_{u'y'}(\kappa)\right]^2}{\left[\sum\limits_{\kappa=\kappa_{min}}^{\kappa_{max}} r_{u'u'}(\kappa)\right]\left[\sum\limits_{\kappa=\kappa_{min}}^{\kappa_{max}} r_{y'y'}(\kappa)\right]}\right\} \qquad (4.8.2)$$

is equal to zero if the system under investigation is linear and the measurements are noise-free. The nonlinearity index (4.8.2) differs from zero if either the process is nonlinear or the measurements are noisy. In (4.8.2) $r(\kappa)$ denotes the discrete time correlation function and κ is the discrete time shift. κ_{min} and κ_{max} has to be chosen in such a way that all the correlation functions should be zero outside the domain. ∎

Proof. Theorem 4.8.1 follows from the discrete time equivalent of (4.8.1). ∎
The minimum and maximum shifting times have to be chosen so that the auto- and cross-correlation functions outside the selected domain are zero. If the test signal is pseudo-random then its auto-correlation function is periodic and the above-mentioned have to be fulfilled only in one period. For pseudo-random test signals with maximum length the distance between two neighboring peaks in the auto-correlation function series has to be bigger than the relative settling time of the process – in the units of the sampling time. Using, e.g., a PRBS, the length of the domain should be less by one than the period of the signal and for a PRTS less by one than the half of the period.

The cross-correlation function becomes zero and thus the nonlinearity index is one if the nonlinear system has a quadratic static characteristic and the mean value of the input signal coincides with its extremum point. (See, e.g., Example 4.8.1)

The linear correlation method needs less computational effort than the linear spectral density method.

Unfortunately, the output auto-correlation function can be affected by the noise even if the noise is independent from the input signal.

TABLE 4.8.1 Nonlinearity indices for linear and nonlinear correlation and dispersion methods

Model	Mean value of the input signal	Order of the time lag	Figure	Nonlinearity indices based on the above functions			
				linear correlation	dispersion	nonlinear cross-correlation	second-order auto-correlation
linear dynamic	1	1	4.8.1a	0.0	0.0	0.0	0.0
	1	2	4.8.2a	0.0	0.0	0.0	0.0
quadratic simple Hammerstein	-1	1	4.8.1b	1.0	1.0	0.579	0.511
	-1	2	4.8.2b	1.0	1.0	0.460	0.524
quadratic simple Hammerstein	0	1	4.8.1c	0.038	0.451	0.259	0.479
	0	2	4.8.2c	0.069	0.452	0.218	0.566
quadratic simple Wiener	-1	1	4.8.1d	1.0	1.0	0.375	0.769
	-1	2	4.8.2d	1.0	1.0	0.309	0.841
quadratic simple Wiener	0	1	4.8.1e	0.047	0.220	0.119	0.669
	0	2	4.8.2e	0.099	0.580	0.140	0.766

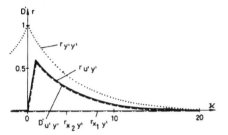

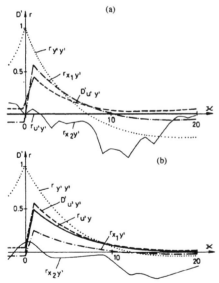

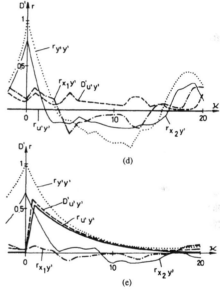

Fig. 4.8.1 Correlation and dispersion functions at different first-order models:

(a) linear model $(u_0 = 1)$;

(b) simple Hammerstein model $(u_0 = -1)$;

(c) simple Hammerstein model $(u_0 = 0)$;

(d) simple Wiener model $(u_0 = -1)$;

(e) simple Wiener model $(u_0 = 0)$.

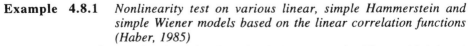

Example 4.8.1 *Nonlinearity test on various linear, simple Hammerstein and simple Wiener models based on the linear correlation functions (Haber, 1985)*

The test was performed for the noise-free simple Hammerstein (Figure 4.2.1a) and simple Wiener (Figure 4.2.1b) models. The cascade models contain different linear dynamic (LD) and nonlinear static (NS) parts. The same static characteristics and linear dynamic terms were simulated as in Example 4.2.1. The static curves and the step responses of the linear dynamic parts were given already in Figure 4.2.1.

The input and output signals were sampled by 2 [s] at the models having a first-order time lag and by 4 [s] at those ones having a second-order one. The test signal was always a PRTS signal of maximum length 80 with amplitude ±2 around the values listed in Table 4.8.1. The minimum switching time was equal to the sampling time.

The nonlinearity test method was applied to the following models:
(a)   linear model (Wiener model with linear static characteristic);
(b)   nonlinear simple Hammerstein model excited in the static extremum;
(c)   nonlinear simple Hammerstein model not excited in the static extremum;
(d)   nonlinear simple Wiener model excited in the static extremum;
(e)   nonlinear simple Wiener model not excited in the static extremum.

The linear cross-correlation functions are plotted in Figures 4.8.1 and 4.8.2 for the listed situations. (The discrete values are drawn by continuous lines as being continuous functions.) The auto-correlation function of the normalized input signal was always a Dirac function. Figure 4.8.1 refers to models having a first-order time lag and Figure 4.8.2 to those having the oscillating time lag. Table 4.8.1 sums up the estimated nonlinearity indices calculated from the linear cross- and auto-correlation functions.

The linear correlation method detected only the linear systems as linear. If the mean value of the test signal coincided with the extremum point of the quadratic curve $(u_0 = -1)$, the linear cross-correlation function was zero and the nonlinearity index was equal to one. At other working points the index was between zero and one.   ∎

4.9 SECOND-ORDER (NONLINEAR) CROSS-CORRELATION METHOD

Assumption 4.9.1   Excite the system by a random signal so that the mean value of the test signal differs from zero, all odd order moments of the normalized input signal are zero and all even order moments exist.   ∎

In other words, all even order auto-correlation functions of the normalized input signal should be zero and all odd order auto-correlation functions must not be zero.

Usual test signals like Gaussian white noise or pseudo-random multi-level test signals with maximum length fulfill all requirements of Assumption 4.9.1.

Theorem 4.9.1   (Billings and Fakhouri, 1978; Haber, 1979, 1985, 1989; Korenberg and Hunter, 1990) Given a test signal satisfying Assumption 4.9.1, then the system is linear then and only then if the second-order (nonlinear) cross-correlation function

$$r_{x_1 y'}(\kappa) = 0 \qquad \forall \kappa$$

is zero for every shifting time for linear systems and differs from zero for nonlinear systems. The multiplier is defined as

$$x_1(k) = \frac{u'^2(k) - \mathrm{E}\{u'^2(k)\}}{\sigma\{u'^2(k)\}} \qquad\qquad (4.9.1)$$

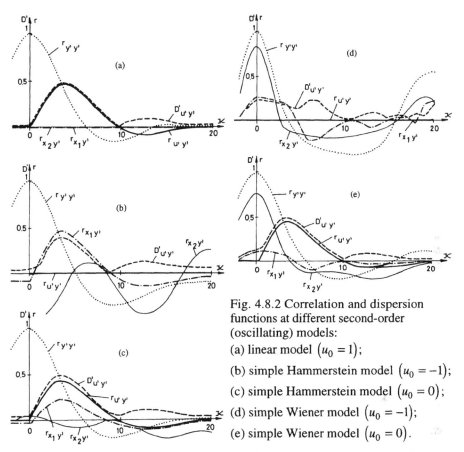

Fig. 4.8.2 Correlation and dispersion functions at different second-order (oscillating) models:

(a) linear model $(u_0 = 1)$;

(b) simple Hammerstein model $(u_0 = -1)$;

(c) simple Hammerstein model $(u_0 = 0)$;

(d) simple Wiener model $(u_0 = -1)$;

(e) simple Wiener model $(u_0 = 0)$.

Proof. (Haber, 1989) The cross-correlation function between the multiplier and the normalized noisy output signal is equal to that of the multiplier and the noise-free output signal $w(k)$ divided by the standard deviation of the noisy output signal, if both the multiplier $x_1(k)$ and additive output noise $e(k)$ have zero mean value then:

$$r_{x_1 y'}(\kappa) = E\{x_1(k - \kappa)y'(k)\} = E\left\{x_1(k - \kappa)\frac{w(k) + e(k) - E\{y(k)\}}{\sigma\{y(k)\}}\right\}$$

$$= \frac{1}{\sigma\{y(k)\}}E\{x_1(k - \kappa)w(k)\}$$

The noise-free system can be described by its Volterra series. The nonlinear cross-correlation function takes the form

$$r_{x_1 y'}(\kappa) = \frac{1}{\sigma\{u'^2(k)\}}\frac{1}{\sigma\{y(k)\}}$$

$$\times \sum_{n=0}^{\infty}\sum_{\kappa_1=0}^{\infty}\ldots\sum_{\kappa_n=0}^{\infty}h_n(\kappa_1, \ldots, \kappa_n)E\left\{[u'^2(k) - E\{u'^2(k)\}]\prod_{i=1}^{n}u(k - \kappa_i)\right\}$$

(4.9.2)

Introduce the following notations:

$$u_0 = E\{u(k)\}$$
$$\sigma_u = \sigma\{u(k)\}$$

and

$$r(\kappa_1 - \kappa_2, \ldots, \kappa_1 - \kappa_n) = E\left\{\prod_{i.1}^{n} u'(k - \kappa_i)\right\} \qquad (4.9.3)$$

Equation (4.9.2) becomes zero if the term

$$
\begin{aligned}
r_n &= E\left\{\left[u'^2(k) - E\{u'^2(k)\}\right]\prod_{i=1}^{n} u(k - \kappa_i)\right\} \\
&= E\left\{\left[u'^2(k) - r(0)\right]\prod_{i=1}^{n}\left[u_0 + \sigma_u u'(k - \kappa_i)\right]\right\}
\end{aligned}
\qquad (4.9.4)
$$

or the sums of the corresponding Volterra kernels are zero for all values of i ($i = 0, \ldots, n$). The sums of the Volterra kernels differ from zero as assumed in Assumption 4.1.1. Therefore it is enough to prove that $r_i = 0$, $i = 0, \ldots, n$. Instead of presenting the complicated formula for the calculation of (4.9.2) at any i, consider (4.9.4) for little n values starting by $i = 0$.

1. *Constant channel*

$$r_0 = E\left\{\left[u'^2(k) - r(0)\right]\right\} = r(0) - r(0) = 0$$

Now it can be seen what was the reason for choosing the multiplier $x_1(k)$ so that its mean value was zero. Otherwise the nonlinear correlation function would not be zero for a constant term.

2. *Linear channel*

$$r_1 = E\left\{\left[u'^2(k) - r(0)\right]\left[u_0 + \sigma_u u'(k - \kappa_1)\right]\right\} = u_0[r(0) - r(0)] + \sigma_u r(0, -\kappa_1) = 0$$

since every even order auto-correlation function of the change of the test signal is equal to zero because of (4.9.3) and Assumption 4.9.1.

3. *Quadratic channel*

$$
\begin{aligned}
r_2 &= E\left\{\left[u'^2(k) - r(0)\right]\prod_{i=1}^{2}\left[u_0 + \sigma_u u'(k - \kappa_i)\right]\right\} \\
&= u_0[r(0) - r(0)] + \sum_{i=1}^{2} u_0 \sigma_u r(0, -\kappa_i) + \sigma_u^2\left[r(0, -\kappa_1, -\kappa_2) - r(0)r(\kappa_1 - \kappa_2)\right] \\
&= \sigma_u^2\left[r(0, -\kappa_1, -\kappa_2) - r(0)r(\kappa_1 - \kappa_2)\right] \neq 0
\end{aligned}
$$

since the odd order auto-correlation functions are not equal to zero for all shifting times. Taking, e.g., a pseudo-random n_r level test signal of maximum length N_p, the components of r_2 are at $\kappa_1 = \kappa_2 = 0$ (Krempl, 1973):

$$r(0,0,0) = \frac{N_p + 1}{N_p} \frac{\left(n_r^2 - 1\right)\left(3n_r^2 - 7\right)}{240}$$

$$r(0) = \frac{N_p + 1}{N_p} \frac{\left(n_r^2 - 1\right)}{12}$$

and thus

$$r_2 \neq 0$$

4. *Cubic channel*

$$r_3 = E\left\{\left[u'^2(k) - r(0)\right]\prod_{i=1}^{3}\left[u_0 + \sigma_u u'(k - \kappa_i)\right]\right\}$$

$$= u_0^3\left[r(0) - r(0)\right] + \sum_{i=1}^{3} u_0^2 \sigma_u r(0, -\kappa_i) + u_0 \sigma_u^2\left[r(0, -\kappa_1, -\kappa_2) - r(0)r(\kappa_1 - \kappa_2)\right]$$

$$+ u_0 \sigma_u^2\left[r(0, -\kappa_2, -\kappa_3) - r(0)r(\kappa_2 - \kappa_3)\right]$$

$$+ u_0 \sigma_u^2\left[r(0, -\kappa_1, -\kappa_3) - r(0)r(\kappa_1 - \kappa_3)\right]$$

$$+ \sigma_u^3\left[r(0, -\kappa_1, -\kappa_2, -\kappa_3) - r(0)r(\kappa_1 - \kappa_2, \kappa_1 - \kappa_3)\right]$$

$$= u_0 \sigma_u^2\left[r(0, -\kappa_1, -\kappa_2) - r(0)r(\kappa_1 - \kappa_2)\right]$$

$$+ u_0 \sigma_u^2\left[r(0, -\kappa_2, -\kappa_3) - r(0)r(\kappa_2 - \kappa_3)\right]$$

$$+ u_0 \sigma_u^2\left[r(0, -\kappa_1, -\kappa_3) - r(0)r(\kappa_1 - \kappa_3)\right]$$

since all even order auto-correlation functions of the normalized input signal are zero for all shifting time values. The remaining terms are identical with those obtained for the quadratic channel multiplied by the mean values of the test signal. Thus the cubic channel can be detected if the mean value of the test signal differs from zero.

5. *Channels of higher degree*
From the foregoing it can be seen that by increasing the degree of the channel by one, the resulting terms will have similar structure as the original one but each term will be multiplied by a given power of the mean value of the input signal and, in addition, a new term, the auto-correlation function of a degree higher by one than that of the investigated nonlinear kernel, appears. Assuming, e.g., a fourth-degree channel the entering term takes the form:

$$\sigma_u^4\left[r(0, -\kappa_1, -\kappa_2, -\kappa_3, -\kappa_4) - r(0)r(\kappa_1 - \kappa_2, \kappa_1 - \kappa_3, \kappa_1 - \kappa_4)\right] \qquad (4.9.5)$$

The expression of (4.9.5) is not equal to zero, since, e.g., at $\kappa_1 = \kappa_2 = \kappa_3 = \kappa_4$ (Krempl, 1973):

$$r(0,0,0,0,0) = \frac{N_p + 1}{N_p} \frac{\left(n_r^2 - 1\right)\left(3n_r^4 - 18n_r^2 + 3n_r\right)}{1344}$$

and thus

$$r_4 \neq 0$$

even when the test signal has a zero mean value $\left(u_0 = 0\right)$.   ∎

We could see that with odd degree nonlinear subsystems it was essential that the mean value of the test signal differed from zero, otherwise the test could not distinguish between linear and nonlinear systems. This fact will be expressed in Theorem 4.9.2 that uses Assumption 4.9.2.

Assumption 4.9.2   Excite the system by a random signal so that the mean value of the test signal and all odd order moments of the normalized input signal are zero and all even order moments exist.   ∎

Theorem 4.9.2   (Billings and Fakhouri, 1978; Billings and Voon, 1983; Korenberg and Hunter, 1990) Given a test signal satisfying Assumption 4.9.2 then the second-order (nonlinear) cross-correlation function $r_{x_1 y'}(\kappa)$ with (4.9.1) is zero if the system satisfying Assumption 4.1.1 is linear or has nonlinear terms of odd degree and $r_{x_1 y'}(\kappa)$ differs from zero if the system satisfying Assumption 4.1.1 has even degree nonlinear terms.   ∎

Proof.   The proof of Theorem 4.9.2 is a consequence of the proof of Theorem 4.9.1. The mean value of the input signal is zero $\left(u_0 = 0\right)$ and the function r_n contains only an auto-correlation function of $(n+1)$-th order which term is equal to zero at any delay time only at odd n values (degrees).   ∎

It can be easily understood that the two cases:
- the detection of an even degree nonlinear system does not need a random or pseudo-random signal with not zero mean value;
- the detection of an odd degree nonlinear system needs a random or pseudo-random signal with not zero mean value,

are related to each other. Namely, the problem of exciting a nonlinear system of an odd degree by a test signal of non-zero mean value can be transformed into a new coordinate system where the test signal takes a zero mean value and the nonlinear system can be described by both odd and even degree channels.

Corollary 4.9.1   As an analytical nonlinearity index

$$\nu = \max_\kappa \left\{ \left| r_{x_1 y'}(\kappa) \right| \right\}$$

can be taken. If the process satisfying Assumption 4.1.1 is linear then the nonlinearity measure is zero even if the measurements are noisy. If Assumption 4.9.1 holds then a non-zero ν detects any nonlinear system and if Assumption 4.9.2 holds then a non-zero ν detects only a nonlinear system that has even degree nonlinear terms.   ∎

If the system is disturbed by noises the cross-correlation function $r_{x_1 y'}(\kappa)$ will not be exactly zero for every shifting time even if the system is linear. Consequently the question arises of how can we check whether the cross-correlation function differs significantly from zero or not. This question is a general problem with the cross-correlation function tests and since in the next section we treat another nonlinearity test that is also based on the zero checking of another cross-correlation function, it seems to be expedient to treat the problem in a general way.

Let $x_1(k)$ and $x_2(k)$ arbitrary stationary stochastic signals. The question is how can we check whether the cross-correlation function $r_{x_1 x_2}(\kappa)$ is zero in a shifting-time interval $[\kappa_{min}, \kappa_{max}]$

$$r_{x_1 x_2}(\kappa) = E\{x_1(k-\kappa)x_2(k)\} \overset{?}{=} 0 \qquad \kappa_{min} \leq \kappa \leq \kappa_{max}$$

Instead of computing the cross-correlation function $r_{x_1 x_2}(\kappa)$ it is expedient

- to normalize, and
- to whiten

both signals $x_1(k)$ and $x_2(k)$. The normalized signals $x_1'(k)$ and $x_2'(k)$ are

$$x_1'(k) = \left[x_1(k) - E\{x_1(k)\}\right]/\sigma\{x_1(k)\}$$
$$x_2'(k) = \left[x_2(k) - E\{x_2(k)\}\right]/\sigma\{x_2(k)\}$$

Owing to the normalization the cross-correlation function at zero shifting time lies between -1 and 1

$$\left| r_{x_1' x_2'}(0) \right| \leq 1$$

Furthermore the auto-correlation functions of the normalized signals are between -1 and 1 for every shifting time

$$\left| r_{x_1' x_1'}(\kappa) \right| \leq 1 \qquad \forall \kappa$$
$$\left| r_{x_2' x_2'}(\kappa) \right| \leq 1 \qquad \forall \kappa$$

To understand the importance of the whitening of the normalized components, assume that $x_1'(k-\kappa)$ and $x_2'(k))$ are independent in the shifting-time domain $[\kappa_{min}, \kappa_{max}]$

$$r_{x_1' x_2'}(\kappa) = E\{x_1'(k-\kappa)x_2'(k)\} = 0 \qquad \kappa_{min} \leq \kappa \leq \kappa_{max}$$

The auto-correlation function of $r_{x_1' x_2'}(\kappa)$ is

$$E_\kappa\left\{r_{x_1' x_2'}(\kappa - \Delta\kappa)r_{x_1' x_2'}(\kappa)\right\}$$
$$= E_\kappa\left\{E_k\{x_1'(k-\kappa+\Delta\kappa)x_2'(k)\}E_k\{x_1'(k-\kappa)x_2'(k)\}\right\}$$

$$= E_\kappa\left\{E_k\left\{x_1'(k-\kappa+\Delta\kappa)x_1'(k-\kappa)\right\}E_k\left\{x_2'(k)x_2'(k)\right\}\right\}$$
$$= E_\kappa\left\{r_{x_1'x_1'}(\Delta\kappa)r_{x_2'x_2'}(0)\right\} = r_{x_1'x_1'}(\Delta\kappa)$$

Here $E_\kappa\{...\}$ denotes the mean value over the shifting time κ and $E_k\{...\}$ denotes the mean value over the discrete time k. We can see that $r_{x_1'x_1'}(\Delta\kappa)$ has the same auto-correlation function as the signal $x_1'(k)$ has if $x_2'(k)$ is a normalized white noise (Box and Jenkins, 1976). Therefore it is practical to whiten both components of the correlation function. The whitening filters can be obtained by identification of a SISO system with zero input signal and the output signals $x_1'(k)$ and $x_2'(k)$, respectively. Denote the resulting filters by $H_{w1}(q^{-1})$ and $H_{w2}(q^{-1})$, respectively. The whitened signals are

$$x_1'^{\mathrm{w}}(k) = H_{w1}^{-1}(q^{-1})x_1'(k)$$

and

$$x_2'^{\mathrm{w}}(k) = H_{w2}^{-1}(q^{-1})x_2'(k)$$

where

$$H_{w1}^{-1}(q^{-1}) = \frac{1}{H_{w1}(q^{-1})}$$

and

$$H_{w2}^{-1}(q^{-1}) = \frac{1}{H_{w2}(q^{-1})}$$

Another way of computing the whitened signals is to filter of the signals by the inverse of the square root of its covariance matrix. Let the vectors x_1' and x_2' contain the signal values $x_1'(k)$ and $x_2'(k)$

$$x_1' = \left[x_1'(1), ..., x_1'(N)\right]^{\mathrm{T}}$$
$$x_2' = \left[x_2'(1), ..., x_2'(N)\right]^{\mathrm{T}}$$

Compute the covariance matrices and factorize them to the product of a matrix and its transpose

$$\mathrm{cov}\{x_1'\} = S_1^{\mathrm{T}}S_1$$
$$\mathrm{cov}\{x_2'\} = S_2^{\mathrm{T}}S_2$$

S_1 and S_2 are also called the square roots of the covariance matrices. The vector of the whitened signals can be obtained by multiplying the signal vectors $x_1'(k)$ and $x_2'(k)$ by the inverse of the square root of their covariance matrix, respectively

$$x_1'^w = \left(S_1^T\right)^{-1} x_1'$$

$$x_2'^w = \left(S_2^T\right)^{-1} x_2'$$

The normalized whitened signals are the elements of $x_1'^w$ and $x_2'^w$

$$x_1'^w = \left[x_1'^w(1), \ldots, x_1'^w(N)\right]^T$$

$$x_2'^w = \left[x_2'^w(1), \ldots, x_2'^w(N)\right]^T$$

(If the signals $x_1(k)$ and $x_2(k)$ have zero mean then the normalization and whitening can be performed in one step, then the covariance matrix of the not normalized signal has to be computed.)

Through the normalization and whitening both components of the correlation function $r_{x_1'^w x_2'^w}(\kappa)$, i.e., $x_1'^w(k)$ and $x_2'^w(k)$ are uncorrelated sequences with zero mean and unit standard deviation.

If at least one of $x_1'^w(k)$ and $x_2'^w(k)$ is a random variable, then the product $x_1'^w(k-\kappa) x_2'^w(k)$ is a random variable. If the number of data pairs is very large then owing to the *Central Limit Theorem* the correlation function at every shifting time has a Gaussian distribution (Bohlin, 1978; Leontaritis and Billings, 1987). If $x_1'^w(k-\kappa)$ and $x_2'^w(k)$ have zero mean, unit standard deviation, and are uncorrelated, then the mean value of the correlation function at every shifting time is zero and its standard deviation is about $1/\sqrt{N}$ (Bendat and Piersol, 1971). Two signals are uncorrelated in the shifting-time domain $[\kappa_{min}, \kappa_{max}]$ at 95% probability level if (Bendat and Piersol, 1971)

$$\left| r_{x_1'^w x_2'^w}(\kappa) \right| \leq \frac{1.95}{\sqrt{N}} \cong \frac{2}{\sqrt{N}}$$

with the assumption that the size of the shifting time domain is much smaller than the number of data $[\kappa_{max} - \kappa_{min} + 1 << N]$.

There exists another test for checking whether the correlation function under investigation is zero in the given shifting time domain.

The sum of normally distributed independent random variables with zero mean and unit standard deviation has a chi-square distribution. Therefore the correlation function can be considered zero in the shifting time domain $\kappa_{min} \leq \kappa \leq \kappa_{max}$ if the statistics

$$\Xi^2 = N \sum_{\kappa=\kappa_{min}}^{\kappa_{max}} r_{x_1'^w x_2'^w}^2(\kappa)$$

is less than the tabulated value $\Xi^2(\kappa_{max} - \kappa_{min} + 1, \alpha)$, where $\kappa_{max} - \kappa_{min} + 1$ is the degree of freedom and $100(1-\alpha)$ percent is the significance level (Bohlin, 1978; Leontaritis and Billings, 1987).

The chi-square test checks a cumulative quantity over the whole shifting time domain in contrast to the test that checks every value of the correlation function whether it falls into twice the standard deviation range of the zero value. Consequently the chi-square test can allow some peaks of the correlation function outside the allowed range, if the other values are small enough. (Billings and Voon, 1986)

As is seen, the nonlinearity test works also under noisy circumstances if the noise disturbing the output has zero mean value and is independent from the input signal.

Example 4.9.1   *Nonlinearity test on various linear, simple Hammerstein and simple Wiener models based on the nonlinear cross-correlation function (Haber, 1985)*

The nonlinearity test was performed for the noise-free simple Hammerstein (Figure 4.2.1a) and simple Wiener (Figure 4.2.1b) models. The cascade models contain different linear dynamic (LD) and nonlinear static (NS) parts. The same static characteristics and linear dynamic terms were simulated as in Example 4.4.1. The static curves and the step responses of the linear dynamic parts were given already in Figure 4.2.1.

The input and output signals were sampled by 2 [s] at the models having a first-order time lag and by 4 [s] at those ones having a second-order one. The test signal was always a PRTS signal of maximum length 80 with amplitude ±2 around the values listed in Table 4.8.1. The minimum switching time was equal to the sampling time. The nonlinearity test method was applied to the following models:
  (a)   linear model (Wiener model with static characteristic);
  (b)   nonlinear simple Hammerstein model, excited in the static extremum;
  (c)   nonlinear simple Hammerstein model, not excited in the static extremum;
  (d)   nonlinear simple Wiener model, excited in the static extremum;
  (e)   nonlinear simple Wiener model, not excited in the static extremum.

The nonlinear cross-correlation functions are plotted in Figures 4.8.1 and 4.8.2 for the listed situations. (The discrete values are drawn by continuous lines as being continuous time functions.) The auto-correlation function of the normalized input signal was always a Dirac function. Figure 4.8.1 refers to models having a first-order time lag and Figure 4.8.2 to those ones having the oscillating time lag. Table 4.8.1 sums up the estimated nonlinearity indices calculated from the second-order (nonlinear) cross-correlation functions.

The nonlinear cross-correlation method detected the linear or the nonlinear feature of the process well. The nonlinearity indices had their greatest values when the mean value of the test signal was in the extremum point of the quadratic surface.   ∎

Example 4.9.2   *Second-order (nonlinear) cross-correlation nonlinearity test for the simple Hammerstein and Wiener models with an odd degree nonlinear steady state characteristics (Haber, 1989)*

To show that an odd degree nonlinear system can be detected nonlinear by the nonlinear cross-correlation test only if the mean value of the test signal differs from zero, two third degree (cubic) nonlinear systems were investigated. Both the simple Hammerstein and the simple Wiener models contained the linear dynamic parts with the transfer function

$$G(s) = \frac{1}{1 + 2 \cdot 0.5 \cdot 10s + 10^2 s^2}$$

and the static cubic characteristics

$$Y = 2 + U + 0.25U^3$$

(As it appears from the above, the quadratic term was deliberately omitted.) The input and output signals were sampled by 4 [s]. The following cases were investigated:

- (a) cubic simple Hammerstein model, the mean value of the test signal was zero;
- (b) cubic simple Hammerstein model, the mean value of the test signal was not zero;
- (c) cubic simple Wiener model, the mean value of the test signal was zero;
- (d) cubic simple Wiener model, the mean value of the test signal was not zero.

The test signal was a pseudo-random quinary (five-level) signal (PRQS) of a maximum length of 124 with amplitude ±2 around the mean value. Figure 4.9.1 shows the nonlinear cross-correlation functions for the listed situations (a)-(d). (The discrete values are drawn by continuous lines as being continuous time functions.) The first-order linear auto-correlation function of the normalized input signal was always a Dirac function. The figures contain the mean values u_0 of the input signal and the computed nonlinearity indices v.

It can be seen that only those cases were detected as nonlinear, where the mean value of the input signal differed from zero.   ∎

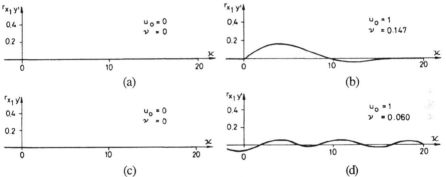

Fig. 4.9.1 Second-order (nonlinear) cross-correlation functions at different second-order systems: (a) cubic simple Hammerstein model $(u_0 = 0)$; (b) cubic simple Hammerstein model $(u_0 \neq 0)$; (c) cubic simple Wiener model $(u_0 = 0)$; (d) cubic simple Wiener model $(u_0 \neq 0)$

4.10 SECOND-ORDER (OUTPUT) AUTO-CORRELATION METHOD

Use again a random or pseudo-random input signal that satisfies Assumption 4.9.1. Then not only can the normalized second-order cross-correlation function detect the linear or nonlinear feature of a process but also the normalized second-order auto-correlation function of the output signal, as stated in Theorem 4.10.1.

Theorem 4.10.1   Excite the system satisfying Assumption 4.1.1 by a test signal satisfying Assumption 4.9.1. The cross-correlation of the normalized square of the normalized output signal

$$x_2(k) = \frac{\left[y'^2(k) - E\{y'^2(k)\}\right]}{\sigma\{y'^2(k)\}} \tag{4.10.1}$$

and of the normalized output signal, called simply the normalized second-order auto-correlation function of the normalized output signal,

$$r_{x_2y'}(\kappa) = r_{(y'^2)'y'}(\kappa) = E\left\{\left(y'^2\right)'(k-\kappa)y'(k)\right\} \tag{4.10.2}$$

is zero for all shifting times if the system is linear and differs from zero if the system is nonlinear.   ■

Before presenting the proof of Theorem 4.10.1 we show that Theorem 4.10.1 is a modification of a theorem of Billings and Voon (1983) presented here as Lemma 4.10.1.

Lemma 4.10.1
(Billings and Voon, 1983; Korenberg and Hunter 1990) Excite the system satisfying Assumption 4.1.1 by a test signal satisfying Assumption 4.9.1. The second-order auto-correlation function of the normalized output signal

$$r_{y'^2y'}(\kappa) = E\left\{y'^2(k-\kappa)y'(k)\right\} \tag{4.10.3}$$

is zero for all shifting times if the system is linear and differs from zero if the system is nonlinear.   ■

Proof.   (Haber 1989) We show the equivalence between Theorem 4.10.1 and Lemma 4.10.1 in respect of detecting the linear or nonlinear feature of a process.

$$r_{(y'^2)'y'}(\kappa) = E\left\{\left(y'^2\right)'(k-\kappa)y'(k)\right\} = \frac{1}{\sigma\{y'^2(k)\}}E\left\{\left[y'^2(k-\kappa) - E\{y'^2(k)\}\right]y'(k)\right\}$$

$$= \frac{1}{\sigma\{y'^2(k)\}}E\left\{y'^2(k-\kappa)y'(k)\right\} = \frac{1}{\sigma\{y'^2(k)\}}r_{y'^2y'}(\kappa)$$

As is seen, both cross-correlation functions (4.10.2) and (4.10.3) are zero or differ from zero simultaneously.   ■

The advantage of using (4.10.2) over (4.10.3) is that the correlation function lies between -1 and 1 at least for the zero shifting time that is not the case if (4.10.3) is used.

Now we deal with the proof of the second-order (output) auto-correlation method.

Proof of Theorem 4.10.1 (Haber 1989) The measured output signal $y(k)$ is assumed to be the sum of the noise-free output signal $w(k)$ and the noise $e(k)$. By using the results of the proof of the equivalency of Theorem 4.10.1 and Lemma 4.10.1 we get

$$r_{(y'^2)'y'}(\kappa) = \frac{1}{\sigma\{y'^2(k)\}}r_{y'^2y'}(\kappa)$$

$$= \frac{1}{\sigma\{y'^2(k)\}} E\left\{\left[\frac{w(k-\kappa)+e(k-\kappa)-E\{y(k)\}}{\sigma\{y(k)\}}\right]^2 \frac{w(k)+e(k)-E\{y(k)\}}{\sigma\{y(k)\}}\right\}$$

$$= \frac{E\left\{\left[w(k-\kappa)+e(k-\kappa)-E\{y(k)\}\right]^2\left[w(k)+e(k)-E\{y(k)\}\right]\right\}}{\sigma\{y'^2(k)\}\sigma^3\{y(k)\}} \quad (4.10.4)$$

The noise-free output signal $w(k)$ can be described by the Volterra operator $V[\ldots]$ consisting of the sum of the homogeneous Volterra operators $V_i[\ldots]$,

$$w(k) = V[u(k)] = \sum_{n=0}^{\infty} V_i[u(k)]$$

$$= \sum_{n=0}^{\infty} \sum_{\kappa_1=0}^{\infty} \cdots \sum_{\kappa_n=0}^{\infty} h_n(\kappa_1, \ldots, \kappa_n)\prod_{i=1}^{n} u(k-\kappa_i)$$

Since the noise term has zero mean value, the mean value of the measured output signal is equal to that of the noise-free output signal

$$E\{y(k)\} = E\{w(k)\}$$

Equation (4.10.4) contains the output signal decreased by its mean value

$$w(k) - E\{w(k)\} = V[u(k)] - E\{V[u(k)]\}$$

$$= \sum_{n=0}^{\infty} \sum_{\kappa_1=0}^{\infty} \cdots \sum_{\kappa_n=0}^{\infty} h_n(\kappa_1, \ldots, \kappa_n)\left[\prod_{i=1}^{n} u_i(k-\kappa_i) - E\left\{\prod_{i=1}^{n} u_i(k-\kappa_i)\right\}\right]$$

$$= V'[u(k)]$$

The symbol $V'[u(k)]$ was introduced for the noise-free output signal decreased by its mean value. The following expression should be checked for zero value

$$v^*(\kappa) = E\left\{\left[V'[u(k-\kappa)]+e(k-\kappa)\right]^2\left[V'[u(k)]+e(k)\right]\right\} \quad (4.10.5)$$

because (4.10.4) is zero if and only if $v^*(\kappa)$ is zero.

We show now that if the noise has a zero mean value then it has no effect on the expression (4.10.5) and therefore it can be neglected in (4.10.5). In order to prove it, express (4.10.5) in detail

$$v^*(\kappa) = E\left\{\left[V'[u(k-\kappa)]+e(k-\kappa)\right]^2\left[V'[u(k-\kappa)]+e(k)\right]\right\}$$

$$= E\left\{\left[\left[V'[u(k-\kappa)]\right]^2 + 2e(k-\kappa)V'[u(k-\kappa)]+e^2(k-\kappa)\right]\left[V'[u(k)]+e(k)\right]\right\}$$

$$= E\Big\{ \big[V'[u(k-\kappa)]\big]^2 V'[u(k)] + 2e(k-\kappa)V'[u(k-\kappa)]V'[u(k)]$$

$$+ e^2(k-\kappa)V'[u(k)] + e(k)\big[V'[u(k-\kappa)]\big]^2$$

$$+ 2e(k-\kappa)e(k)V'[u(k-\kappa)] + e^2(k-\kappa)e(k)\Big\}$$

$$= E\Big\{ \big[V'[u(k-\kappa)]\big]^2 V'[u(k)] \Big\} \tag{4.10.6}$$

because all terms that contain the noise term have zero expected value. In detail it means that the terms that contain the noise in linear form are zero because the noise is independent of the second term in the products:

$$E\big\{ 2e(k-\kappa)V'[u(k-\kappa)]V'[u(k)] \big\} = 0$$

$$E\Big\{ e(k)\big[V'[u(k-\kappa)]\big]^2 \Big\} = 0$$

The terms where the noise occur in quadratic form are also zero because they are multiplied by the noise-free output signal reduced by its mean value $V'[u(k-\kappa)]$ and the noise is independent of the noise-free process output

$$E\big\{ e^2(k-\kappa)V'[u(k)] \big\} = 0$$

$$E\big\{ 2e(k-\kappa)e(k)V'[u(k-\kappa)] \big\} = 0$$

Finally the term which is a cubic function of the noise is also zero because all even order auto-correlation functions of the noise are zero

$$E\big\{ e^2(k-\kappa)e(k) \big\} = r_e(0,\kappa) = 0$$

According to (4.10.5) and (4.10.6) further on we have to prove that the second-order auto-correlation function of the noise-free output signal

$$v^*(\kappa) = E\Big\{ \big[V'[u(k-\kappa)]\big]^2 V'[u(k)] \Big\} \tag{4.10.7}$$

is zero if and only if the process is linear.

We shall call $v^*(\kappa)$ the modified nonlinearity measure because it is zero if and only if the second-order auto-correlation function of the normalized measured output signal (4.10.3) is zero.

Instead of expressing (4.10.7) for any nonlinear degree let us investigate (4.10.7) for small n values starting from $n = 0$.

1. *Constant channel:*

Now

$$V'[u(k)] = V'_0[u(k)] = h_0 - h_0 = 0$$

and the modified nonlinearity measure is zero.

2. *Linear system:*
The noise-free output signal decreased by its mean value is

$$V'[u(k)] = h_0 + V_1[u(k)] - [h_0 + E\{V_1[u(k)]\}]$$

$$= \sum_{\kappa_1=0}^{\infty} h_1(\kappa_1)\Delta u(k-\kappa_1) = V_1[\Delta u(k)]$$

and thus the modified nonlinearity measure is

$$v^*(\kappa) = E\{[V_1[\Delta u(k-\kappa)]]^2 V_1[\Delta u(k)]\} = E\{V_1^3(\Delta u(k-\kappa), \Delta u(k-\kappa), \Delta u(k))\}$$

$$= \sum_{\kappa_1=0}^{\infty} \sum_{\kappa_2=0}^{\infty} \sum_{\kappa_3=0}^{\infty} h_1(\kappa_1)h_1(\kappa_2)h_1(\kappa_3)r_{\Delta u}(\kappa_1-\kappa_2, \kappa_1+\kappa-\kappa_3) = 0$$

(4.10.8)

because every even order auto-correlation function of the stochastic component of the input signal is zero.

$\Delta u(\kappa)$ is the stochastic part of the test signal, i.e., its deviation around its steady state value u_0, see (4.4.1). ∎

In (4.10.8) we applied the symbol of the n-linear Volterra operator (Schetzen, 1980):

$$V_n(u_1(k),...,u_n(k)) = \sum_{\kappa_1=0}^{\infty} \cdots \sum_{\kappa_n=0}^{\infty} h_n(\kappa_1,...,\kappa_n)\prod_{i=1}^{n} u_i(k-\kappa_i)$$

We shall denote the product of the i-linear and j-linear Volterra operators by $V_i V_j$

$$V_i(u_1(k),...,u_i(k))V_j(u_{i+1}(k),...,u_{i+j}(k))$$

$$= (V_iV_j)(u_1(k),...,u_i(k), u_{i+1}(k),...,u_{i+j}(k))$$

$$= \sum_{\kappa_1=0}^{\infty} \cdots \sum_{\kappa_{i+j}=0}^{\infty} h_{i+j}(\kappa_1,...,\kappa_{i+j})\prod_{\ell=1}^{i+j} u_\ell(k-\kappa_\ell)$$

If the operators in the products are the same, we shall denote it by a power, e.g.,

$$V_i(u_1(k),...,u_i(k))V_i(u_{i+1}(k),...,u_{i+i}(k))$$

$$= V_i^2(u_1(k),...,u_i(k), u_{i+1}(k),...,u_{i+i}(k))$$

(The symbol $V_1^n(...)$ denotes that the nth order kernel is the product of n equal first-order kernels.)

3. *Quadratic system:*

The noise-free output signal decreased by its mean value is

$$V'[u(k)] = V_1[\Delta u(k)] + V_2[u(k)] - E\{V_2[u(k)]\}$$

$$= V_1[\Delta u(k)] + V_2[\Delta u(k)] + 2V_2[\Delta u(k), u_0] - E\{V_2[\Delta u(k)]\}$$

and thus the modified nonlinearity measure is

$$v^*(k) = E\left\{\left[V_1[\Delta u(k-\kappa)] + V_2[u(k-\kappa)] + 2V_2(\Delta u(k-\kappa), u_0) - \left[E\{V_2[\Delta u(k)]\}\right]^2\right.\right.$$

$$\left.\left.\times\left[V_1[\Delta u(k)] + V_2[\Delta u(k)]\right] + 2V_2(\Delta u(k), u_0) - E\{V_2[\Delta u(k)]\}\right]\right\}$$

Instead of expressing all terms occurring one can see that even the highest power term in the input signal

$$E\left\{\left[V_2[\Delta u(k-\kappa)]\right]^2 V_2[\Delta u(k)]\right\}$$

$$= E\left\{\left(V_2^2 V_2\right)\left(\Delta u(k-\kappa), \Delta u(k-\kappa), \Delta u(k-\kappa), \Delta u(k), \Delta u(k)\right)\right\}$$

$$= \left(V_2^2 V_2\right) r_{\Delta u}(0, 0, 0, \kappa, \kappa) \neq 0$$

differs from zero because the odd order auto-correlation functions of the stochastic component of the input signal are generally not zero. Here we assumed that the sum of the Volterra kernels is finite and differs from zero as stated in Assumption 4.1.1.

4. *Cubic system:*
A cubic system generally also contains quadratic forms and the test would detect the process as nonlinear. Here we shall investigate a cubic system that does not contain a quadratic part

$$V_0(k) = V[u(k)] = h_0 + V_1[u(k)] + V_3[u(k)]$$

The noise-free output signal decreased by its mean value is

$$V'[u(k)] = V_1[\Delta u(k)] + V_3[u(k)] - E\{V_3[u(k)]\}$$

$$= V_1[\Delta u(k)] + V_3[\Delta u(k)] + 3V_3(\Delta u(k), \Delta u(k), u_0)$$

$$+ 3V_3(\Delta u(k), u_0, u_0) + V_3(u_0, u_0, u_0) - E\{V_3(\Delta u(k), \Delta u(k), \Delta u(k))\}$$

$$- 3E\{V_3(\Delta u(k), \Delta u(k), u_0)\} - 3E\{V_3(\Delta u(k), u_0, u_0)\} - E\{V_3(u_0, u_0, u_0)\}$$

$$= V_1[\Delta u(k)] + V_3[\Delta u(k)] + 3V_3(\Delta u(k), \Delta u(k), u_0) + 3V_3(\Delta u(k), u_0, u_0)$$

$$- E\{V_3[u(k)]\} - 3E\{V_3(\Delta u(k), \Delta u(k), u_0)\} - 3E\{V_3(\Delta u(k), u_0, u_0)\} \qquad (4.10.9)$$

In (4.10.9)

$$E\{V_3(\Delta u(k), u_0, u_0)\} = 0$$

because the expected value of a linear function of a signal with zero mean value is zero and

$$E\{V_3[\Delta u(k)]\} = E\left\{\sum_{\kappa_1=0}^{\infty}\sum_{\kappa_2=0}^{\infty}\sum_{\kappa_3=0}^{\infty} \Delta u(k-\kappa_1)\Delta u(k-\kappa_2)\Delta u(k-\kappa_3)\right\}$$

$$= \sum_{\kappa_1=0}^{\infty} \sum_{\kappa_2=0}^{\infty} \sum_{\kappa_3=0}^{\infty} E\{\Delta u(k-\kappa_1)\Delta u(k-\kappa_2)\Delta u(k-\kappa_3)\} = 0$$

because the expected value of the product of three arbitrary delayed stochastic signals with zero mean value is zero.

The dimension of a Volterra kernel in an n-linear operator is decreased by as much as the constant value u_0 in the argument is. Denote the reduced order operators by star superscripts $(*)$,

$$V_3(\Delta u(k), \Delta u(k), u_0) = u_0 V_2^*(\Delta u(k), \Delta u(k)) = u_0 V_2^*[u(k)]$$
$$V_3(\Delta u(k), u_0, u_0) = u_0^2 V_1^*(\Delta u(k)) = u_0^2 V_1^*[\Delta u(k)]$$

Thus (4.10.9) can be rewritten as

$$V'[u(k)] = V_1[\Delta u(k)] + V_3[\Delta u(k)] + 3u_0 V_2^*[\Delta u(k)]$$
$$+ 3u_0^2 V_1^*[\Delta u(k)] - 3u_0 E\{V_2^*[\Delta u(k)]\}$$
$$= V_1^{**}[\Delta u(k)] + 3u_0 V_2^*[\Delta u(k)] + V_3[\Delta u(k)] - 3u_0 E\{V_2^*[\Delta u(k)]\}$$

where

$$V_1^{**}[\Delta u(k)] = V_1[\Delta u(k)] + 3u_0^2 V_1^*[\Delta u(k)] \qquad (4.10.10)$$

is a linear Volterra operator of the stochastic component of the input signal (and it depends on the mean value of the input signal).

The modified nonlinearity measure is

$$v^*(\kappa) = E\{[V_1^{**}[\Delta u(k-\kappa)] + 3u_0 V_2^*[\Delta u(k-\kappa)] + V_3[\Delta u(k-\kappa)]$$
$$- 3u_0 [E\{V_2^*[\Delta u(k)]\}]^2 [V_1^{**}[\Delta u(k)]] + 3u_0 V_2^*[\Delta u(k)] + V_3[\Delta u(k)]$$
$$- 3u_0 E\{V_2^*[\Delta u(k)]\}]\} \qquad (4.10.11)$$

Since the mean value of the input signal differs from zero both odd and even degree Volterra operator functions of the stochastic component of the input signal occur in (4.10.11). The modified nonlinearity measure will not be zero and therefore the process will be detected nonlinear. Here we assumed that the sum of the Volterra kernels is finite and differs from zero as stated in Assumption 4.1.1.

5. *Channels of higher degree:*
From the foregoing it can be seen that the modified nonlinearity measure generally differs from zero except the case when the mean value of the input signal is zero and the process has only odd degree subsystems or the nonlinear subsystems of different degree are of differentiating type. To avoid a nonlinear system being detected linear the system has to satisfy Assumption 4.1.1.   ∎

Theorem 4.10.2   Given a test signal satisfying Assumption 4.9.2, then the cross-correlation function $r_{x_2 y'}(\kappa)$ of the multiplier $x_2(k)$ (4.10.1) and of the normalized output signal is zero if the system is linear or has nonlinear terms of odd degree and $r_{x_2 y'}(\kappa)$ differs from zero if the system satisfying Assumption 4.1.1 has even degree nonlinear terms.   ∎

Theorem 4.10.2 is a modification of a theorem of Billings and Voon (1983) presented here as Lemma 4.10.2.

Lemma 4.10.2
(Billings and Voon, 1983; Korenberg and Hunter, 1990) Given a test signal satisfying Assumption 4.9.2 then the second-order (output) auto-correlation function $r_{y'^2 y'}(\kappa)$ is zero if the system is linear or has nonlinear terms of odd degree and $r_{y'^2 y'}(\kappa)$ differs from zero if the system satisfying Assumption 4.1.1 has even degree nonlinear terms. ∎

Proof of the equivalency of Theorem 4.10.2 and Lemma 4.10.2. The equivalence between Theorem 4.10.2 and Lemma 4.10.2 in respect of detecting the linear or nonlinear feature of a process can be seen similar to the mutual validity of Theorem 4.10.1 and Lemma 4.10.1. (Both cross-correlation functions (4.10.2) and (4.10.3) are zero or differ from zero simultaneously.)   ∎

Now we deal with the proof of Theorem 4.10.2 (and simultaneously with the proof of Lemma 4.10.2).

Proof of Theorem 4.10.2. It has already been shown that the test detects the even degree nonlinear terms. We deal now with the odd degree components. First consider the lowest odd degree nonlinear subsystem, the cubic one.

1. *Cubic system:*

If the mean value of the test signal is zero

$$u_0 = 0$$

then (4.10.11) can be further reduced to

$$v^*(\kappa) = E\left\{\left[V_1[\Delta u(k-\kappa)] + V_3[\Delta u(k-\kappa)]\right]^2 \left[V_1[\Delta u(k)] + V_3[\Delta u(k)]\right]\right\} \quad (4.10.12)$$

because from (4.10.10)

$$V_1^{**}[\Delta u(k)] = V_1[\Delta u(k)] \qquad \text{if} \quad u_0 = 0$$

Having performed the multiplications it can be seen that the expected values of only the odd degree operators have to be calculated and they are all zero because of the corresponding feature of the stochastic component of the input signal.

2. *Channels of higher odd degree:*

If the degree is not restricted to three then – as a generalization of (4.10.12) – we obtain

$$v^*(\kappa) = E\left\{\left[\sum_{\ell=0}^{\infty} V_{2\ell+1}[\Delta u(k-\kappa)]\right]^2 \left[\sum_{\ell=0}^{\infty} V_{2\ell+1}[\Delta u(k)]\right]\right\}$$

All the terms in the first bracket will be even degree functions of the stochastic component of the input signal after having performed the raising to the second power. These terms are multiplied by odd degree functions of the stochastic component of the input signal which means that the modified nonlinearity measure contains expected values of odd degree Volterra operator functions of a stochastic signal with zero mean value and is therefore equal to zero. Consequently the nonlinear system with odd degree subsystems cannot be detected nonlinear by the test under investigation. ∎

Corollary 4.10.1   As an analytical nonlinearity index

$$v = \max_{\kappa}\left\{\left|r_{x_2 y'}(\kappa)\right|\right\}$$

can be taken. If the process satisfying Assumption 4.1.1 is linear then the nonlinearity measure is zero even the measurements are noisy. If Assumption 4.9.1 holds then a non-zero v detects any nonlinear system and if Assumption 4.9.2 holds then a non-zero v detects only a nonlinear system having even degree nonlinear terms. ∎

As is seen, the second-order (output) auto-correlation test also works under noisy circumstances if the noise disturbing the output has zero mean value and it is independent from the input signal.

The cross-correlation function $r_{x_2 y}(\kappa)$ can be checked for being zero or differing significantly from zero for a given shifting time domain by the methods treated at the end of the last section.

Example 4.10.1 *Nonlinearity test on various linear, simple Hammerstein and simple Wiener models based on the second-order (output) auto-correlation function (Haber, 1985)*

The nonlinearity test was performed for the noise-free simple Hammerstein (Figure 4.2.1a) and simple Wiener (Figure 4.2.1b) models. The cascade models contain different linear dynamic (LD) and nonlinear static (NS) parts. The same static characteristics and linear dynamic terms were simulated as in Example 4.2.1. The static curves and the step responses of the linear dynamic parts were given already in Figure 4.2.1.

The input and output signals were sampled by 2 [s] at the models having a first-order time lag and by 4 [s] at those ones having a second-order one. The test signal was always a PRTS signal of maximum length 80 with amplitude ±2 around the values listed in Table 4.8.1. The minimum switching time was equal to the sampling time.

The nonlinearity test method was applied at the following models:
(a)   linear model (Wiener model with linear static characteristic);
(b)   nonlinear simple Hammerstein model, excited in the static extremum;
(c)   nonlinear simple Hammerstein model, not excited in the static extremum;
(d)   nonlinear simple Wiener model, excited in the static extremum;
(e)   nonlinear simple Wiener model, not excited in the static extremum.

The second-order auto-correlation functions of the normalized output signal are plotted in Figures 4.8.1 and 4.8.2 for the listed situations. (The discrete values are drawn by a continuous line as being continuous time functions.) The auto-correlation function of the normalized input signal was always a Dirac function. Figure 4.8.1 refers to the models having a first-order time lag and Figure 4.8.2 to those ones having the oscillating time lag. Table 4.8.1 sums up the estimated nonlinearity indices calculated from the second-order (output) auto-correlation functions.

The second-order (output) auto-correlation method detected the linear or the nonlinear feature of the process well. The nonlinearity indices had their greatest values when the mean value of the test signal was in the extremum point of the quadratic surface.   ∎

Example 4.10.2 *Second-order (output) auto-correlation nonlinearity test for simple Hammerstein and Wiener models with an odd degree nonlinear steady state characteristics (Haber, 1989)*

To show that an odd degree nonlinear system can be detected nonlinear by the second-order auto-correlation tests only if the mean value of the test signal differs from zero, two third-degree (cubic) nonlinear systems were investigated. Both the simple Hammerstein and the simple Wiener models contained the linear dynamic parts with the transfer function

$$G(s) = \frac{1}{1 + 2 \cdot 0.5 \cdot 10s + 10^2 s^2}$$

and the static cubic characteristics

$$Y = 2 + U + 0.25U^3$$

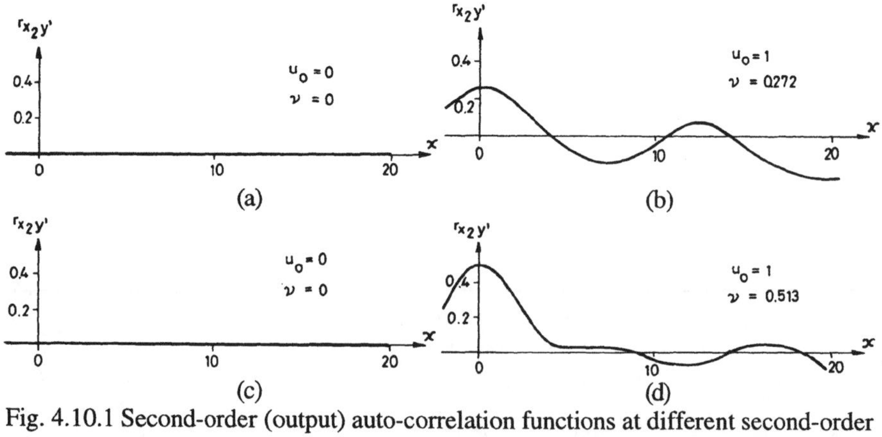

Fig. 4.10.1 Second-order (output) auto-correlation functions at different second-order systems: (a) cubic simple Hammerstein model $(u_0 = 0)$; (b) cubic simple Hammerstein model $(u_0 \neq 0)$; (c) cubic simple Wiener model $(u_0 = 0)$; (d) cubic simple Wiener model $(u_0 \neq 0)$

(As it appears from the above, the quadratic term was deliberately omitted.) The input and output signals were sampled by 4 [s]. The following cases were investigated:

(a)   cubic simple Hammerstein model: the mean value of the test signal was zero;
(b)   cubic simple Hammerstein model: the mean value of the test signal was not zero;
(c)   cubic simple Wiener model: the mean value of the test signal was zero;
(d)   cubic simple Wiener model: the mean value of the test signal was not zero.

The test signal was a pseudo-random quinary (five-level) signal (PRQS) of a maximum length of 124 with amplitude ± 2 around the mean value. Figure 4.10.1 shows the higher order auto-correlation functions for the listed situation (a)-(d). (The discrete values are drawn by continuous lines as being continuous time functions.) The first-order linear auto-correlation function of the normalized input signal was always a Dirac function. The figures contain the mean values u_0 of the input signal and the computed nonlinearity indices v.

It can be seen that only those cases were detected as nonlinear, where the mean value of the input signal differed from zero.   ■

Remarks:
1. Moustafa and Emara-Shabaik (1992) showed that a nonlinear system excited by a white noise of Gaussian distribution is of even degree if and only if the second order auto-correlation function of the output signal

$$r_{y^2 y}(\kappa) = E\left\{y^2(k), y(k+\kappa)\right\}$$ (4.10.13)

   differs for all shifting times from zero. The second order auto-correlation function of the output signal (4.10.13) is very similar to (4.10.2), which is also a second-order auto-correlation function but defined on the normalized output signal.
2. The second-order auto-correlation function is a special kind of the third-order cumulant of the output signal. Some nonlinearity tests were defined based on the cumulants.
3. Emara-Shabaik and Moustafa (1994) showed that a nonlinear system excited by a white noise of Gaussian distribution is of even degree if the third-order cumulant of the output signal

$$r_y(\kappa_1, \kappa_2) = E\left\{y(k), y(k+\kappa_1)y(k+\kappa_2)\right\}$$ (4.10.14)

   differs for all shifting time combinations from zero. As is seen, (4.10.13) is a special case of (4.10.14) with $\kappa_1 = 0$ and $\kappa_2 = \kappa$.
4. Emara-Shabaik and Moustafa (1994) showed that a system excited by a white noise of Gaussian distribution is nonlinear if and only if the fourth-order cumulant of the output signal

$$r_y(\kappa_1, \kappa_2, \kappa_3) = E\left\{y(k), y(k+\kappa_1)y(k+\kappa_2)y(k+\kappa_3)\right\}$$ (4.10.15)

   differs for all shifting time combinations from zero.
5. As the calculation of the third-order cumulants is easier than that of the fourth-order cumulant, Emara-Shabaik and Moustafa (1994) recommended the following procedure for testing the nonlinearity of a process:
   - First check for an even degree nonlinearity with the third-order cumulants of (4.10.14).
   - If all third-order cumulants are zero, then the system is either linear or has an odd degree nonlinearity.
   - The fourth-order cumulant (4.10.15) shows whether the system is linear or not.

4.11 SECOND-ORDER (OUTPUT) SPECTRAL DENSITY METHOD

Subba Rao and Gabr (1979) suggested the two-dimensional function

$$v_q(\omega_1, \omega_2) = \frac{S_{yy}(j\omega_1, j\omega_2)^2}{S_{yy}(j\omega_1)S_{yy}(j\omega_2)S_{yy}(j(\omega_1 + \omega_2))} \tag{4.11.1}$$

for the nonlinearity test if the system was excited by a white noise with zero mean value. In (4.11.1) $S_{yy}(j\omega)$ is the Fourier transform of the auto-correlation function of the output signal, i.e., the auto-spectral density function and $S_{yy}(\omega_1, \omega_2)$ is the two-dimensional Fourier transform of the second-order auto-correlation function of the output signal

$$S_{yy}(j\omega_1, j\omega_2) = \int_{\tau_1=-\infty}^{\infty} \int_{\tau_2=-\infty}^{\infty} r_{yy}(\tau_1, \tau_2) \, e^{-j\omega_1\tau_1} e^{-j\omega_2\tau_2} d\tau_1 d\tau_2 \tag{4.11.2}$$

where $r_{yy}(\tau_1, \tau_2)$ is the second-order auto-correlation function of the output signal

$$r_{yy}(\tau_1, \tau_2) = E\{y(t)y(t + \tau_1)y(t + \tau_2)\}$$

Assumption 4.11.1   The process is excited by white noise with zero mean value. The third-order cumulant of the test signal is zero

$$\mu_3 = E\{u^3(t)\} = 0$$ ∎

Theorem 4.11.1   (Subba Rao and Gabr, 1979) Given an input signal satisfying Assumption 4.11.1, then

$$v_q(\omega_1, \omega_2) = \text{const} \qquad \forall\omega_1 \quad \text{and} \quad \forall\omega_2$$

if the process is linear or has odd degree nonlinear components. ∎

Proof.   (Haber, 1989) If the input signal is white noise with zero mean value then the auto-spectral density function of the output signal is

$$S_{yy}(j\omega) = \sigma_u^2 |G(j\omega)|^2$$

The spectral density functions in the denominator of (4.11.1) also have the following forms:

$$S_{yy}(j\omega_1) = \sigma_u^2 |G(j\omega_1)|^2 \tag{4.11.3}$$

$$S_{yy}(j\omega_2) = \sigma_u^2 |G(j\omega_2)|^2 \tag{4.11.4}$$

$$S_{yy}(j(\omega_1 + \omega_2)) = \sigma_u^2 |G(j(\omega_1 + \omega_2))|^2 \tag{4.11.5}$$

The output signal of a linear process is defined by the convolution

$$y(t) = \int_{t_1=0}^{\infty} g(t_1)u(t-t_1)dt_1$$

The second-order auto-correlation function is

$$r_{yy}(\tau_1, \tau_2) = E\{y(t)y(t+\tau_1)y(t+\tau_2)\}$$

$$= E\left\{ \int_{t_1=0}^{\infty} g(t_1)u(t-t_1)dt_1 \int_{t_2=0}^{\infty} g(t_2)u(t+\tau_1-t_2)dt_2 \int_{t_3=0}^{\infty} g(t_3)u(t+\tau_2-t_3)dt_3 \right\}$$

$$= \int_{t_1=0}^{\infty} g(t_1) \int_{t_2=0}^{\infty} g(t_2) \int_{t_3=0}^{\infty} g(t_3)E\{u(t-t_1)u(t+\tau_1-t_2)u(t+\tau_2-t_3)\}dt_3dt_2dt_1$$

$$= \mu_3 \int_{t_1=0}^{\infty} g(t_1)g(t_1+\tau_1)g(t_1+\tau_2)dt_1$$

because the second-order auto-correlation function of the white noise input signal is

$$E\{u(t-t_1)u(t+\tau_1-t_2)u(t+\tau_2-t_3)\} = \begin{cases} \mu_3 & \text{if } t_2 = t_1+\tau_1 \text{ and } t_3 = t_1+\tau_2 \\ 0 & \text{otherwise} \end{cases}$$

The second-order auto-spectral density function can be calculated by (4.11.2)

$$S_{yy}(j\omega_1, j\omega_2) = \mu_3 \int_{\tau_1=0}^{\infty} \int_{\tau_2=0}^{\infty} \int_{t_1=0}^{\infty} g(t_1)g(t_1+\tau_1)g(t_1+\tau_2)dt_1\, e^{-j\omega_1\tau_1} e^{-j\omega_2\tau_2} d\tau_1 d\tau_2$$

$$= \mu_3 \int_{t_1=0}^{\infty} g(t_1) \int_{\tau_1=0}^{\infty} \int_{\tau_2=0}^{\infty} g(t_1+\tau_1)g(t_1+\tau_2) \qquad (4.11.6)$$

$$\times e^{-j\omega_1(\tau_1+t_1)} e^{-j\omega_2(\tau_2+t_1)} d\tau_1 d\tau_2 e^{j\omega_1 t_1} e^{j\omega_2 t_1} dt_1$$

Introduce the new variables

$$t_3 = t_1 + \tau_1$$
$$t_4 = t_1 + \tau_2$$

in (4.11.6) and notice that

$$g(t_3) = 0 \qquad t_3 < 0$$
$$g(t_4) = 0 \qquad t_4 < 0$$

By means of them the second-order auto-spectral density function becomes

$$S_{yy}(j\omega_1, j\omega_2) = \mu_3 \int\limits_{t_1=0}^{\infty} g(t_1) \left[\int\limits_{t_3=0}^{\infty} g(t_3) e^{-j\omega_1 t_3} dt_3\right] \left[\int\limits_{t_4=0}^{\infty} g(t_4) e^{-j\omega_2 t_4} dt_4\right] e^{j\omega_1 t_1} e^{j\omega_2 t_2} dt_1$$

$$= \mu_3 \int\limits_{t_1=0}^{\infty} g(t_1) G(j\omega_1) G(j\omega_2) e^{j\omega_1 t_1} e^{j\omega_2 t_2} dt_1 = \mu_3 G(j\omega_1) G(j\omega_2) G(-j(\omega_1 + \omega_2))$$

$$(4.11.7)$$

Now substitute (4.11.3), (4.11.4), (4.11.5) and (4.11.7) into (4.11.1)

$$v_q(\omega_1, \omega_2) = \frac{\mu_3^2 |G(j\omega_1)|^2 |G(j\omega_2)|^2 |G(-j(\omega_1+\omega_2))|^2}{\sigma_u^6 |G(j\omega_1)|^2 |G(j\omega_2)|^2 |G(j(\omega_1+\omega_2))|^2} = \frac{\mu_3^2}{\sigma_u^6} = const$$

It can be seen that $v_q(\omega_1, \omega_2)$ is zero if the process is linear and the third-order cumulant is zero. The third-order cumulant, and as well as the skewness if the signal has zero mean value, is zero if the amplitude distribution of the test signal is symmetric. ■

The computation of $v_q(\omega_1, \omega_2)$ is very time consuming. It needs many more calculations than the coherence function since $v_q(\omega_1, \omega_2)$ is a two-dimensional and the coherence function is only a one-dimensional function of the frequency.

Corollary 4.11.1   A nonlinearity index can be defined by checking the zero value and randomness of the deviations

$$\Delta v_d(\omega_1^*, \omega_2^*) = v_d(\omega_1^*, \omega_2^*)$$

$$- \frac{1}{(\omega_{1max}^* - \omega_{1min}^* + 1)(\omega_{2max}^* - \omega_{2min}^* + 1)} \sum_{\omega_{1min}^*}^{\omega_{1max}^*} \sum_{\omega_{2min}^*}^{\omega_{2max}^*} v_d(\omega_1^*, \omega_2^*)$$

in the frequency domain $\omega_{1min}^* \le \omega_1^* \le \omega_{1max}^*$; $\omega_{2min}^* \le \omega_2^* \le \omega_{2max}^*$, where ω^* denotes discrete frequency points. Given an input signal satisfying Assumption 4.11.1, then if the deviations $\Delta v_q(\omega_1^*, \omega_2^*)$ are random values then the process is linear or has odd degree nonlinear components.   ■

A further disadvantage of the method is that the spectral density function of the output signal can be influenced by the noise even if the noise is independent of the input signal.

4.12 DISPERSION METHOD

Assumption 4.12.1   Excite the system by a random or pseudo-random signal so that all odd order moments of the normalized input signal are zero and the even order moments exist.   ■

The functional relation between two variables cannot always be detected by the linear cross-correlation function. See, for example, a static quadratic function

$$y(k) = u^2(k)$$

and let the input signal be a zero mean white noise Gaussian signal. Then the cross-correlation function is equal to zero for every time shift

$$r_{uy}(\kappa) = E\{u(k-\kappa)u^2(k)\} = 0 \qquad \forall \kappa \qquad \text{if} \quad E\{u(k)\} = 0$$

Another function, the cross-dispersion function can, however, be defined which is not zero in this case

$$D_{uy}(\kappa) = \text{var}\{E_u\{E_y\{y(k)|u(k-\kappa)\}\}\} = E_u\{[E_y\{y(k)|u(k-\kappa)\} - E_y\{y(k)\}]^2\} \quad (4.12.1)$$

The indices in $E_u\{...\}$ and $E_y\{...\}$ denote the signal according to which the expected value has to be computed.

The cross-dispersion function is the variance of the conditional mean value of the output signal assuming the given input signal is valid. Since the variance is always positive, the cross-dispersion function is always positive contrary to the cross-correlation function. Take now two special cases:

1. *No relation between input and output signals*

    In this case the mean value of the output signal is equal to that of the conditional value, since the input signal has no effect on the output signal and the cross-dispersion function is equal to zero.

2. *The input–output relation is a static deterministic function with a possible time delay*

    If

    $$y(k) = f(u(k-\kappa))$$

    where $f(...)$ denotes a static function and κ is the time delay, then the cross-dispersion function is equal to the variance of the output signal

    $$D_{uy}(\kappa) = \text{var}\{f(u(k-\kappa))\}$$

To set comparable measures it is expedient to normalize the cross-dispersion function to the variance of the output signal. The normalized cross-dispersion function $D'_{uy}(\kappa)$ is defined as the (positive) square root of the function

$$D'_{uy}(\kappa) = \sqrt{\frac{D_{uy}(\kappa)}{\text{var}\{y(k)\}}} \qquad\qquad (4.12.2)$$

The easiest way of computing the normalized cross-dispersional function is when the normalized input and output sequences are considered

$$D'_{uy}(\kappa) = \sqrt{\frac{D_{uy}(\kappa)}{\text{var}\{y(k)\}}} = \sqrt{\frac{E\{[E\{y(k)|u(k-\kappa)\} - \bar{y}]^2\}}{\sigma_y^2}}$$

$$= \sqrt{\frac{\mathrm{E}\left\{\left[\mathrm{E}\left\{\left(\sigma_y y'(k) + \bar{y}\right)\left(\sigma_u u'(k-\kappa) + \bar{u}\right)\right\} - \bar{y}\right]^2\right\}}{\sigma_y^2}}$$

$$= \sqrt{\mathrm{E}\left\{\left[\mathrm{E}\left\{y'(k)|u'(k-\kappa)\right\}\right]^2\right\}} = D'_{u'y'}(\kappa) \tag{4.12.3}$$

It follows from the above that for any process

$$0 \le D'_{u'y'}(\kappa) \le 1$$

is valid. The normalized cross-dispersion function is zero if no input–output relation exists

$$D'_{u'y'}(\kappa) = 0 \qquad \forall \kappa$$

and it is equal to one at the shifting time equal to the delay time d of a static deterministic relation

$$D'_{u'y'}(\kappa) = \begin{cases} 1 & \kappa = d \\ 0 & \kappa \ne d \end{cases}$$

There is an obvious similarity between the normalized cross-correlation and cross-dispersion functions. The absolute value of the normalized cross-correlation function is between zero and one

$$0 \le \left| r_{u'y'}(\kappa) \right| \le 1 \qquad \forall \kappa$$

and it is equal to one at

$$\left| r_{u'y'}(\kappa) \right| = \begin{cases} 1 & \kappa = d \\ 0 & \kappa \ne d \end{cases}$$

if a static linear relation exists between the input and the delayed output signals. On the basis of the above a theorem can be formulated.

Theorem 4.12.1   If the test signal satisfies Assumption 4.12.1 then a static noisy system with a possible delay time

$$y(k) = f\big(u(k-\kappa)\big) + e(k) \tag{4.12.4}$$

is linear if and only if the normalized cross-dispersion function is equal to the absolute value of the normalized cross-correlation function

$$D'_{u'y'}(\kappa) = \left| r_{u'y'}(\kappa) \right| \qquad \forall \kappa$$

The process is nonlinear if the two functions differ from each other. ($f(...)$ is a static function.)   ∎

Proof. Both the cross-correlation and the cross-dispersion functions consist of expected values. The noises disturbing the output signal are filtered through the averaging process. Further on the theorem relies on the previously mentioned facts:
- the cross-dispersion function is unity for any deterministic functional relation and the cross-correlation function is zero only when the relation is linear; and
- both functions are zero if the output signal does not depend on the input signal.

∎

Rajbman (e.g., Rajbman and Chadeev, (1980)) introduced the absolute degree of nonlinearity as

$$v_1(\kappa) = \sqrt{D_{u'y'}^{'2}(\kappa) - r_{u'y'}^2(\kappa)}$$ (4.12.5)

and the relative degree of nonlinearity as

$$v_2(\kappa) = \sqrt{\frac{D_{u'y'}^{'2}(\kappa) - r_{u'y'}^2(\kappa)}{D_{u'y'}^{'2}(\kappa)}} = \sqrt{1 - \frac{r_{u'y'}^2(\kappa)}{D_{u'y'}^{'2}(\kappa)}}$$ (4.12.6)

Two similar definitions can be introduced, such as, e.g.,

$$v_3(\kappa) = D_{u'y'}'(\kappa) - \left|r_{u'y'}(\kappa)\right|$$ (4.12.7)

and

$$v_4(\kappa) = 1 - \frac{\left|r_{u'y'}(\kappa)\right|}{D_{u'y'}'(\kappa)}$$ (4.12.8)

It can be shown that the square of the absolute degree of nonlinearity means physically the normalized variance of the output signal around the best fitted linear model. This is stated in the following theorem.

Theorem 4.12.2   If the test signal satisfies Assumption 4.12.1 then a nonlinearity index can be defined as

$$v(\kappa) = D_{u'y'}^{'2}(\kappa) - r_{u'y'}^2(\kappa) \qquad \forall \kappa$$ (4.12.9)

The measure (4.12.9) is equal to

$$v(\kappa) = \min_{\hat{c}_1} E\left\{\left[E\{y'(k)|u'(k-\kappa)\} - \hat{c}_1 u'(k-\kappa)\right]^2\right\}$$ (4.12.10)

which is equal to

$$v(\kappa) = E\left\{\left[E\{y'(k)|u'(k-\kappa)\} - r_{u'y'}(\kappa)u'(k-\kappa)\right]^2\right\}$$ (4.12.11)

The index is zero for all shifting times for linear systems only.   ∎

Proof.   The nonlinearity indices of (4.12.10) and (4.12.11) are special cases of (4.3.11) and (4.3.10) when only the functional relation

$$\hat{w}'(k) = y'(k-\kappa)|u'(k-\kappa) = f'\big(u'(k-\kappa)\big)$$

is investigated between the normalized input and output signals. ($f'(...)$ is a static function.) It was shown in the proof of Theorem 4.3.3 that (4.12.10) and (4.12.11) are identical to each other. It remains to show that (4.12.9) and (4.12.11) are identical:

$$v(\kappa) = E\Big\{\big[E\{y'(k)|u'(k-\kappa)\} - r_{u'y'}(\kappa)u'(k-\kappa)\big]^2\Big\}$$

$$= E\Big\{\big[E\{y'(k)|u'(k-\kappa)\}\big]^2\Big\} - 2r_{u'y'}(\kappa)E\big\{u'(k-\kappa)E\{y'(k)|u'(k-\kappa)\}\big\}$$ ∎

$$+ r_{u'y'}^2(\kappa)E\big\{u'^2(k-\kappa)\big\} = D_{u'y'}^{\prime 2}(\kappa) - r_{u'y'}^2(\kappa)$$

Example   4.12.1 *Calculation of the nonlinearity measure of a static quadratic process excited by a white noise signal with not zero mean value. (Continuation of Example 4.3.1) (Haber, 1989)*

The normalized cross-correlation function was given by (4.3.2)

$$r_{u'y'}(0) = \frac{\sqrt{2}u_0}{\sqrt{2u_0^2 + \sigma_u^2}}$$

For a deterministic relation the cross-dispersion function is equal to one

$$D_{u'y'}'(\kappa) = 1$$

The square of the nonlinearity index (4.12.6) is

$$v_2^2(0) = 1 - \frac{r_{u'y'}^2(0)}{D_{u'y'}^{\prime 2}(0)} = 1 - \frac{2u_0^2}{2u_0^2 + \sigma_u^2} = \frac{\sigma_u^2}{2u_0^2 + \sigma_u^2}$$

which is equal to the nonlinearity index (4.3.29) calculated in Example 4.3.1.   ∎

Corollary 4.12.1   To decide whether a process is linear or not the nonlinearity index has to be calculated for every delay κ between the input and output signals. The process is linear if and only if the nonlinearity index is zero for all shifting times.   ∎

A cumulative index that is zero if and only if the process is linear can be defined by several ways, e.g.,

1.   The overall nonlinearity index is the greatest one of those assigned to the different shifting times in the shifting time domain $\kappa_{min} \le \kappa \le \kappa_{max}$ (Haber, 1985)

$$v = \max_\kappa v(\kappa) \qquad \kappa_{min} \le \kappa \le \kappa_{max} \tag{4.12.12}$$

or

$$v_i = \max_\kappa v_i(\kappa) \qquad \kappa_{min} \le \kappa \le \kappa_{max} \qquad i = 1, 2, 3, 4 \tag{4.12.13}$$

2. The overall nonlinearity index is the average value of those assigned to the different shifting times in the shifting time domain $\kappa_{min} \leq \kappa \leq \kappa_{max}$

$$v = \frac{1}{\kappa_{max} - \kappa_{min} + 1} \sum_{\kappa_{min}}^{\kappa_{max}} v(\kappa) \qquad\qquad (4.12.14)$$

or

$$v_i = \frac{1}{\kappa_{max} - \kappa_{min} + 1} \sum_{\kappa_{min}}^{\kappa_{max}} v_i(\kappa) \qquad i = 1, 2, 3, 4 \qquad (4.12.15)$$

3. Rajbman and Chadeev (1980) recommended also to use the average value of $v_1^2(\kappa)$ in the domain $\kappa_{min} \leq \kappa \leq \kappa_{max}$

$$v = \frac{1}{\kappa_{max} - \kappa_{min} + 1} \sum_{\kappa_{min}}^{\kappa_{max}} v_1^2(\kappa)$$

$$= \frac{1}{\kappa_{max} - \kappa_{min} + 1} \sum_{\kappa_-}^{\kappa_-} D_{u'y'}^{\prime 2}(\kappa) - r_{u'y'}^2(\kappa) \qquad (4.12.16)$$

They interpreted the square of (4.12.16) $\sqrt{v}$ as the nonlinearity degree of the system.

4. Várlaki et al., (1985) recommended normalizing (4.12.16) by the average value of the square of the partial cross-dispersion functions assigned to the different shifting times

$$v = \frac{\dfrac{1}{\kappa_{max} - \kappa_{min} + 1} \displaystyle\sum_{\kappa_{min}}^{\kappa_{max}} D_{u'y'}^{\prime 2}(\kappa) - r_{u'y'}^2(\kappa)}{\dfrac{1}{\kappa_{max} - \kappa_{min} + 1} \displaystyle\sum_{\kappa_{min}}^{\kappa_{max}} D_{u'y'}^{\prime 2}(\kappa)}$$

$$= \frac{\displaystyle\sum_{\kappa_{min}}^{\kappa_{max}} \left(D_{u'y'}^{\prime 2}(\kappa) - r_{u'y'}^2(\kappa) \right)}{\displaystyle\sum_{\kappa_{min}}^{\kappa_{max}} D_{u'y'}^{\prime 2}(\kappa)} = 1 - \frac{\displaystyle\sum_{\kappa_{min}}^{\kappa_{max}} r_{u'y'}^2(\kappa)}{\displaystyle\sum_{\kappa_{min}}^{\kappa_{max}} D_{u'y'}^{\prime 2}(\kappa)} \qquad (4.12.17)$$

The computation of the cross-dispersion function is far more complicated than that of the correlation function. The range between the minimum and maximum values of the input signal has to be divided into small windows to calculate the empirical conditional expected value in the formula

$$D_{u'y'}(\kappa) = E_{u'}\left\{ \left[E_{y'}\left\{ y'(k) \middle| u'(k - \kappa) \right\} \right]^2 \right\}$$

Unfortunately this has to be done for every shifting time between κ_{min} and κ_{max}. The situation is somewhat better for test signals with few amplitude values (e.g., PRBS or PRTS).

The recommendations about the whitening of the input signal before performing a cross-correlation test and the suggested statistical tests for determining whether a cross-correlation function is zero or not, made at the end of section 4.9, can be applied for the cross-dispersion function based tests as well.

The output noises of zero mean value are filtered by the averaging process of the calculation of the conditional mean value.

Example   4.12.2 *Nonlinearity test on various linear, simple Hammerstein and simple Wiener models based on cross-correlation and cross-dispersion functions (Haber, 1985)*

The nonlinearity test method was performed for the noise-free simple Hammerstein (Figure 4.2.1a) and simple Wiener (Figure 4.2.1b) models. The cascade models contain different linear dynamic (LD) and nonlinear static (NS) parts. The same static characteristics and linear dynamic terms were simulated as in Example 4.2.1. The static curves and step responses of the linear dynamic parts were given already in Figure 4.2.1.

The input and output signals were sampled by 2 [s] at the models having a first-order time lag and by 4 [s] at those ones having a second-order one. The test signal was always a PRTS signal of maximum length 80 with amplitude ±2 around the values listed in Table 4.8.1. The minimum switching time was equal to the sampling time.

The nonlinearity test method was applied at the following models:
(a)   linear model (Wiener model with linear static characteristic);
(b)   nonlinear simple Hammerstein model, excited in the static extremum;
(c)   nonlinear simple Hammerstein model, not excited in the static extremum;
(d)   nonlinear simple Wiener model, excited in the static extremum;
(e)   nonlinear simple Wiener model, not excited in the static extremum.

The linear cross-correlation and cross-dispersion functions are plotted in Figures 4.8.1 and 4.8.2 for the listed situations. (The discrete values are drawn by continuous lines as being continuous time functions.) The auto-correlation function of the normalized input signal was always a Dirac function. Figure 4.8.1 refers to models having a first-order time lag and Figure 4.8.2 to those ones having the oscillating time lag. Table 4.8.1 sums up the estimated nonlinearity indices calculated from the linear cross-correlation and cross-dispersion functions.

The dispersion nonlinearity index was defined as the maximal deviation between the normalized dispersion and the absolute value of the normalized cross-correlation functions. It was one at the quadratic systems excited near the extremum point and this measure decreased by going away from there. At linear systems it became zero as the absolute values of the normalized linear cross-correlation and cross-dispersion functions coincided. Unfortunately, the computed cross-dispersion function did not tend to zero at large shifting times, contrary to the cross-correlation function, which fact leaded to a high nonlinearity index at large shifting times.   ∎

4.13 METHOD BASED ON THE COMPARISON OF THE VARIANCES OF THE NOISE-FREE MODEL OUTPUT SIGNALS OF THE IDENTIFIED BEST LINEAR AND OF AN ALTERNATIVE NONLINEAR MODEL

Theorem 4.13.1   Várlaki *et al.*, (1985) recommended the following analytical nonlinearity index

$$v = 1 - \frac{\text{var}\{y(k)\} - \text{var}\{y(k) - \hat{w}_{\text{lin}}(k)\}}{\text{var}\{y(k)\} - \text{var}\{y(k) - \hat{w}_{\text{nlin}}(k)\}} \qquad (4.13.1)$$

where $\hat{w}_{\text{lin}}(k)$ and $\hat{w}_{\text{nlin}}(k)$ are the noise-free output signals of the identified best linear and of an alternative nonlinear model, respectively. The nonlinearity measure falls between zero and one

$$0 \le v \le 1$$

If the process is linear the index is zero and if the process is nonlinear it differs from zero.   ∎

Proof.   If the process is linear then

$$\hat{w}_{\text{lin}}(k) = \hat{w}_{\text{nlin}}(k)$$

and $v = 1 - 1 = 0$. If the process is nonlinear then

$$\text{var}\{y(k) - \hat{w}_{\text{nlin}}(k)\} < \text{var}\{y(k) - \hat{w}_{\text{lin}}(k)\}$$

i.e.,

$$\text{var}\{y(k)\} - \text{var}\{y(k) - \hat{w}_{\text{nlin}}(k)\} > \text{var}\{y(k)\} - \text{var}\{y(k) - \hat{w}_{\text{lin}}(k)\}$$

and the nonlinearity index falls between zero and one.   ∎

Lemma 4.13.1
(Várlaki *et al.*, 1985) The nonlinearity index (4.13.1) is equivalent to the following expression

$$v = 1 - \frac{\text{var}\{\hat{w}_{\text{lin}}(k)\}}{\text{var}\{\hat{w}_{\text{nlin}}(k)\}} \qquad (4.13.2)$$

Proof.   Assume an additive noise $e(k)$ at the output of the process. The measured output signal $y(k)$ can be given by an estimated noise-free model output signal $\hat{w}(k)$ and an additive noise term $e(k)$

$$y(k) = \hat{w}(k) + e(k)$$

where the mean value of the noise is zero and for stationary, ergodic noise processes

$$E\{e(k)\} = \overline{e(k)} = 0$$

Equation (4.13.1) contains the following expression both in the denominator and in the numerator

$$\mathrm{var}\{y(k)\} - \mathrm{var}\{y(k) - \hat{w}(k)\}$$

$$= \mathrm{E}\left\{[y(k) - \bar{y}]^2\right\} - \mathrm{E}\left\{[y(k) - \bar{y} + \bar{y} - \hat{w}(k)]^2\right\}$$

$$= \mathrm{E}\left\{[y(k) - \bar{y}]^2\right\} - \mathrm{E}\left\{[y(k) - \bar{y}]^2\right\} - \mathrm{E}\left\{[\bar{y} - \hat{w}(k)]^2\right\}$$

$$-2\mathrm{E}\left\{[y(k) - \bar{y}][\bar{y} - \hat{w}(k)]\right\} = -\mathrm{E}\left\{[\hat{w}(k) - \bar{y}]^2\right\} + 2\mathrm{E}\left\{[\hat{w}(k) - \bar{y}]^2\right\}$$

$$-2\mathrm{E}\left\{e(k)[\bar{y} - \hat{w}(k)]\right\} = \mathrm{E}\left\{[\hat{w}(k) - \bar{y}]^2\right\} = \mathrm{var}\{\hat{w}(k)\} \qquad (4.13.3)$$

if the noise term is uncorrelated with the noise-free model output signal and the mean value of the model output signal is equal to the mean value of the measured output signal.

Applying (4.13.3) to the identified best linear and nonlinear models, we obtain (4.13.2) from (4.13.1).   ∎

In Example 4.3.1 the nonlinearity index for a quadratic static function was calculated. We show that (4.13.2) results in the same measure.

Example 4.13.1 *Calculation of the nonlinearity measure of a static quadratic process excited by a white noise signal with not zero mean value. (Continuation of Examples 4.3.1 and 4.12.1.) (Haber, 1989)*

The variance of the best linear model was given by (4.3.35)

$$\mathrm{var}\{\hat{w}'_{\mathrm{lin}}(k)\} = \hat{c}_1^2 \sigma_u^2 = 4c^2 u_0^2 \sigma_u^2 \qquad (4.13.4)$$

The output signal of the best nonlinear model coincides with the noise-free output signal. Its variance was given by (4.3.28)

$$\mathrm{var}\{\hat{w}'_{\mathrm{nlin}}(k)\} = 2c^2 \sigma_u^2 \left[2u_0^2 + \sigma_u^2\right] \qquad (4.13.5)$$

The nonlinearity index (4.13.2) is then

$$v = 1 - \frac{\mathrm{var}\{\hat{w}'_{\mathrm{lin}}(k)\}}{\mathrm{var}\{\hat{w}'_{\mathrm{nlin}}(k)\}} = 1 - \frac{4c^2 u_0^2 \sigma_u^2}{2c^2 \sigma_u^2 \left[2u_0^2 + \sigma_u^2\right]} = \frac{\sigma_u^2}{2u_0^2 + \sigma_u^2}$$

which is equal to the nonlinearity index (4.3.29) calculated in Example 4.3.1.

The nonlinearity index of (4.13.2) has a close relation to the relative degree of nonlinearity (4.12.6). This relation is shown now for a static system with a possible delay time.

Lemma 4.13.2

(Haber, 1989) Assume that the input signal satisfies Assumption 4.12.1. Then for a static system with a possible delay time and output noise

$$y(k) = f\big(u(k - \kappa)\big) + e(k) \qquad (4.13.6)$$

where $f(...)$ is a static function and $e(k)$ is a white noise, the nonlinearity index (4.13.2) is equal to the square of the relative degree of nonlinearity (4.12.6).

Proof.   The parameters of the best fitting linear model can be estimated from the equation

$$y(k) = \hat{c}_0(\kappa) + \hat{c}_1(\kappa)u(\kappa) + e(k) \tag{4.13.7}$$

Substitute the normalized input and output values into (4.13.7)

$$\sigma_y y'(k) + \bar{y} = \hat{c}_0(\kappa) + \hat{c}_1(\kappa)\left[\sigma_u u'(\kappa) + \bar{u}\right] + e(k) \tag{4.13.8}$$

thus

$$y'(k) = \frac{\sigma_u}{\sigma_y}\hat{c}_1(\kappa)u'(\kappa) + \frac{\hat{c}_0(\kappa) + \hat{c}_1(\kappa)\bar{u} - \bar{y}}{\sigma_y} + e(k) \tag{4.13.9}$$

In (4.13.9) the estimate of the constant term is

$$\hat{c}_0(\kappa) = \bar{y} - \hat{c}_1(\kappa)\bar{u} \tag{4.13.10}$$

and the proportional term can be determined by computing the cross-correlation

$$\frac{\sigma_u}{\sigma_y}\hat{c}_1(\kappa) = \hat{c}'_1(\kappa) = r_{u'y'}(\kappa) \tag{4.13.11}$$

The computed noise-free output signal of the best linear model is

$$\hat{w}_{\text{lin}}(k) = \hat{c}_0(\kappa) + \hat{c}_1(\kappa)u(\kappa) \tag{4.13.12}$$

Substituting the estimated parameters from (4.13.10) and (4.13.11) into (4.13.12), the noise-free model output signal is

$$\hat{w}_{\text{lin}}(k) = \bar{y} + \sigma_y r_{u'y'}(\kappa)u'(\kappa) \tag{4.13.13}$$

Taking the expected value of both sides of (4.13.13) shows that the mean value of the measured output signal and that of the best linear model are equal

$$\overline{\hat{w}_{\text{lin}}(k)} = \bar{y} \tag{4.13.14}$$

The variance of the best linear model is

$$\text{var}\left\{\hat{w}^2_{\text{lin}}(k)\right\} = \text{E}\left\{\left[\sigma_y r_{u'y'}(\kappa)u'(\kappa)\right]^2\right\} = \sigma_y^2 r_{u'y'}^2(\kappa) \tag{4.13.15}$$

If the best nonlinear model completely describes the noise-free process model, then the variance of the noise-free output signal is

$$\text{E}\left\{\left[w^2(\kappa) - \bar{w}\right]^2\right\} = \text{E}\left\{\left[\text{E}\left\{y(k)|u(k-\kappa)\right\} - \bar{y}\right]^2\right\} = \sigma_y^2 D'^2_{u'y'}(\kappa) \tag{4.13.16}$$

Here we used (4.12.3) and that the mean value of the noisy and noise-free output signals are equal

$$\overline{w} = \overline{y}$$

Substituting (4.13.15) and (4.13.16) into (4.13.2), we obtain for the nonlinearity index

$$v = 1 - \frac{\sigma_y^2 r_{u'y'}^2(\kappa)}{\sigma_y^2 D_{u'y'}'^2(\kappa)} = 1 - \frac{r_{u'y'}^2(\kappa)}{D_{u'y'}'^2(\kappa)} \qquad (4.13.17)$$

                                                                                        ■

The nonlinearity measure introduced by Várlaki *et al.*, (1985) eliminates the effect of the noises superposed to the output signal. This can be seen from (4.13.2) since the variances of the noise-free model output signals are compared. Indirectly the relations of the definitions (4.13.1) and (4.13.2) to (4.13.6) support this statement, since both the cross-correlation and cross-dispersion function filter the noises.

The method requires the identification of the best linear and an alternative nonlinear model. The parameter estimation algorithms require more computations than the time domain and correlation methods.

4.14 METHOD BASED ON THE IDENTIFICATION OF ORTHOGONAL SUBSYSTEMS

Assumption 4.14.1   Assume that the noise-free process can be approximated by an *n*-degree nonlinear model for the given input signal

$$w(k) = \sum_{i=0}^{n} y_i(k)$$

where

$$E\{y_i(k)y_j(k)\} = 0 \qquad \text{if } i \neq j$$

Theorem 4.14.1   The nonlinearity index

$$v = 1 - \frac{E\{y_1^2(k)\}}{\sum_{i=1}^{n} E\{y_i^2(k)\}} \qquad (4.14.1)$$

is equivalent to (4.13.2). Consequently if the process is linear then the nonlinearity measure is zero, and if the process is nonlinear the measure differs from zero.

Proof.   The mean value of the square of the noise-free output signal is

$$E\{w^2(k)\} = \sum_{i=0}^{n} E\{y_i^2(k)\} \qquad (4.14.2)$$

Assume that an additive noise term is at the output of the process. Then the measured output signal is

$$y(k) = w(k) + e(k)$$

The mean of the square of the output signal is

$$E\{y^2(k)\} = E\{w^2(k)\} + E\{e^2(k)\} \tag{4.14.3}$$

because

$$E\{w(k)e(k)\} = 0$$

if the noise is independent of the process input signal and has a zero mean value.
The variance of the measured output signal is

$$\text{var}\{y(k)\} = E\left\{[y(k) - E\{y(k)\}]^2\right\} = E\{y^2(k)\} - [E\{y(k)\}]^2 \tag{4.14.4}$$

The output signal of the degree-0 subsystem is the mean value of both the noisy and the noise-free output signals

$$E\{y(k)\} = E\{w(k) + e(k)\} = E\{w(k)\} = E\left\{\sum_{i=0}^{n} y_i(k)\right\} = y_0 \tag{4.14.5}$$

It concludes from (4.14.2) to (4.14.5) that the variance of the measured output signal is

$$\text{var}\{y(k)\} = E\{e^2(k)\} + \sum_{i=1}^{n} E\{y_i^2(k)\} = E\{e^2(k)\} + \text{var}\{\hat{w}_{\text{nlin}}(k)\} \tag{4.14.6}$$

Approximate the process first by a linear and then by an alternative nonlinear model. On the basis of Equations (4.14.2) and (4.14.3) the variances of the noise-free model outputs are

$$\text{var}\{\hat{w}_{\text{lin}}(k)\} = E\left\{[[y_0 + y_1(k)] - [\bar{y}_0 + \bar{y}_1]]^2\right\}$$
$$= E\left\{[y_1(k) - \bar{y}_1]^2\right\} = E\{y_1^2(k)\} \tag{4.14.7}$$

and if the fitted nonlinear model is perfect then

$$\text{var}\{\hat{w}_{\text{nlin}}(k)\} \cong \text{var}\{w(k)\} = E\left\{\left[\sum_{i=0}^{n} y_i(k) - \sum_{i=0}^{n} \bar{y}_i(k)\right]^2\right\}$$
$$= E\left\{\left[\sum_{i=1}^{n} y_i(k)\right]^2\right\} = \sum_{i=1}^{n} E\{y_i^2(k)\} \tag{4.14.8}$$

Having substituted (4.14.7) and (4.14.8) into the formula of the nonlinearity degree (4.13.2), we obtain (4.14.1). ∎

The nonlinearity index lies between 0 and 1, as shown already in Section 4.13.
It is relatively easy to compute the numerator of the fraction of (4.14.1) but it is more difficult to calculate the denominator of it .

Using (4.14.6) and (4.14.7) Equation (4.14.1) can be rewritten as

$$v = 1 - \frac{\mathrm{var}\{\hat{w}_{\mathrm{lin}}(k)\}}{\mathrm{var}\{y(k)\} - \mathrm{var}\{e(k)\}} \tag{4.14.9}$$

If the system is not very noisy then (4.14.9) can be approximated by

$$v \cong 1 - \frac{\mathrm{var}\{\hat{w}_{\mathrm{lin}}(k)\}}{\mathrm{var}\{y(k)\}} \tag{4.14.10}$$

which index can be much more easily computed than (4.14.9) since $\mathrm{var}\{y(k)\}$ can be calculated without any parameter estimation.

The above simple form of the nonlinearity measure was introduced by Schetzen (1980). The analytical index v of (4.14.10) is zero if and only if
- the system is linear and
- there is no additive noise.

The advantage of this measure lies in its easy computation. The disadvantage is that one cannot distinguish between noise and nonlinear effects, similarly to the usage of the coherence function (Várlaki, Terdik and Lototsky, 1985).

The output signal can be separated to orthogonal terms, e.g., by means of the Fourier analysis. That was applied in Section 4.6 under the name of the *frequency method*. The harmonics have no DC component and the average value of their square is the effective value of the harmonics

$$\mathrm{E}\{y_i^2(t)\} = \frac{1}{T}\int_{t=0}^{T} y_i^2(t)dt = y_{i\mathrm{eff}} = \frac{Y_i^2}{2}$$

(Y_i is the amplitude of the ith harmonic.) Observe that for a periodic test signal the nonlinearity measure (4.14.1) becomes (4.6.2) and (4.14.9) becomes (4.6.3) since

$$\mathrm{var}\{y(t)\} = y_{\mathrm{eff}} - y_0^2$$

If the noise is filtered from the output signal or its variance is negligible compared to the variance of the noise-free signal then the approximating formula (4.6.6) can be applied, which is an equivalent form of (4.14.10).

Schetzen (1980) has shown that if the input signal has zero mean value then there is no need to identify the first-degree (orthogonal) subsystem but it is enough to compute the cross-correlation function between the input and the noisy output signals. This will be shown now.

Assume first that the input signal is a white noise. Then the variance of the output signal of the linear submodel is

$$\mathrm{var}\{\hat{w}_{\mathrm{lin}}(k)\} = \mathrm{var}\{y_1(k)\} = \left[\sum_{\kappa=0}^{\infty} h(\kappa)u(k-\kappa)\right]^2$$

$$= \sum_{\kappa_1=0}^{\infty}\sum_{\kappa_2=0}^{\infty} h(\kappa_1)h(\kappa_2)u(k-\kappa_1)u(k-\kappa_2)$$

$$= \sigma_u^2 \sum_{\kappa_1=0}^{\infty} h^2(\kappa_1) = \frac{1}{\sigma_u^2} \sum_{\kappa_1=0}^{\infty} r_{uy}^2(\kappa_1)$$

since the relation between the weighting function series and the cross-correlation series is

$$r_{uy}(\kappa) = \sigma_u^2 h(\kappa)$$

Thus the analytical nonlinearity index in the case of a white noise input signal is

$$v = 1 - \frac{\sum_{\kappa_1=0}^{\infty} r_{uy}^2(\kappa_1)}{\mathrm{var}\{u(k)\}\left[\mathrm{var}\{y(k)\} - \mathrm{var}\{e(k)\}\right]}$$

which can be approximated by

$$v \cong 1 - \frac{\sum_{\kappa_1=0}^{\infty} r_{uy}^2(\kappa_1)}{\mathrm{var}\{u(k)\}\mathrm{var}\{y(k)\}}$$

if the effect of the noise can be neglected.

If the input signal is a color noise then it has to be whitened first. Denote the identified linear whitening filter by $H_w(q^{-1})$ and the white noise source signal by $u^w(k)$. Then

$$u(k) = H_w(q^{-1})u^w(k) \qquad (4.14.11)$$

and

$$u^w(k) = \frac{1}{H_w(q^{-1})}u(k) = H_w^{-1}(q^{-1})u(k)$$

The relation between $u(k)$ and $u^w(k)$ can also be seen in Figure 4.14.1.

The variance of the output signal of the linear submodel is now

$$\mathrm{var}\{\hat{w}_{\mathrm{lin}}(k)\} = \mathrm{var}\{y_1(k)\} = \frac{1}{\sigma_{u^w}^2} \sum_{\kappa_1=0}^{\infty} r_{u^w y}^2(\kappa_1)$$

The unmeasurable values σ_{u^w} and $r_{u^w y}(\kappa)$ can be replaced by the measureable ones

$$\sigma_{u^w}^2 = \frac{\sigma_u^2}{\sum_{\kappa_1=0}^{\infty} h_w^2(\kappa_1)} \qquad (4.14.12)$$

and

$$r_{u^w y}(\kappa) = \sum_{\kappa_1=0}^{\infty} h_w^{-1}(\kappa_1) r_{uy}(\kappa_1 + \kappa) \qquad\qquad (4.14.13)$$

■

The results of the above are summarized in the following theorem.

Assumption 4.14.2   Apply a random or pseudo-random test signal $u(k)$ with a zero mean. All odd order moments of the signal should be zero and all even order ones should differ from zero.

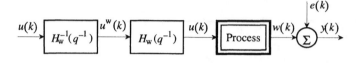

Fig. 4.14.1 Whitening filter $H_w^{-1}(q^{-1})$ of the input signal $u(k)$ and the computed white source signal $u^w(k)$

Theorem 4.14.2   (Schetzen, 1980) If the input signal satisfies Assumption 4.14.2 then a nonlinearity index can be defined as

$$\nu = 1 - \frac{\sum\limits_{\kappa_1=0}^{\infty} r_{u^w y}(\kappa_1)}{\mathrm{var}\{u^w(k)\}\big[\mathrm{var}\{y(k)\} - \mathrm{var}\{\varepsilon(k)\}\big]}$$

where $u^w(k)$ is the whitened input sequence with variance (4.14.12), and $r_{u^w y}(\kappa_1)$ can be calculated according to (4.14.13). The whitening filter $H_w(q^{-1})$ is defined by (4.14.11). $\varepsilon(k)$ is the computed residual. The nonlinearity index is zero if the process is linear and it differs from zero if the process has even degree nonlinear terms.   ■

Like other nonlinearity tests, a process having only odd degree subsystems cannot be detected as nonlinear if the input signal has only a stochastic component with zero mean value. This fact can easily be shown for a static cubic process having no quadratic channel.

Example 4.14.1 *Nonlinearity measure of a static cubic system having no quadratic channel (Haber, 1989)*
According to (4.14.8) and (4.14.1) the process would be detected as linear if the cubic subsystem had no contribution to the variance of the output signal. Assume that the input signal is Gaussian white noise with standard deviation σ_u. The Hermite polynomials are orthogonal to the above input signal. The constant, linear, and cubic orthogonal subsystems are as follows:

$$y_0 = c_0 \cdot 1$$
$$y_1(k) = c_1 u(k)$$
$$y_3(k) = c_3\big[u^3(k) - 3\sigma_u^2 u(k)\big]$$

where c_i are any constants that are multiplied by the Hermite polynomials of the input signal (Schetzen, 1980). The mean value of the square of the output signal of the cubic channel is

$$\mathrm{E}\left\{y_3^2(k)\right\} = c_3^2 \mathrm{E}\left\{\left[u^3(k) - 3\sigma_u^2 u(k)\right]^2\right\}$$

$$= c_3^2 \mathrm{E}\left\{u^6(k)\right\} - 6\sigma_u^2 \mathrm{E}\left\{u^4(k)\right\} + 9\sigma_u^4 \mathrm{E}\left\{u^2(k)\right\}$$

$$= c_3^2\left[9\sigma_u^6 - 18\sigma_u^6 + 9\sigma_u^6\right] = 0$$

if the input signal is a Gaussian white noise with zero mean value, as stated before. ∎

4.15 METHOD BASED ON THE IDENTIFICATION OF THE FIRST-ORDER URYSON MODEL

Várlaki et al., (1985) recommend fitting a first-order Uryson series model to the input–output data. If the model has no significant nonlinear channels then the system is detected as linear. The idea behind this suggestion is the following (Gallman and Narendra, 1971):
- The Uryson model is a general valid description of the nonlinear dynamic systems. Experience has shown, however, that a wide class of polynomial systems can be well approximated even by the first-order Uryson model.
- If the input signal is a Gaussian white noise then the model components are orthogonal to each other and the parameter estimation does not require any matrix inversion.

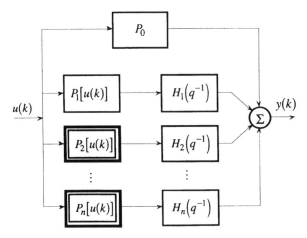

Fig. 4.15.1 First-order Uryson model

Figure 4.15.1 shows the scheme of the first-order Uryson model. $P_i[u(k)]$ denotes static polynomials of ith degree. $H_i(q^{-1})$ are pulse transfer functions of linear dynamic processes whose input signals are the outputs of the corresponding static nonlinear blocks. The zero subscript denotes the constant channel. One of the possibilities is to define the polynomials as powers of the input signal

$$P_i[u(k)] = u^i(k)$$

There are known several ways for the parameter estimation of the system (Gallman and Narendra, 1971):

- application of direct search method (gradient procedure) for the unknown parameters;
- iterative estimation of the parameters of the pulse transfer functions;
- correlation technique if special – orthogonal – polynomials are applied.

If the input signal is a Gaussian white noise with zero mean and standard deviation σ_u then an evident choice of the static polynomials is the application of the Hermite polynomials, because then the linear dynamic blocks can easily be identified independently from each other by the correlation method. The orthonormal Hermite polynomials are defined by (e.g., Schetzen, 1980)

$$P_0[u(k)] = P_0 = 1$$

$$P_{i+1}[u(k)] = \frac{1}{\sqrt{i+1}\,\sigma_u^{i+1}}\left[u(k)P_i[u(k)] - \sigma_u^2\frac{dP_i[u(k)]}{du(k)}\right] \qquad i \geq 1$$

Let us denote the output signals of the ith degree Hermite polynomial by $v_i(k)$

$$v_i(k) = P_i[u(k)]$$

As the Hermite polynomial was defined, $v_i(k)$ is known in the knowledge of the input signal. The noise-free output signal of the process can be approximated by

$$w(k) = h_0 + \sum_{i=1}^{n}\sum_{\kappa=0}^{m} h_i(\kappa)v_i(k-\kappa)$$

where n is the degree of the process and m is the number of the terms of the weighting function series taken into account. The measured output signal may also contain noise

$$y(k) = w(k) + e(k) = h_0 + \sum_{i=1}^{\infty}\sum_{\kappa=0}^{\infty} h_i(\kappa)v_i(k-\kappa) + e(k)$$

The Hermite polynomials of different degrees are orthogonal to each other, thus

$$r_{v_i v_j}(\kappa) = E\{P_i[u(k-\kappa)]P_j[u(k)]\} = \begin{cases} 0 & \text{if } i \neq j \\ \delta(\kappa) & \text{if } i = j \end{cases} \tag{4.15.1}$$

i.e., the effect of the additive noise can be filtered by the correlation technique.

Compute the cross-correlation function between the signal $v_i(k)$ and the measured output signal

$$r_{v_i y}(\kappa) = E\{v_i(k-\kappa)y(k)\} = E\{v_i(k-\kappa)w(k)\} + E\{v_i(k-\kappa)e(k)\} = h_i(\kappa) \tag{4.15.2}$$

The pulse transfer function of any channel can be estimated as

$$\hat{h}_i(\kappa) = E\{P_i[u(k-\kappa)]y(k)\} \qquad i \geq 1$$

and the coefficient of the degree-0 polynomial is

$$\hat{h}_0 = E\{P_0[u(k)]y(k)\} = E\{y(k)\} = \bar{y}$$

The process is linear if all pulse transfer functions are zero except that in the linear channel

$$v = 0 \qquad \text{if } \hat{h}_i(\kappa) = 0 \qquad \text{for } i \geq 2 \quad \text{and } \forall \kappa$$

A straightforward consequence is the following theorem.

Assumption 4.15.1   The test signal is Gaussian white noise with zero mean.

Theorem 4.15.1   If the input signal satisfies Assumption 4.15.1 then the following nonlinearity index can be defined

$$v = 1 - \frac{\displaystyle\sum_{\kappa=0}^{m} \hat{h}_1^2(\kappa)}{\displaystyle\sum_{i=1}^{n} \sum_{\kappa=0}^{m} \hat{h}_i^2(\kappa)} \tag{4.15.3}$$

If the process is linear then the measure is zero and if the process is nonlinear the measure differs from zero.   ∎

Proof.   If the process is linear then the higher degree kernels $h_i(\kappa)$, $i > 1$, are zero, and $v = 0$. If the process is nonlinear then some of the $h_i(\kappa)$, $i > 1$ are not zero. Then the denominator has some positive terms more than the numerator, consequently the fraction is less than one and $v > 0$.   ∎

Corollary 4.15.1   The nonlinearity index (4.15.3) can be rewritten as

$$v = \frac{\displaystyle\sum_{i=2}^{n} \sum_{\kappa=0}^{m} \hat{h}_i^2(\kappa)}{\displaystyle\sum_{i=1}^{n} \sum_{\kappa=0}^{m} \hat{h}_i^2(\kappa)} \tag{4.15.4}$$

or

$$v = \frac{1}{1 + \dfrac{\displaystyle\sum_{\kappa=0}^{m} \hat{h}_1^2(\kappa)}{\displaystyle\sum_{i=2}^{n} \sum_{\kappa=0}^{m} \hat{h}_i^2(\kappa)}} \tag{4.15.5}$$

∎

(4.15.5) is identical to the result of Várlaki *et al.*, (1985), who derived it from the nonlinearity measure (4.13.1).

The nonlinearity measure (4.15.5) falls between 0 and 1, as stated already about (4.14.1) in Section 4.14.

Corollary 4.15.2   Consider the reciprocal of the fraction term in the denominator of (4.15.5)

$$\mu = \frac{\displaystyle\sum_{i=2}^{n}\sum_{\kappa=0}^{m}\hat{h}_i^2(\kappa)}{\displaystyle\sum_{\kappa=0}^{m}\hat{h}_1^2(\kappa)} \qquad\qquad (4.15.6)$$

(4.15.6) is equal to

$$\mu = \frac{\nu}{1-\nu} = \begin{cases} 0 & \text{if } \nu = 0 \\ \infty & \text{if } \nu = 1 \end{cases}$$

Consequently the process is linear if the always positive measure μ is zero. This is easy to understand from (4.15.6) because in this case all higher degree channels should be zero. ∎

The linear feature can be checked also by the chi-square test. It is known that the estimated parameters have a Gaussian distribution if the noise is white and Gaussian and a least squares estimation method was applied. The estimates of the weighting function series are bias-free and the standard deviation of each parameter is equal to the standard deviation of the output noise σ_e because of (4.15.1)

$$\sigma\big(\hat{h}_i(\kappa)\big) = \sigma_\varepsilon$$

The standard deviation of the residuals σ_ε can be calculated as

$$\sigma_\varepsilon = \sqrt{\frac{1}{N-n_\theta}\big[y(k)-\hat{w}_{\text{nlin}}(k)\big]^2}$$

where the number of parameters n_θ is

$$n_\theta = 1 + n(m+1)$$

Since the model components $v_i(k-\kappa)$ are orthogonal to each other, the estimates of the weighting function series are independent, and the sum of the normalized weighting function series of higher degree channels has a chi-square distribution

$$\Xi^2 = \sum_{i=2}^{n}\sum_{\kappa=0}^{m}\frac{\hat{h}_i^2(\kappa)}{\sigma\big\{\hat{h}_i(\kappa)\big\}} = \frac{1}{\sigma_\varepsilon}\sum_{i=2}^{n}\sum_{\kappa=0}^{m}\hat{h}_i^2(\kappa) \qquad\qquad (4.15.7)$$

The process is linear with probability $100(1-\alpha)$ percent if the calculated value Ξ^2 of

(4.15.7) is less than the tabulated value $\Xi^2[(m+1)(n-1),\alpha]$.

Várlaki et al., (1985) recommended using the F-test for deciding whether the process is linear or not. According to their method, first a linear model has to be fitted to the measured input and output records

$$\hat{w}_{\text{lin}}(k) = \hat{h}_0 P_0[u(k)] + \sum_{\kappa=0}^{m} \hat{h}_1(\kappa) P_1[u(k-\kappa)]$$

$$= \hat{h}_0 + \sum_{\kappa=0}^{m} \hat{h}_1(\kappa) \frac{u(k-\kappa)}{\sigma_u}$$

(4.15.8)

The model output $\hat{w}_{\text{lin}}$ in (4.15.8) has to be subtracted from the measured output signal

$$y^*(k) = y(k) - \hat{w}_{\text{lin}}(k)$$

and then a nonlinear first-order Uryson model has to be fitted to the input and to the modified output signal series. Denote the estimated weighting function series by $\hat{h}_i^*(\kappa)$. If the process is linear then the estimates of the nonlinear terms are zero. Their zero value can be checked either by the chi-square test or by the F-test. To apply the F-test, form the ratio of the normalized sum of squares of the parameters of the nonlinear and linear terms

$$F = \frac{\dfrac{1}{(n-1)(m+1)} \displaystyle\sum_{i=2}^{n} \sum_{\kappa=0}^{m} \dfrac{1}{\sigma_\varepsilon} \hat{h}_i^{*2}(\kappa)}{\dfrac{1}{(m+1)} \displaystyle\sum_{\kappa=0}^{m} \dfrac{1}{\sigma_\varepsilon} \hat{h}_i^{*2}(\kappa)} = \frac{1}{(n-1)} \frac{\displaystyle\sum_{i=2}^{n} \sum_{\kappa=0}^{m} \hat{h}_i^{*2}(\kappa)}{\displaystyle\sum_{\kappa=0}^{m} \hat{h}_i^{*2}(\kappa)}$$

(4.15.9)

If F is less than the F distribution value $F[(n-1)(m+1), m+1, \alpha]$ then the process is linear with a probability of $100(1-\alpha)$ percent. Consequently an analytic nonlinearity index can be defined as

$$\nu = \frac{F - F[(n-1)(m+1), m+1, \alpha]}{F[(n-1)(m+1), m+1, \alpha]}$$

(4.15.10)

For nonlinear systems the measure (4.15.10) is greater than zero.

Comparing (4.15.6) with (4.15.9), we see that

$$(n-1)F = \mu$$

i.e., the F-test applied to the system – not reduced by the estimated linear model output – is equivalent to the nonlinearity test based on the comparison of the variances of the noise-free model output signals of the best linear and nonlinear identified models if a first-order Uryson model approximation was applied.

Both measures (4.15.6) and (4.15.9) are zero if the process is linear and are greater than zero if the process is nonlinear. The F-test gives a numerical threshold for the decision with a given probability.

An additive independent output noise does not affect the nonlinearity index because the cross-correlation functions between the outputs of the Hermite polynomials and the independent noise are zero, see (4.15.2).

4.16 METHOD BASED ON PARAMETER ESTIMATION AND F-TEST OF LINEAR AND SIMPLE NONLINEAR STRUCTURES

This method is based on a fast parameter estimation of the system first by a linear and then by a simple nonlinear structure. The assumed linear order should not be less than that of the process and the delay times should approximately be equal to that of the process. These parameters are generally known *a priori* from physical relations. The advantage of the method is that the applied nonlinear structure does not have to coincide with that of the real system, i.e., a simple model being linear in the parameters is suitable for this purpose. This model may be for example the generalized Hammerstein model, the parametric Volterra model or the bilinear model, etc. If the difference between the loss functions V_{lin} and V_{nlin} of the identification by linear and nonlinear models, respectively, is significant, then the process is a nonlinear one.

Proposition 4.16.1   Naka (1975) introduced the nonlinearity index as

$$\nu = \frac{V_0 - V_{\text{nlin}}}{V_0 - V_{\text{lin}}} \qquad (4.16.1)$$

where V_0 is a large value $\left(V_0 > V_{\text{lin}} > V_{\text{nlin}}\right)$. This index is equal to one for linear systems and it is greater than one for nonlinear ones.   ■

The disadvantage of the formula (4.16.1) is that the assumption of a more complex model usually leads to smaller loss functions, but the difference might be not significant. The F-test (e.g., Gustavsson, 1972) is suitable for a better nonlinearity test.

Theorem 4.16.1   Denote the number of parameters by $n_{\theta\text{lin}}$ for the linear model and by $n_{\theta\text{nlin}}$ for the nonlinear one. The difference between the number of parameters is

$$\Delta n_{\theta} = n_{\theta\text{nlin}} - n_{\theta\text{lin}}$$

The system is nonlinear if the value

$$F = \frac{V_{\text{lin}} - V_{\text{nlin}}}{V_{\text{nlin}}} \frac{N - n_{\theta\text{nlin}}}{n_{\theta\text{nlin}} - n_{\theta\text{lin}}} \qquad (4.16.2)$$

is greater than the corresponding $F(N, \Delta n_{\theta}, \alpha)$ distribution value at the prescribed probability level $100(1 - \alpha)$ percent .   ■

Proposition 4.16.2   As an analytical nonlinearity index take

$$\nu = \frac{F - F(N, \Delta n_{\theta}, \alpha)}{F(N, \Delta n_{\theta}, \alpha)} \qquad (4.16.3)$$

The measure of (4.16.3) is greater than zero for nonlinear systems.   ∎

This criterion is sufficient but not necessary for detecting the nonlinear feature of the system, because the chosen approximating nonlinear structure might not give significantly better identification results than the linear model. The so-called parametric Volterra model (Haber and Keviczky, 1974) seems to be such a nonlinear structure that gives significantly better results than the linear assumption in many practical cases. The reason is that the model approximates the Volterra series with an infinite long memory by means of few parameters. An even simpler model is the generalized Hammerstein model that can also be tried to fit the measured input–output records.

Theorem 4.16.2   (Haber, 1989) If the measured process output is the sum of the noise-free output signal of the estimated model and a white noise, then it can be shown that the nonlinearity measure (4.13.1), which was based on the difference of the variances of the model output signals of the best linear and an alternative nonlinear model, is only zero if (4.16.2) is zero.   ∎

In other words the F-test gives a measure of the nonlinearity under statistical circumstances because a probability is coordinated to the nonlinearity measure, which was not the case with the formula (4.13.1).

Proof.   The nonlinearity index of (4.13.1) can be rewritten as

$$v = 1 - \frac{\text{var}\{y(k)\} - \text{var}\{\varepsilon_{\text{lin}}(k)\}}{\text{var}\{y(k)\} - \text{var}\{\varepsilon_{\text{nlin}}(k)\}},$$ (4.16.4)

where $\varepsilon_{\text{lin}}(k)$ is the prediction error of the best linear and $\varepsilon_{\text{nlin}}(k)$ is that of the alternative nonlinear model. Rearranging (4.16.4) we obtain

$$v = \frac{\text{var}\{\varepsilon_{\text{lin}}(k)\} - \text{var}\{\varepsilon_{\text{nlin}}(k)\}}{\text{var}\{y(k)\} - \text{var}\{\varepsilon_{\text{nlin}}(k)\}}$$ (4.16.5)

The variances of the residuals are the normalized loss functions of the identification

$$V_{\text{lin}} = \frac{1}{2}\sum_{k=1}^{N}\varepsilon_{\text{lin}}^2(k) = \frac{N - n_{\theta\text{lin}}}{2}\,\text{var}\{\varepsilon_{\text{lin}}(k)\}$$ (4.16.6)

$$V_{\text{nlin}} = \frac{1}{2}\sum_{k=1}^{N}\varepsilon_{\text{nlin}}^2(k) = \frac{N - n_{\theta\text{nlin}}}{2}\,\text{var}\{\varepsilon_{\text{nlin}}(k)\}$$ (4.16.7)

Using (4.16.6) and (4.16.7) Equation (4.16.5) becomes

$$v = \frac{\dfrac{2}{N - n_{\theta\text{lin}}}V_{\text{lin}} - \dfrac{2}{N - n_{\theta\text{nlin}}}V_{\text{nlin}}}{\text{var}\{y(k)\} - \dfrac{2}{N - n_{\theta\text{nlin}}}V_{\text{nlin}}} \cong \frac{V_{\text{lin}} - V_{\text{nlin}}}{\dfrac{N}{2}\text{var}\{y(k)\} - V_{\text{nlin}}}$$ (4.16.8)

Observe that both (4.16.8) and (4.16.2) have the same numerator factors, i.e., the process is linear only if both nonlinear measures (4.13.1) and (4.16.2) are equal to zero or can be considered zero with a given probability.   ■

Remarks:
1. The noises can be filtered by an appropriate parameter estimation algorithm.
2. Gooijer and Kumar (1992) used the Volterra series model as an alternative to the linear model.

Example 4.16.1 *Nonlinearity test of the simple Wiener model based on parameter estimation of simple, linear and nonlinear structures (Haber and Zierfuss, 1987)*

The method based on the parameter estimation of simple linear and nonlinear structures was used for the nonlinear simple Wiener model having a first-order lag term. All the parameters were the same as used in Examples 4.2.1 and 4.4.1, i.e., the transfer function of the linear part is

$$\frac{V(s)}{U(s)} = \frac{1}{1+10s}$$

and the equation of the nonlinear static term is

$$y(k) = 2 + v(k) + 0.5v^2(k)$$

The test signal was a PRTS with maximum length 26, amplitude ±2 and mean value zero. The sampling time was 2 [s] and the minimum switching time of the PRTS signal was 5-times more. $N = 130$ data pairs were used for the identification from $k = 131$ to $k = 260$. The parameters of the following structures were estimated:

- *second-order linear model (L2):*

  $$\hat{y}(k) = (1.398 \pm 0.6697)\hat{y}(k-1) - (0.4735 \pm 0.05563)\hat{y}(k-2)$$
  $$+(0.1607 \pm 0.04092) + (0.3312 \pm 0.01561)u(k-1)$$
  $$-(0.1794 \pm 0.02882)u(k-2)$$

- *second-order quadratic generalized Hammerstein model (H2):*

  $$\hat{y}(k) = (1.354 \pm 0.06704)\hat{y}(k-1) - (0.4367 \pm 0.5541)\hat{y}(k-2)$$
  $$+(0.1195 \pm 0.03747) + (0.2133 \pm 0.02702)u(k-1)$$
  $$-(0.1068 \pm 0.03118)u(k-2) + (0.05870 \pm 0.1153)u^2(k-1)$$
  $$-(0.02832 \pm 0.01224)u^2(k-2)$$

- *second-order quadratic parametric Volterra model (V2):*

  $$\hat{y}(k) = (1.458 \pm 0.02812)\hat{y}(k-1) - (0.5242 \pm 0.02325)\hat{y}(k-2)$$
  $$+(0.1281 \pm 0.01553) + (0.2228 \pm 0.01121)u(k-1)$$
  $$-(0.1556 \pm 0.01308)u(k-2) + (0.211 \pm 0.005024)u^2(k-1)$$
  $$-(0.07344 \pm 0.005405)u^2(k-2) + (0.08429 \pm 0.03487)u(k-1)u(k-2)$$

TABLE 4.16.1 Calculation of the nonlinearity measures at the identification of the simple Wiener model by linear, generalized Hammerstein and parametric Volterra models

Model	Order	Standard deviation of the residual	Loss function	F-value	Nonlinearity index
linear (L2)	2	0.167	3.6126		0.0
quadratic generalized Hammerstein (H2)	2	0.150	2.925	$\dfrac{3.626-2.925}{2.925}\dfrac{130-7}{7-5}=14.74$	$\dfrac{14.74-3.07}{3.07}=3.80$
quadratic parametric Volterra (V2)	2	0.062	0.500	$\dfrac{3.626-0.5}{0.5}\dfrac{130-8}{8-5}=254.25$	$\dfrac{254.25-2.7}{2.7}=93.17$

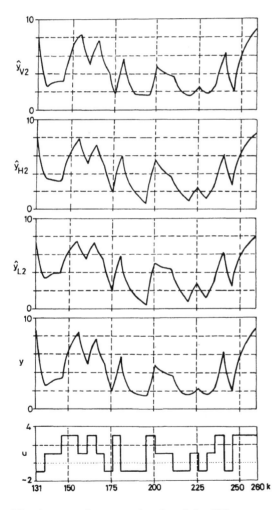

Fig. 4.16.1 The measured input u and output y signals of the first-order quadratic Wiener model and the computed output signals of the estimated second-order linear $\hat{y}_{L2}$, generalized Hammerstein $\hat{y}_{H2}$ and parametric Volterra models $\hat{y}_{V2}$

The input and output signals of the Wiener model and the computed output signals based on the identified models are drawn in Figure 4.16.1. The nonlinear character of the investigated system is well seen by the better output matching of the approximating nonlinear models than the linear one. The calculation of the nonlinearity measures is given in Table 4.16.1. The F-values were computed according to (4.16.2) and the nonlinearity indices according to (4.16.3). The $F(N, \Delta n_\theta, \alpha)$ values are $F(130, 2, 0.05) = 3.07$ for the generalized Hammerstein model and $F(130, 3, 0.05) = 2.7$ for the parametric Volterra model.

The nonlinear feature of the process was well detected.   ∎

4.17 METHOD BASED ON RESIDUAL ANALYSIS OF THE BEST FITTING LINEAR MODEL

It is quite easy to fit the best linear model to the input–output data. In practice it is often enough to assume that the order of the linear parametric model is equal to three or four and assume the dead time at the minimum derived from *a priori* (e.g., physical) knowledge. Repeated parameter estimation with different orders and dead times in a small range also leads quickly to the best fitting linear model. It is expedient to use more effective methods than the least squares method if the measured data are noisy. After having obtained the parameters of the best fitting linear process (and noise) model the residuals (the estimated source noise record) $\varepsilon(k)$ can be computed.

If the process is linear then the cross-correlation function between any nonlinear function of the input signal $g_i^u(u(k))$ and the residual is zero for all shifting times

$$r_{g_i^u, \varepsilon}(\kappa) = 0 \qquad \forall \kappa$$

To decide whether the correlation function is zero or not, it is expedient to normalize its components. Different recommendations for the components of the cross-correlation function and therefore the nonlinearity tests are known. The following conditions have to be fulfilled if the process is linear, i.e., the process is nonlinear if any of them is not valid:

- Billings and Fakhouri (1978), Haber (1979, 1985), Billings and Voon (1986):

$$r_{x_1 \varepsilon'}(\kappa) = 0 \qquad \forall \kappa = 0$$

where $\varepsilon'(k)$ is the normalized residual

$$\varepsilon'(k) = \frac{\varepsilon(k) - E\{\varepsilon(k)\}}{\sigma\{\varepsilon(k)\}} \approx \frac{\varepsilon(k)}{\sigma\{\varepsilon(k)\}}$$

and $x_1(k)$ is identical with the multiplier of (4.9.1) used in the second-order (nonlinear) cross-correlation test method,

$$x_1(k) = \frac{u'^2(k) - E\{u'^2(k)\}}{\sigma\{u'^2(k)\}}$$

- Billings and Voon (1986) also suggest a similar test, where the square of the residual is considered

$$r_{(x_1^2)'(\varepsilon^2)'}(\kappa) = 0 \qquad \forall \kappa$$

- Leontaritis and Billings (1987) recommend building the cross-correlation function for residual analysis from any nonlinear function

$$g_i^{uy\varepsilon}\left(u(k),u(k-1),\ldots,y(k-1),y(k-2),\ldots,\varepsilon(k-1),\varepsilon(k-2),\ldots\right) \qquad (4.17.1)$$

and from the residual $\varepsilon(k)$. The model is linear if and only if

$$r_{\left(g_i^{uy\varepsilon}\right)'\varepsilon'}(\kappa)=0 \qquad \kappa\ge 0 \qquad (4.17.2)$$

A negative shifting time is not allowed in (4.17.2), since the auto-correlation function of the residuals does not vanish for zero shifting time. That is the reason also for computing the nonlinear function (4.17.1) from $y(k-1)$ and not from $y(k)$, since the actual value of the measured output signal can be affected by the noise $e(k)$;

- On the basis of theoretical considerations and simulations Leontaritis and Billings (1987) recommend as a good choice of the multiplier the square of the measured output signal

$$g_i^{uy\varepsilon}(\ldots)=y^2(k-1)$$

It is expedient, however, to normalize both components in the cross-correlation function. The recommended test for testing the linearity of the residuals is therefore

$$r_{x_2\varepsilon'}(\kappa)=0 \qquad \kappa>0$$

where

$$x_2(k)=\frac{y'^2(k)-\mathrm{E}\left\{y'^2(k)\right\}}{\sigma\left\{y'^2(k)\right\}}$$

The above test is equivalent to the nonlinearity test of Billings and Voon (1983) applied now to the residuals and not to the process output, as in (4.10.2).

Above and further correlation function based model validity tests will be presented in Chapter 6 in details.

Example 4.17.1 *Nonlinearity test of the simple Wiener model based on fitting a good approximating linear model and analyzing the residuals (Haber and Zierfuss, 1987)*

The simple Wiener model was the same as in Example 4.16.1, i.e., the linear dynamic part had the transfer function

$$\frac{V(s)}{U(s)}=\frac{1}{1+10s}$$

and the equation of the nonlinear static term was

$$y(k)=2+v(k)+0.5v^2(k)$$

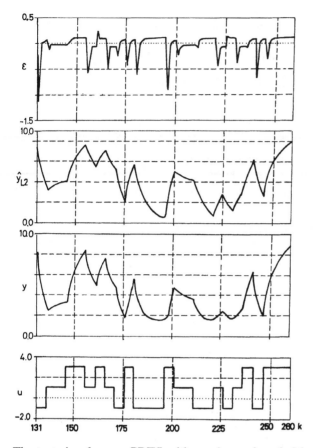

Fig. 4.17.1 The measured
input u and output y signals
of the noise-free simple
Wiener model, the computed
output signal $\hat{y}$ and the
residuals ε of the fitted linear
model

The test signal was a PRTS with maximum length 26, amplitude ± 2 and mean value
zero. The sampling time was 2 [s] and the minimum switching time of the PRTS
signal was 5-times more. $N = 26 \cdot 5 = 130$ data pairs were used for the identification
from $k = 131$ to $k = 260$. Both noise-free and disturbed cases were simulated. In the
latter case a white noise with zero mean and standard deviation 0.2 was added to the
noise-free output signal. (It corresponded to a 10 percent noise to signal ratio.)

It could be seen from the measurements' records that the process had no dead time.
From fitting of linear models with different orders, the second-order model showed a
good fitting and its estimated parameters were all significant. In the noise-free case the
least squares method, and in the noisy case the extended matrix method with ten
iterations, was applied for the parameter's estimation.

Let us first investigate the noise-free simulation. The measured input and output
signals, the computed output signal of the estimated linear model, and the computed
residuals are seen in Figure 4.17.1. The estimated parameters and their standard
deviations were already given in Example 4.16.1 by (4.16.9). Figure 4.17.2 shows the
auto-correlation function of the residuals and the cross-correlation functions
$r_{u'\varepsilon'}(\kappa), r_{x_1\varepsilon'}(\kappa), r_{x_2\varepsilon'}(\kappa)$. The correlation functions can be considered to be zero if their

values lie inside the 95% confidence limits around zero. This threshold is $\pm 2/\sqrt{130}$

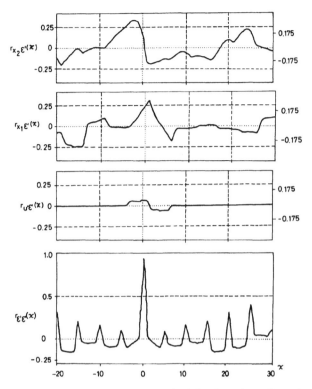

Fig. 4.17.2 Auto-correlation function and different cross-correlation functions of the residuals at the fitting of a linear model to the noise-free simple Wiener model

$= \pm 0.175$, which is also marked in the plots. Based on the plot $r_{\varepsilon'\varepsilon'}(\kappa)$ and, even better, based on the plot of $r_{u'\varepsilon'}(\kappa)$ the linear approximation seems to be satisfactory. The nonlinear feature is then clearly shown by the cross-correlation functions $r_{x_1\varepsilon'}(\kappa)$ and $r_{x_2\varepsilon'}(\kappa)$, which exceed the zero confidence limits ± 0.175 significantly.

The same procedure has been performed for the noisy simulations. The results of the parameter estimations are as follows

$$\hat{y}(k) = -(0.02587 \pm 0.5536)\hat{y}(k-1) + (0.6875 \pm 0.4526)\hat{y}(k-2)$$
$$+(0.6415 \pm 0.1979) + (0.3825 \pm 0.02393)u(k-1)$$
$$+(0.3277 \pm 0.2151)u(k-2) + (0.8412 \pm 0.5583)\varepsilon(k-1)$$
$$+(0.05077 \pm 0.074)\varepsilon(k-2)$$

The measured input and output signals, the computed output signal of the estimated process model (without the noise model), and the computed residuals are seen in Figure 4.17.3. Figure 4.17.4 shows the auto-correlation function of the residuals and the cross-correlation functions $r_{u'\varepsilon'}(\kappa)$, $r_{x_1\varepsilon'}(\kappa)$, $r_{x_2\varepsilon'}(\kappa)$. On the basis of the plots of the correlation functions $r_{\varepsilon'\varepsilon'}(\kappa)$ and $r_{u'\varepsilon'}(\kappa)$ the linear approximation seems to be satisfactory. The nonlinear feature is clearly seen from $r_{x_2\varepsilon'}(\kappa)$, which function exceeds the zero confidence limits ± 0.175 significantly.

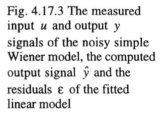

Fig. 4.17.3 The measured
input u and output y
signals of the noisy simple
Wiener model, the computed
output signal $\hat{y}$ and the
residuals ε of the fitted
linear model

4.18 CONCLUSIONS

Sixteen different nonlinearity test methods were reviewed. Analytical indices have been
presented for the nonlinearity tests. The effect of the additive noise was also
investigated. Several simulation runs were performed where the linear and nonlinear
characters of known linear and nonlinear dynamic models were correctly detected.

It is already known that the normalized dispersion and cross-correlation functions hardly
differ from each other if the process is slightly nonlinear, e.g., the excitation is far from
the extremum of a quadratic curve. It was an interesting observation that the absolute
value of the nonlinear cross-correlation function was approximately proportional to the

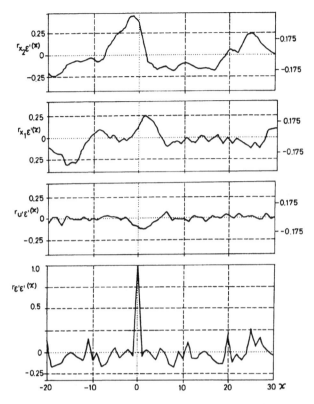

Fig. 4.17.4 Auto-correlation function and different cross-correlation functions of the residuals at the fitting of a linear model to the noisy simple Wiener model

deviation between the normalized dispersion function and the absolute value of the linear cross-correlation function and furthermore the coefficient of proportionality was generally bigger than one

$$\left|r_{x_1 y'}(\kappa)\right| \cong c(\kappa)\left[D'_{u'y'}(\kappa) - \left|r_{u'y'}(\kappa)\right|\right] \qquad c(\kappa) > 1$$

This means that the nonlinear cross-correlation method is a much more sensitive nonlinearity test method than the dispersion method.

According to the authors' – perhaps subjective – opinion, the following methods proved most suitable for detecting the linear or nonlinear feature of a process:

- the time domain test;
- second-order (nonlinear) cross-correlation method;
- second-order (output) auto-correlation method;
- method based on parameter estimation of simple linear and nonlinear structures;
- method based on residual analysis of the best fitting linear model.

The steady state and the output average methods are also good procedures but they detect only the steady state nonlinear character of a process.

The time domain test is the easiest to perform; however, it needs repeated experiments with a new test signal being proportional to that applied in the previous test. This test is applied very often in the industry (see e.g., Thomson *et al.*, 1996).

The correlation methods need some more computation than the time domain test. The main drawback of the method is, however, that the process have to be excited for a relatively long period by a stationary stochastic test signal.

Both the time domain and correlation tests require special test signals (including repeated experiments). The main advantage of the methods based on parameter estimation of only linear or of linear and simple nonlinear models is that they do not require special test signals. (Of course, the identifiability criteria should be fulfilled.) It means that the latter methods can be applied even when the other methods cannot be used.

The dispersion method is instrumental among the nonlinearity tests, because the method based on the comparison of the variances of the noise-free model output signals of the identified best linear and alternative nonlinear models is a consequence of it, and from the latter one several nonlinearity test procedures such as

- the frequency method;
- the method based on the identification of orthogonal subsystems;
- the method based on parameter estimation and F-test

can be derived.

Both the method based on the identification of the first-order Uryson model and the frequency method are special representations of the method based on the identification of orthogonal subsystems.

The second-order (nonlinear) cross-correlation and second-order (output) auto-correlation methods detect only those systems nonlinear whose components of different degrees are not of differentiating or integrating type.

A nonlinear system with odd degree nonlinear components may be detected as linear by the second-order (nonlinear) cross-correlation and second-order (output) auto-correlation methods if the mean value of the stochastic test signal is zero.

From the last two remarks it follows that nonlinearity tests are to be applied with caution.

As it has already been mentioned in the introduction

- the nonlinearity test cannot be extrapolated outside the input signal domain investigated;
- the test should be performed with all possible test signals for which types it was defined in the input signal domain.

4.19 REFERENCES

Allgöver, F. (1995a). Definition and computation of a nonlinearity measure, *IFAC Symp. on Nonlinear Control System Design,* (Tahoe City: CA, USA), pp. 257-262.

Allgöwer, F. (1995b). *Nonlinearity Measure – A Tool for the Analysis and Synthesis of Nonlinear Control Systems* (in German), (In: Design of nonlinear control systems, Editor: S. Engell), Publishing House Oldenburg, (Munich: Germany), pp. 309-331.

Bendat, J.S. and A.G. Piersol (1971). *Random Data: Analysis and Measurement Procedures,* Wiley Interscience, (New York: USA).

Bendat, J.S. and A.G. Piersol (1980). *Engineering Applications of Correlation and Spectral Analysis,* J. Wiley and Sons, (New York: USA).

Bendat, J.S. (1990). *Nonlinear System Analysis and Identification from Random Data,* J. Wiley and Sons, (New York: USA).

Billings, S.A. and S.Y. Fakhouri (1978). Theory of separable processes with applications to the identification of nonlinear systems. *Proc. IEE,* Part D, Vol. 129, 9, pp. 1051-1058.

Billings, S.A. and W.S.F. Voon (1983). Structure detection and model validity tests in the identification of nonlinear systems, *IEE Proc.,* Part D, Vol. 130, 4, pp. 193-199.

Billings, S.A. and W.S.F. Voon (1986). Correlation based model validity tests for nonlinear models, *Int. Journal of Control,* Vol. 44, 1 , pp. 235-244.

Bohlin, T. (1978). Maximum-power validation of models without higher-order fitting, *Automatica,* Vol. 14, pp. 137-146.

Box G.E.P. and G.M. Jenkins (1976). *Time Series Analysis - Forecasting and Control,* Holden-Day, (Oakland: CA, USA).

Cox, D.R. and D.V. Hinkley (1974). *Theoretical Statistics,* Chapman and Hall, (London: UK).

Draper, N.R. and H. Smith (1966). *Applied Regression Analysis,* John Wiley and Sons, (New York: USA).

Emara-Shabaik, H.E. and K.A.F. Moustafa (1994). Characterization of dynamic system nonlinearities via probabilistic approach, *Int. Journal of Systems Science,* Vol. 25, 3, pp. 603–611.

Gallman, P.G. and K.S. Narendra (1971). Identification of nonlinear systems using an Uryson model, *Report Ct-38,* Department of Engineering and Applied Science, Yale University , (New Haven: USA).

Gardiner, A.B. (1966). Elimination of the effect of nonlinearities on process cross-correlations, *Electronics Letters,* Vol. 2, pp. 164.

Gardiner, A.B. (1973). Identification of processes containing single-valued nonlinearities, *Int. Journal of Control,* Vol. 5, pp. 1029-1039.

Gasser, T. (1975). Goodness-of-fit tests for correlated data, *Biometrica,* Vol. 62, 3, pp. 563-570.

Gooijer, J.G. and K. Kumar (1992). Some recent developments in nonlinear time series modelling, testing and forecasting, *Int. Journal of Forecasting,* Vol. 8, pp. 135–156.

Gustavsson, I. (1972). Comparison of different models for identification of industrial processes, *Automatica,* Vol. 8, pp. 127-142.

Haber, R. (1979). Parametric identification of nonlinear dynamic systems based on correlation functions, *Prepr. 5th IFAC Symp. on Identification and System Parameter Estimation,* (Darmstadt: FRG), pp. 515-522.

Haber, R. (1985). Nonlinearity tests for dynamic processes. *Prepr. 7th IFAC Symp. on Identification and System Parameter Estimation,* (York: UK), pp. 409-413.

Haber, R. (1989). Different nonlinearity test methods for dynamic processes. *Report FHK-AV-PLT-89/1,* Laboratory of Process Control, Faculty of Plant and Process Engineering, Cologne Polytechnic, (Köln: Germany).

Haber, R. and L. Keviczky (1974). Nonlinear structures for system identification, *Periodica Polytechnica - Electrical Engineering,* Vol. 18, 4, pp. 394-404.

Haber, R. and R. Zierfuss (1987). Investigation of different model validity test methods. *Report TUV-IMPA-87/1,* Institute of Machine- and Process Automation, Technical University of Vienna, (Vienna: Austria).

Kaminskas, V. and A. Rimidis (1982). System structure identification, *Prepr. 6th IFAC Symposium on Identification and System Parameter Estimation,* (Washington, D.C.: USA), pp. 104-109.

Kennedy, J.B. and A.M. Neville (1976). *Basic Statistical Methods for Engineers and Scientists,* Thomas Y. Crowell Company, (New York: USA).

Korenberg, M.J. and I.W. Hunter (1990). The identification of nonlinear biological systems: Wiener kernel approach. *Annals of Biomedical Engineering,* Vol. 18, pp. 629-654.

Krempl, R. (1973). Application of three-level pseudo-random signals for parameter estimation of nonlinear systems, *Prepr. 3rd IFAC Symposium on Identification and System Parameter Estimation,* (Hague: The Netherlands), pp. 835-838.

Leontaritis, I.J. and S.A. Billings (1987). Model selection and validation methods for nonlinear systems, *Int. Journal of Control,* Vol. 45, 1, pp. 311-341.

Moustafa, K.A.F. and H.E. Emara-Shabaik (1992). Nonlinearity detection of weakly nonlinear dynamic systems using cumulants, *Int. Journal of Systems Science,* Vol. 23, 7, pp. 1167–1177.

Naka, Ken-Ichi (1975). Identification of a function and structure in central nervous system, *Prepr. 1st Symposium on Testing and Identification of Nonlinear Systems,* (Pasadena: USA), pp. 205-220.

Peebles, P.Z. (1980). *Probability, Random Variables and Random Signal Principles,* McGraw-Hill Book Company, (New York: USA).

Rajbman, N.S. and V.M. Chadeev (1980). *Identification of Industrial Processes,* North Holland Publ. Co., (Amsterdam: The Netherlands), pp. 435.

Saboke, J. (1985). Dynamic behavior of tool machines (in German), *Messen-Prüfen-Automatisieren,* Vol. 5, pp. 534-543.

Schetzen, M. (1980). *The Volterra and Wiener Theories of Nonlinear Systems,* Wiley Interscience, (New York: USA).

Subba Rao, T. and M.M. Gabr (1979). A test for linearity of stationary time series, *Technical Report 105,* Department of Mathematics (Statistics), Institute of Science and Technology, University of Manchester, (Manchester: UK).

Szücs, B., E. Monos and F. Csáki (1975). New aspects of blood pressure control, *Prepr. 6th IFAC World Congress,* (Boston: USA), 54, pp. 5.1-5.10.

Thomson, M., S.P. Schooling and M. Soufian (1995). The practical application of a nonlinear identification methodology, *Control Engineering Practice,* Vol. 4, 3, pp. 295–306.

Várlaki, P., G. Terdik and V.A. Lototsky (1985). Tests for linearity and bilinearity of dynamic systems, *Prepr. 7th IFAC Symposium on Identification and System Parameter Estimation,* (York: UK), pp. 427-432.

5. Structure Identification

5.1 INTRODUCTION

In the last decade several methods have been elaborated for the identification of nonlinear dynamic systems. Most of the methods assume that the structure of the system is given *a priori*. Therefore they are in reality parameter estimation algorithms, and structure identification is thus usually performed by repeated parameter estimation. However, in nonlinear system theory several methods are known to determine the structure of a system. (The words *structure identification, structure selection* and *structure determination* are used as synonyms in this chapter.)

The structure identification algorithms are organized according to the classes of nonlinear dynamic models and to the kind of experiment performed on the unknown process. The classification of the methods in this chapter follows the system recommended by Haber and Unbehauen (1990). The following methods are reviewed.

1. *Structure identification of block oriented models:*
   - based on estimated Volterra kernels;
   - frequency method;
   - from impulse or step responses;
   - based on the estimated Wiener–Hammerstein model.
2. *Structure identification of cascade models:*
   - correlation analysis;
   - frequency method;
   - from impulse or step responses.
3. *Structure identification quasi-linear models with signal dependent parameters:*
   - from normal operating data;
   - from step responses.
4. *Structure identification of models linear-in-parameters:*
   - all possible regressions;
   - regression analysis using non-orthogonal model components;
   - regression analysis using orthogonal model components.
5. *Group method of data handling (GMDH)*

Generally the models may have polynomial nonlinear terms, which may cause a continuous nonlinear steady state characteristic and/or a continuous function of dynamic parameter (e.g., time constant) on any signal (e.g., input or output signal). For simplification most models are only considered up to quadratic form. Extensions to higher degree nonlinear models can be done easily. (Not all methods can be applied if only a polynomial description is used. The correlation analysis applied to the cascade models can be applied for any single valued static nonlinearity. The models linear-in-parameters can be built up from any set of known functions. The submodels of the GMDH can also have structures other than the polynomial.)

If the nonlinear process has a polynomial steady state characteristic with finite degree the output signal can be separated as the outputs of several different parallel (constant, linear, quadratic, cubic, etc.) submodels. The structure identification and parameter estimation of the individual submodels are much easier than those of the whole model. Therefore this method is presented before the other structure identification methods are considered.

5.2 SEPARATION OF THE SYSTEM'S RESPONSE TO PARALLEL CHANNELS OF DIFFERENT NONLINEAR DEGREE

Gardiner (1966, 1968) developed a method for the separation of the output signal of nonlinear dynamic systems to subsystems attributable to the parallel channels of different degrees of nonlinearity. The advantage of the technique is that homogeneous subsystems may be obtained whose identification is simpler. Thus the identification of the linear dynamic and nonlinear static parts of a cascade system does not require any long and error-accumulating iteration procedure, as after determination of the linear part the parameters of the static nonlinearity can be calculated by simple regression.

The output signal $y(k)$ of a degree-n system may be given by the formula

$$y(k) = \sum_{i=0}^{n} V_i[u(k)] \tag{5.2.1}$$

where $u(k)$ is the input signal and $V_i[u(k)]$ is the output of a degree-i subsystem – characterized, e.g., by its degree-i homogeneous Volterra kernel (Figure 5.2.1).

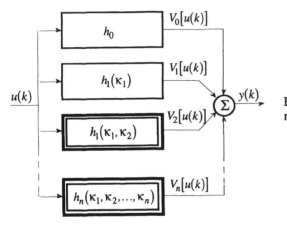

Fig. 5.2.1 Parallel separation of a nonlinear system

Theorem 5.2.1 (Extension of Gardiner, 1966, 1968) Assuming that the experiments are reproducible excite the degree-n system $N \geq (n+1)$-times in such a way that the amplitude of the ith input signal series is γ_i-times greater than the first one $\{u_1(k) \equiv u(k)\}$

$$u_j(k) = \gamma_j u_1(k) = \gamma_j u(k), \qquad j = 2, 3, \ldots, n+1, \qquad \gamma_1 = 1, \ \gamma_{j_1} \neq \gamma_{j_2}, \ j_1 \neq j_2 \tag{5.2.2}$$

Arrange the output signals of the different tests belonging to the same time point k into the vector

$$y = [y_1(k), \ldots, y_{n+1}(k)]^{\mathrm{T}} \tag{5.2.3}$$

and arrange the outputs of the subsystems of different degrees to the input signal $u_1(k) \equiv u(k)$ into the vector $V[u(k)]$

$$V[u(k)] = \left[V_0[u(k)], V_1[u(k)], \ldots, V_n[u(k)]\right]$$ (5.2.4)

Introduce the matrix Γ,

$$\Gamma = \begin{bmatrix} 1 & \gamma_1 & \cdots & \gamma_1^n \\ 1 & \gamma_2 & \cdots & \gamma_2^n \\ \vdots & \vdots & \ddots & \vdots \\ 1 & \gamma_{N+1} & \cdots & \gamma_{N+1}^n \end{bmatrix} \qquad \gamma_1 = 1$$ (5.2.5)

The output signal $y_1(k) \equiv y(k)$ can be separated into its components belonging to the parallel subsystems of different nonlinear degrees for any time points

$$V[u(k)] = \Gamma^* y(k) \qquad 1 \le k \le N$$ (5.2.6)

where Γ^* is the pseudo-inverse of Γ (extension of Gardiner, 1966, 1968)

$$\Gamma^* = \left[\Gamma^{\mathrm{T}}\Gamma\right]^{-1}\Gamma^{\mathrm{T}}$$ (5.2.7)

Proof.   From (5.2.1) and (5.2.2) we obtain

$$y_j(k) = \sum_{i=0}^{n} V_i\left[\gamma_j[u(k)]\right] = \sum_{i=0}^{n} \gamma_j^i V_i[u(k)]$$ (5.2.8)

Collect the measured data in time points k into a matrix form

$$\begin{bmatrix} y_1(k) \\ \vdots \\ y_N(k) \end{bmatrix} = \begin{bmatrix} 1 & \gamma_1 & \cdots & \gamma_1^n \\ \vdots & \ddots & \vdots & \vdots \\ \vdots & \vdots & \ddots & \vdots \\ 1 & \gamma_N & \cdots & \gamma_N^n \end{bmatrix} \begin{bmatrix} V_0[u(k)] \\ \vdots \\ V_n[u(k)] \end{bmatrix} \qquad \gamma_1 = 1$$ (5.2.9)

or introducing the corresponding matrices

$$y(k) = \Gamma V[u(k)]$$ (5.2.10)

The vector $V[u(k)]$ can be obtained from (5.2.10) by means of the least squares method (5.2.6) with (5.2.7).   ∎

Remarks:

1. Gardiner (e.g., 1973b) recommended exciting the process only $N = (n+1)$-times. Then (5.2.6) reduces to

$$V[u(k)] = \Gamma^{-1} y(k)$$

which means that the pseudo-inverse becomes a simple inverse $\left(\Gamma^* = \Gamma^{-1}\right)$.

2. The matrix Γ is called the Vandermonde matrix and its inverse exists if $\gamma_{i_1} \neq \gamma_{i_2}$, $i_1 \neq i_2$.

3. Sometimes it may be reasonable to perform more measurements than $(n+1)$. This may be required because of the noisy measurements or that the degree of the nonlinearity is unknown and the model has to be fitted to more records measured in the investigated domain with different amplitudes. In this case the components of the output signal can be computed by using the LS method (5.2.6) with (5.2.7).

4. The time series of the output signals' components $V_i[u(k)]$ can be obtained from the output signal series $y_i(k)$ (denoted by y_i) by performing the multiplication (5.2.5) for each time:

$$
\begin{bmatrix} V_0[u(1)] & \cdots & V_0[u(N)] \\ \vdots & \ddots & \vdots \\ V_n[u(1)] & \cdots & V_0[u(N)] \end{bmatrix} = \Gamma^* \begin{bmatrix} y_1(1) & \cdots & y_1(N) \\ \vdots & \ddots & \vdots \\ y_N(1) & \cdots & y_N(N) \end{bmatrix}
$$

In the sequel some practical advice is given to simplify the use of Gardiner's method.

Lemma 5.2.1
(Gardiner, 1973b) The dimension of the matrix to be inverted can be decreased by half if the test signals are pairwise the inverse of each other:

$$y_{2j} = -y_{2j-1} \qquad j = 1, \dots, N/2 \qquad\qquad (5.2.11)$$

The number of excitations should be even and $N \geq n+1$ has to be satisfied.
Define the pairwise sums and differences of the measured output signals as

$$y_{2j-1}^*(k) = \frac{y_{2j-1}(k) + y_{2j}(k)}{2} \qquad j = 1, \dots, N/2 \qquad\qquad (5.2.12)$$

and

$$y_{2j}^*(k) = \frac{y_{2j-1}(k) + y_{2j}(k)}{2y_{2j-1}(k)} \qquad j = 1, \dots, N/2 \qquad\qquad (5.2.13)$$

The even degree components of the output signal $y(k) \equiv y_1(k)$ can be calculated by the formula

$$
\begin{bmatrix} V_0[u(k)] \\ V_2[u(k)] \\ \vdots \\ V_v[u(k)] \end{bmatrix} = \Gamma'^* \begin{bmatrix} y_1^*(k) \\ y_3^*(k) \\ \vdots \\ y_{N-1}^*(k) \end{bmatrix} \qquad\qquad (5.2.14)
$$

and the odd degree components of the output signal can be calculated by

$$
\begin{bmatrix} V_0[u(k)] \\ V_2[u(k)] \\ \vdots \\ V_v[u(k)] \end{bmatrix} = \mathbf{\Gamma'}^* \begin{bmatrix} y_1^*(k) \\ y_3^*(k) \\ \vdots \\ y_N^*(k) \end{bmatrix}
$$
(5.2.15)

v and μ are the largest even and odd numbers, respectively, which are less than the degree n.

In both equations (5.2.14) and (5.2.15) $\mathbf{\Gamma'}^*$ is the pseudo-inverse of $\mathbf{\Gamma'}$

$$
\mathbf{\Gamma'} = \begin{bmatrix} 1 & \gamma_1^2 & \gamma_1^4 & \cdots \\ 1 & \gamma_3^2 & \gamma_3^4 & \cdots \\ \vdots & \vdots & \ddots & \vdots \\ 1 & \gamma_{N/2}^2 & \gamma_{N/2}^4 & \cdots \end{bmatrix}
$$
(5.2.16)

Proof.   Write the first four rows in (5.2.9) in detail

$$
\begin{bmatrix} y_1(k) \\ y_2(k) \\ y_3(k) \\ y_4(k) \\ \vdots \\ y_{N'}(k) \end{bmatrix} = \begin{bmatrix} 1 & \gamma_1 & \gamma_1^2 & \gamma_1^3 & \cdots & (\gamma_1)^n \\ 1 & -\gamma_1 & \gamma_1^2 & -\gamma_1^3 & \cdots & (-\gamma_1)^n \\ 1 & \gamma_3 & \gamma_3^2 & \gamma_3^3 & \cdots & (\gamma_3)^n \\ 1 & -\gamma_3 & \gamma_3^2 & -\gamma_3^3 & \cdots & (-\gamma_3)^n \\ \vdots & \vdots & \vdots & \vdots & \ddots & \vdots \\ 1 & -\gamma_{N/2} & \gamma_{N/2}^2 & -\gamma_{N/2}^3 & \cdots & (-\gamma_{N/2})^n \end{bmatrix} \begin{bmatrix} V_0[u(k)] \\ V_1[u(k)] \\ \cdots \\ \cdots \\ \vdots \\ V_n[u(k)] \end{bmatrix}
$$
(5.2.17)

Replace the output signals by their pairwise sums and differences

$$
\begin{bmatrix} y_1(k)+y_2(k) \\ y_1(k)-y_2(k) \\ y_3(k)+y_4(k) \\ y_3(k)-y_4(k) \\ \vdots \\ \cdots \end{bmatrix} = \begin{bmatrix} 2 & 0 & 2\gamma_1^2 & 0 & \cdots \\ 0 & 2\gamma_1 & 0 & 2\gamma_1^3 & \cdots \\ 2 & 0 & 2\gamma_3^2 & 0 & \cdots \\ 0 & 2\gamma_2 & 0 & 2\gamma_3^3 & \cdots \\ \vdots & \vdots & \vdots & \vdots & \cdots \\ \cdots & \cdots & \cdots & \cdots & \cdots \end{bmatrix} \begin{bmatrix} V_0[u(k)] \\ V_1[u(k)] \\ \cdots \\ \cdots \\ \vdots \\ V_n[u(k)] \end{bmatrix}
$$
(5.2.18)

Using (5.2.12) and (5.2.13) we obtain

$$
\begin{bmatrix} y_1^*(k) \\ y_2^*(k) \\ y_3^*(k) \\ y_4^*(k) \\ \vdots \\ \cdots \end{bmatrix} = \begin{bmatrix} 1 & 0 & \gamma_1^2 & 0 & \cdots \\ 0 & 1 & 0 & \gamma_1^3 & \cdots \\ 1 & 0 & \gamma_3^2 & 0 & \cdots \\ 0 & 1 & 0 & \gamma_3^3 & \cdots \\ \vdots & \vdots & \vdots & \vdots & \cdots \\ \cdots & \cdots & \cdots & \cdots & \cdots \end{bmatrix} \begin{bmatrix} V_0[u(k)] \\ V_1[u(k)] \\ \cdots \\ \cdots \\ \vdots \\ V_n[u(k)] \end{bmatrix}
$$
(5.2.19)

Equation (5.2.19) separates into two smaller equations

$$
\begin{bmatrix} y_1^*(k) \\ y_3^*(k) \\ \vdots \\ \cdots \end{bmatrix} = \begin{bmatrix} 1 & \gamma_1^2 & \gamma_1^4 & \cdots \\ 1 & \gamma_3^2 & \gamma_3^4 & \cdots \\ \vdots & \vdots & \vdots & \vdots \\ \cdots & \cdots & \cdots & \cdots \end{bmatrix} \begin{bmatrix} V_0[u(k)] \\ V_2[u(k)] \\ \vdots \\ V_{n-1}[u(k)] \end{bmatrix}
\tag{5.2.20}
$$

and

$$
\begin{bmatrix} y_2^*(k) \\ y_4^*(k) \\ \vdots \\ \cdots \end{bmatrix} = \begin{bmatrix} 1 & \gamma_1^2 & \gamma_1^4 & \cdots \\ 1 & \gamma_3^2 & \gamma_3^4 & \cdots \\ \vdots & \vdots & \vdots & \vdots \\ \cdots & \cdots & \cdots & \cdots \end{bmatrix} \begin{bmatrix} V_1[u(k)] \\ V_3[u(k)] \\ \vdots \\ V_n[u(k)] \end{bmatrix}
\tag{5.2.21}
$$

Now the two new Vandermonde matrices are equal to each other and the output signals of the parallel channels of even and odd degree can be computed separately. ∎

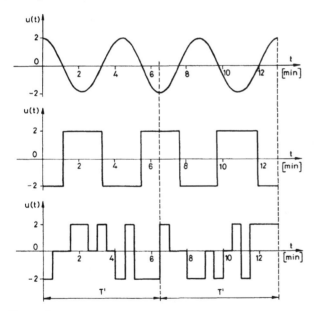

Fig. 5.2.2 Inverse-repeat test signals: (a) sinusoid; (b) square; (c) PRTS signal with maximum length 26 (Test signal (c) is generated by three ternary shift registers. All registers are fed back, the feedback coefficients are 1, -1, -1)

Remarks:

1. There are several test signals that have an inverse repeat character, that is to say, the second half of a period is the inverse of the first half if the mean value is zero. Applying them as a test signal, only $(n+1)/2$ tests have to be performed with different amplitudes, and the first and second half of the runs at each amplitude can be used as pairwise inverse excitations. Figure 5.2.2 presents three different test signals:
   - sinusoid;
   - square wave;
   - pseudo-random multi-level signal (PRMS).

2. The pulse transfer function between the input signal $u(k)$ and the outputs of the homogeneous nonlinear channels $V_i[u(k)]$ can be determined more easily than in the case when only the whole – measured – output signal $y(k)$ would exist. The elements of $\{V_0[u(k)]\}$ are equal to the constant term, linear identification may be performed between $\{V_1[u(k)]\}$ and $\{u(k)\}$, the operator $V_2[\ldots]$ may be approximated by a homogeneous quadratic model, and their parameters have to be fitted to the data sequences $\{u(k)\}$ and $\{V_2[u(k)]\}$.
3. Although the method can be used for any input signal, it is, however, more suitable to the fast evaluation of step responses of different amplitudes for nonlinear systems.

If the output signal has to be separated into the sum of the outputs of the odd and the even degree subsystems, then this can be done from two measurements as shown in Lemma 5.2.2.

Lemma 5.2.2
(Schetzen, 1965a) The output signals of the sums of all odd and even degree subsystems $y_o(k)$ and $y_e(k)$, respectively, can be calculated from the measured output signal if the experiment is repeated by the inverse ((-1)-times) of the original input signal

$$y_o(k) = \frac{1}{2}\left[y(u(k)) - y(-u(k))\right] \tag{5.2.22}$$

$$y_e(k) = \frac{1}{2}\left[y(u(k)) + y(-u(k))\right] \tag{5.2.23}$$

Proof.   The output signal to the input signal $u(k)$ is

$$y(k) = y(u(k)) = \sum_{i=0}^{n} y_i(k) = \sum_{i=0}^{n} V_i[u(k)]$$

and to the inverse of the input signal

$$y(-u(k)) = \sum_{i=0}^{n} V_i[-u(k)] = \sum_{i=0}^{n} (-1)^i V_i[u(k)]$$

(5.2.22) and (5.2.23) result from summing and subtracting the two output signals, respectively.   ∎
      The scheme of the separation is shown in Figure 5.2.3.

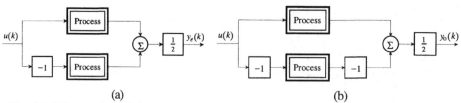

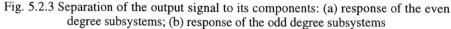
(a)   (b)
Fig. 5.2.3 Separation of the output signal to its components: (a) response of the even degree subsystems; (b) response of the odd degree subsystems

Example 5.2.1 *Separation of the step responses of a simple Hammerstein model into responses attributable to different powers of nonlinearity (Haber et al., 1986)*

The simple Hammerstein model under investigation has the quadratic static part

$$V = 2 + U + 0.5U^2$$

and the linear dynamic transfer function

$$\frac{Y(s)}{V(s)} = \frac{1}{1 + 10s}$$

as seen also in Figure 5.2.4. The process is excited by unit steps

$$u_1(t) = 1 \cdot 1(t) \qquad u_2(t) = 2 \cdot 1(t) \qquad u_3(t) = 4 \cdot 1(t)$$

respectively. The measured responses $y_1(t)$, $y_2(t)$ and $y_3(t)$ are plotted as continuous lines in Figure 5.2.5. Some values of the systems responses, as well as of the not measurable inner variable $v(t)$ are listed in Table 5.2.1.

TABLE 5.2.1 Step responses of a simple Hammerstein model to different excitations and the responses of the constant, linear and quadratic homogeneous channels to the unit step

t	$u(t)$: $0 \to 1$		$u(t)$: $0 \to 2$		$u(t)$: $0 \to 4$		$V_0[u(t)]$	$V_1[u(t)]$	$V_2[u(t)]$
	$v(t)$	$y(t)$	$v(t)$	$y(t)$	$v(t)$	$y(t)$			
0	3.5	2.0	6.0	2.0	14.0	2.0	2.0	0.0	0.0
2.5	3.5	2.332	6.0	2.885	14.0	4.654	2.0	0.221	0.110
5	3.5	2.590	6.0	3.574	14.0	6.722	2.0	0.394	0.197
10	3.5	2.948	6.0	4.528	14.0	9.585	2.0	0.632	0.316
15	3.5	3.165	6.0	5.108	14.0	11.322	2.0	0.777	0.388
20	3.5	3.297	6.0	5.459	14.0	12.376	2.0	0.865	0.432
25	3.5	3.377	6.0	5.672	14.0	13.015	2.0	0.918	0.459
30	3.5	3.425	6.0	5.801	14.0	13.403	2.0	0.950	0.475
40	3.5	3.473	6.0	5.927	14.0	13.780	2.0	0.982	0.491
50	3.5	3.490	6.0	5.973	14.0	13.919	2.0	0.993	0.497

According to Gardiner's method the responses $V_0[u(k)]$, $V_1[u(k)]$ and $V_2[u(k)]$, attributable to the constant, linear and quadratic channels, respectively, can be calculated from

$$V(k) = \begin{bmatrix} 1 & 1 & 1 \\ 1 & 2 & 4 \\ 1 & 4 & 16 \end{bmatrix}^{-1} y(k) = \begin{bmatrix} 2.667 & -2.0 & 0.333 \\ -2.0 & 2.5 & -0.5 \\ 0.333 & -0.5 & 0.167 \end{bmatrix} y(k)$$

The two resulting time functions are plotted as dotted lines in Figure 5.2.5 and some values are listed in Table 5.2.1.

The correctness of the method can easily be checked by computing the responses of the different channels in Figure 5.2.6, which is a parallel representation of the simple (cascade) Hammerstein model of Figure 5.2.4.

After having obtained the output signals of the channels, the parallel subsystems can be identified, since the quasi-input signals of the dynamic parts $u(t)$ and $u^2(t)$ are known. In this case they are unit steps, thus a grapho-analytical method is suitable for the determination of the gain and time constants of the channels. The constant term is the steady state value of $V_0[u(k)]$.

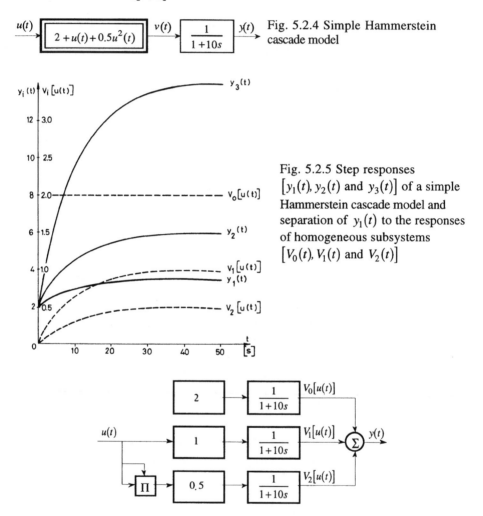

$$u(t) \rightarrow \boxed{2 + u(t) + 0.5u^2(t)} \xrightarrow{v(t)} \boxed{\frac{1}{1+10s}} \xrightarrow{y(t)}$$

Fig. 5.2.4 Simple Hammerstein cascade model

Fig. 5.2.5 Step responses $[y_1(t), y_2(t) \text{ and } y_3(t)]$ of a simple Hammerstein cascade model and separation of $y_1(t)$ to the responses of homogeneous subsystems $[V_0(t), V_1(t) \text{ and } V_2(t)]$

Fig. 5.2.6 Parallel representation of the simple Hammerstein cascade model of Figure 5.2.1

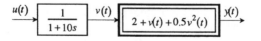

$$u(t) \rightarrow \boxed{\frac{1}{1+10s}} \xrightarrow{v(t)} \boxed{2 + v(t) + 0.5v^2(t)} \xrightarrow{y(t)}$$

Fig. 5.2.7 Simple Wiener cascade model

Example 5.2.2   *Separation of the step response of a simple Wiener model into responses attributable to different powers of nonlinearity (Haber et al., 1986)*

The simple Wiener model under investigation has the linear dynamic transfer function

$$\frac{V(s)}{U(s)} = \frac{1}{1+10s}$$

and the quadratic static part

$$Y = 2 + V + 0.5V^2$$

as can be also seen in Figure 5.2.7. The process is excited by the unit steps

$$u_1(t) = 1 \cdot 1(t) \qquad u_2(t) = 2 \cdot 1(t) \qquad u_3(t) = 4 \cdot 1(t)$$

respectively. The measured responses $y_1(t)$, $y_2(t)$ and $y_3(t)$ are plotted as continuous lines in Figure 5.2.8. Some values of the system responses, as well as of the non-measurable inner variable $v(t)$, are listed in Table 5.2.2.

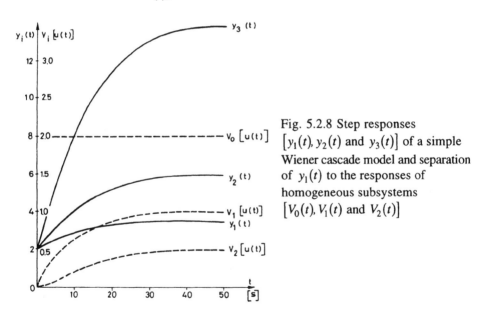

Fig. 5.2.8 Step responses $[y_1(t), y_2(t)$ and $y_3(t)]$ of a simple Wiener cascade model and separation of $y_1(t)$ to the responses of homogeneous subsystems $[V_0(t), V_1(t)$ and $V_2(t)]$

According to Gardiner's method the responses $V_0[u(k)]$, $V_1[u(k)]$ and $V_2[u(k)]$, attributable to the constant, linear and quadratic channels, respectively, can be calculated from

$$V(k) = \begin{bmatrix} 1 & 1 & 1 \\ 1 & 2 & 4 \\ 1 & 4 & 16 \end{bmatrix}^{-1} y(k) = \begin{bmatrix} 2.667 & -2.0 & 0.333 \\ -2.0 & 2.5 & -0.5 \\ 0.333 & -0.5 & 0.167 \end{bmatrix} y(k)$$

The resulting time functions are plotted as dotted lines in Figure 5.2.8 and some values are listed in Table 5.2.2.

TABLE 5.2.2 Step responses of a simple Wiener model to different excitations and the responses of the constant, linear, and quadratic homogeneous channels to the unit step

t	$u(t): 0 \to 1$		$u(t): 0 \to 2$		$u(t): 0 \to 4$		$V_0[u(t)]$	$V_1[u(t)]$	$V_2[u(t)]$
	$v(t)$	$y(t)$	$v(t)$	$y(t)$	$v(t)$	$y(t)$			
0	0.0	0	0.0	2.0	0.0	2.0	2.0	0.0	0.0
2.5	0.221	0.246	0.442	2.540	0.885	3.326	2.0	0.221	0.024
5	0.394	2.471	0.787	3.097	1.574	4.813	2.0	0.394	0.077
10	0.632	2.832	1.264	4.063	2.528	7.725	2.0	0.632	0.199
15	0.777	3.079	1.554	4.761	3.107	9.936	2.0	0.777	0.302
20	0.865	3.238	1.730	5.225	3.459	11.440	2.0	0.865	0.374
25	0.918	3.339	1.836	5.521	3.672	12.412	2.0	0.918	0.421
30	0.950	3.402	1.904	5.717	3.809	13.061	2.0	0.950	0.451
40	0.982	3.464	1.963	5.891	3.927	13.636	2.0	0.982	0.482
50	0.993	3.486	1.986	5.960	3.973	13.865	2.0	0.993	0.493

The correctness of the method can be verified by computing the responses of the different channels in Figure 5.2.9, which is a parallel representation of the simple (cascade) Wiener model of Figure 5.2.7.

After having obtained the output signals of the channels the parallel subsystems can be identified. The constant term is equal to $V_0[u(t)]$ for every t, and the transfer function of the linear part can be estimated from the linear channel whose input and output signals are known. The plots of Figure 5.2.8 show a close relation between $V_1[u(t)]$ and $V_2[u(t)]$. Static regression between the sequences $\{V_1[u(k)]\}$ and $\{V_2[u(k)]\}$ leads to the relation

$$V_2[u(t)] = 0.5V_1^2[u(t)]$$

Thus the quadratic channel, and therefore the simple Wiener model itself, is identified.

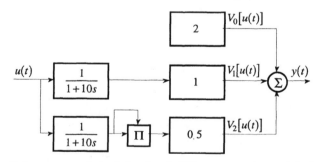

Fig. 5.2.9 Parallel representation of the simple Wiener cascade model of Figure 5.2.7

Example 5.2.3   *Separation of the step response of a simple Wiener–Hammerstein cascade model into responses attributable to different powers of nonlinearity*

The simple Wiener–Hammerstein cascade model under investigation has the linear dynamic part with transfer function

$$\frac{V_1(s)}{U(s)} = \frac{1}{1+5s}$$

before the quadratic static part

$$V_2 = 2 + V_1 + 0.5V_1^2$$

and the second linear dynamic term with the transfer function

$$\frac{Y(s)}{V_2(s)} = \frac{1}{1+10s}$$

behind the quadratic static part. The cascade model is seen in Figure 5.2.10. The process is excited by unit steps

$$u_1(t) = 1 \cdot 1(t) \qquad u_2(t) = 2 \cdot 1(t) \qquad u_3(t) = 4 \cdot 1(t)$$

respectively. The measured responses $y_1(t)$, $y_2(t)$ and $y_3(t)$ are plotted by continuous lines in Figure 5.2.11. Some values of the system's responses, as well as of the non-measurable inner variables $v_1(t)$ and $v_2(t)$ are listed in Table 5.2.3.

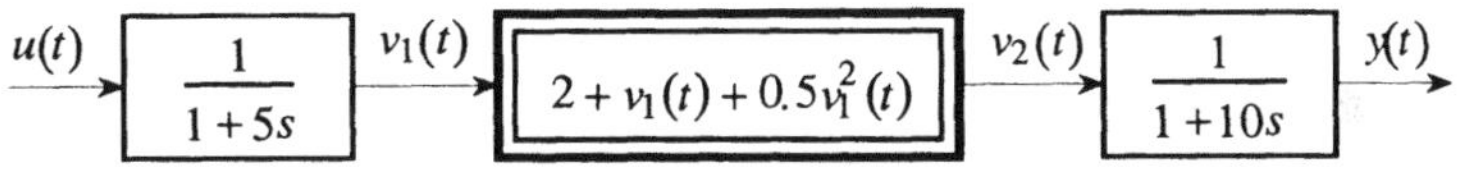

Fig. 5.2.10 Simple Wiener–Hammerstein cascade model

TABLE 5.2.3 Step responses of a simple Wiener–Hammerstein cascade model to different excitations and the responses of the constant, linear, and quadratic homogeneous channels to the unit step

t	$u(t)$: 0 → 1			$u(t)$: 0 → 2			$u(t)$: 0 → 4			$V_0[u(t)]$	$V_1[u(t)]$	$V_2[u(t)]$
	$v_1(t)$	$v_2(t)$	$y(t)$	$v_1(t)$	$v_2(t)$	$y(t)$	$v_1(t)$	$v_2(t)$	$y(t)$			
0	0.0	2.000	2.000	0.0	2.000	2.000	0.0	2.000	2.000	2.0	0.0	0.0
2.5	0.393	2.471	2.056	0.787	3.097	2.126	1.574	4.812	2.309	2.0	0.049	0.001
5	0.632	2.832	2.192	1.264	4.063	2.457	2.528	7.725	3.208	2.0	0.155	0.004
10	0.865	3.238	2.542	1.729	5.224	3.367	3.459	11.440	5.869	2.0	0.400	0.142
15	0.950	3.402	2.856	1.900	5.706	4.215	3.801	13.024	8.445	2.0	0.604	0.252
20	0.982	3.464	3.086	1.963	5.891	4.847	3.927	13.636	10.396	2.0	0.748	0.338
25	0.993	3.487	3.240	1.987	5.960	5.274	3.973	13.866	11.727	2.0	0.843	0.397
30	0.998	3.495	3.340	1.995	5.985	5.550	3.990	13.950	12.589	2.0	0.903	0.436
40	1.000	3.499	3.440	1.999	5.997	5.831	3.999	13.993	13.470	2.0	0.964	0.476
50	1.000	3.500	3.478	2.000	6.000	5.937	4.000	13.999	13.803	2.0	0.987	0.491
60	1.000	3.500	3.492	2.000	6.000	5.977	4.000	14.000	13.927	2.0	0.995	0.497
70	1.000	3.500	3.497	2.000	6.000	5.991	4.000	14.000	13.973	2.0	0.998	0.499
80	1.000	3.500	3.499	2.000	6.000	5.997	4.000	14.000	13.990	2.0	0.999	0.500
90	1.000	3.500	3.500	2.000	6.000	6.000	4.000	14.000	13.996	2.0	1.000	0.500
100	1.000	3.500	3.500	2.000	6.000	6.000	4.000	14.000	13.999	2.0	1.000	0.500

According to Gardiners' method the responses $V_0[u(k)]$, $V_1[u(k)]$, and $V_2[u(k)]$ to the unit step $1(t)$, and attributable to the constant, linear, and quadratic channels, respectively, can be calculated from

$$V(k) = \begin{bmatrix} 1 & 1 & 1 \\ 1 & 2 & 4 \\ 1 & 4 & 16 \end{bmatrix}^{-1} y(k) = \begin{bmatrix} 2.667 & -2.0 & 0.333 \\ -2.0 & 2.5 & -0.5 \\ 0.333 & -0.5 & 0.167 \end{bmatrix} y(k)$$

The resulting time functions are plotted by dotted lines in Figure 5.2.11 and some values are listed in Table 5.2.3.

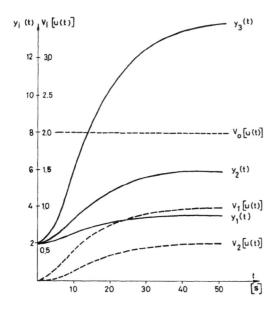

Fig 5.2.11 Step responses $[y_1(t), y_2(t) \text{ and } y_3(t)]$ of a simple Wiener–Hammerstein cascade model and separation of $y_1(t)$ to the responses of homogeneous subsystems $[V_0(t), V_1(t) \text{ and } V_2(t)]$

The correctness of the method can be verified by simulating the responses of the different channels in Figure 5.2.12, which is a parallel representation of the simple Wiener–Hammerstein cascade model of Figure 5.2.10.

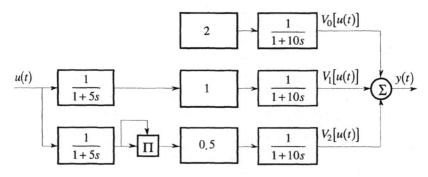

Fig. 5.2.12 Parallel representation of the simple Wiener–Hammerstein cascade model of Figure 5.2.10

After having obtained the output signals of the channels the parallel subsystems can be identified. The constant term is equal to the steady state value of $V_0[u(k)]$. The evaluation of the step response $V_1[u(k)]$ leads to the transfer function

$$G_1(s) = \frac{1}{(1+5s)(1+10s)}$$

The evaluation of the step response $V_2[u(k)]$ as the response of a linear system leads to

$$G_2(s) = \frac{0.5}{(1+5s)(1+2.5s)(1+10s)}$$

The common time constants in $G_1(s)$ and $G_2(s)$ show the time constants of the linear dynamic parts. Since $T_2 = 10$ [s] stands *alone* but $T_1 = 5$ [s] with its half $T_1/2 = 2.5$ [s], this shows that the transfer function with the time constant $T_1 = 5$ [s] is before and that $T_2 = 10$ [s] is after the quadratic static term. (This problem will be dealt with in Section 5.6 in detail, when structure identification of cascade models based on frequency and input step excitations will be treated.) It is enough, now to check that the square of the step response of the linear dynamic term before the quadratic static term is

$$v_1(t) = [1 - \exp(-0.2t)]^2 = 1 - 2\exp(-0.2t) + \exp(-0.4t)$$

which indicates a *quasi-linear* transfer function of the form

$$\frac{\mathcal{L}\{v_1(t)\}}{1/s} = \frac{1}{(1+5s)(1+2.5s)}$$

5.3 CASCADE STRUCTURE IDENTIFICATION FROM PARALLEL CHANNELS OF DIFFERENT NONLINEAR DEGREE

In Section 5.2 we have seen a method for separating a cascade system into parallel channels of different nonlinear degree. The essence of this procedure is that:
- it makes the determination of the highest degree of the nonlinearity possible;
- generally it is easier to identify the individual channels separately than the whole cascade system.

After having identified the parallel channels, the structure and parameters of the cascade system have to be reconstructed. That is the topic of this Section.
First it should be remarked that:
- every cascade system has a parallel equivalent;
- not every parallel subsystem has a cascade equivalent.

This statement will be illustrated by an example.

Example 5.3.1 *Cascade system containing three linear dynamic and two quadratic static elements*
Figure 5.3.1 shows a cascade system containing three linear dynamic and two quadratic static elements.

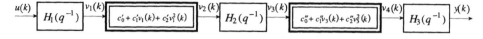

Fig. 5.3.1 Cascade system containing three linear dynamic and two quadratic static
elements

The cascade system can be considered as two cascaded simple Wiener models followed by
a linear dynamic term. In Figure 5.3.2 the simple Wiener models are drawn by their
equivalent generalized Wiener models. The separation of Figure 5.3.2 into parallel
channels has to be executed carefully because of the multiplier before c_2''. (It multiplies
the linear and quadratic channels of the generalized Wiener model at the input of the
system.) Finally, Figure 5.3.3 is the parallel representation of the cascade system.

The transformation of cascade to parallel can be calculated analytically. The output
signal of the first linear filter is

$$v_1(k) = H_1(q^{-1})u(k)$$

The output signal of the first simple Wiener model is

$$v_2(k) = c_0' + c_1'\left[H_1(q^{-1})u(k)\right] + c_2'\left[H_1(q^{-1})u(k)\right]^2$$

The internal signal in the second simple Wiener model is

$$v_3(k) = H_2(q^{-1})\left\{c_0' + c_1'H_1(q^{-1})u(k) + c_2'\left[H_1(q^{-1})u(k)\right]^2\right\}$$

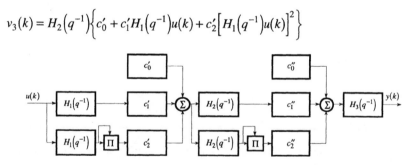

Fig. 5.3.2 Detailed scheme of the cascade system containing three linear dynamic and
two quadratic static elements

The output signal of the second simple Wiener model is

$$v_4(k) = c_0'' + c_1''v_3(k) + c_2''v_3(k)$$

$$= c_0'' + c''H_2(q^{-1})\left\{c_0' + c_1'H_1(q^{-1})u(k) + c_2'\left[H_1(q^{-1})u(k)\right]^2\right\}$$

$$+ c_2''\left[H_2(q^{-1})\left\{c_0' + c_1'H_1(q^{-1})u(k) + c_2'\left[H_1(q^{-1})u(k)\right]^2\right\}\right]^2$$

The output signal of the cascade system is

$$y(k) = H_3(q^{-1})v_4(k)$$

$$= H_3(q^{-1})c_0'' + H_3(q^{-1})c_1''H_2(q^{-1})\left\{c_0' + c_1'H_1(q^{-1})u(k) + c_2'\left[H_1(q^{-1})u(k)\right]^2\right\}$$

$$+H_3(q^{-1})c_2''\left[H_2(q^{-1})\left\{c_0' + c_1'H_1(q^{-1})u(k) + c_2'\left[H_1(q^{-1})u(k)\right]^2\right\}\right]^2$$

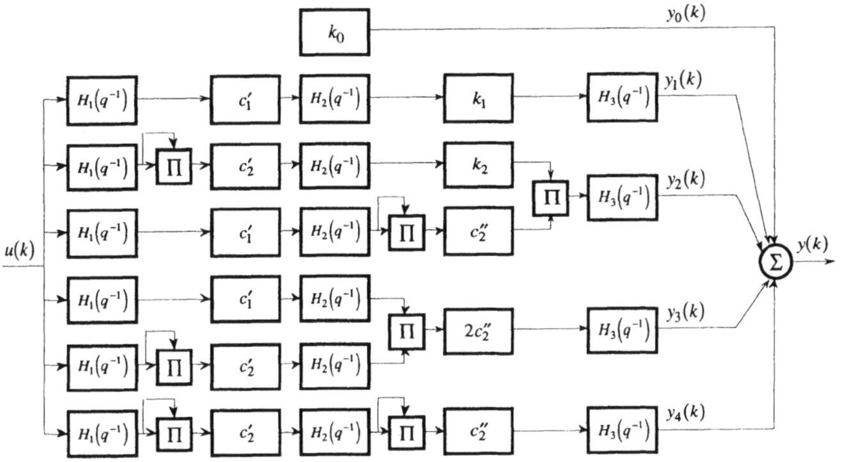

Fig. 5.3.3 Parallel representation of the cascade system containing three linear
dynamic and two quadratic static elements:

$$(k_0 = H_0\left\{c_0' + c_1''H_2(1)c_0' + c_2''\left[H_2(1)c_0'\right]^2\right\}, \quad k_1 = k_2 = c_1'' + 2c_2''H_2(1)c_0')$$

Elementary calculations lead to

$$y(k) = y_0(k) + y_1(k) + y_2(k) + y_3(k) + y_4(k)$$

where

$$y_0(k) = k_0$$

$$y_1(k) = H_3(q^{-1})c_1'H_1(q^{-1})H_3(q^{-1})k_1$$

$$y_2(k) = H_3(q^{-1})\left\{k_2H_2(q^{-1})c_2'\left[H_1(q^{-1})u(k)\right]^2 + c_2''\left[H_2(q^{-1})c_1'H_1(q^{-1})u(k)\right]^2\right\}$$

$$y_3(k) = H_3(q^{-1})c_2''2\left\{H_2(q^{-1})c_1'H_1(q^{-1})u(k)\right\}\left\{H_2(q^{-1})c_2'\left[H_1(q^{-1})u(k)\right]^2\right\}$$

$$y_4(k) = H_3(q^{-1})c_2''\left\{H_2(q^{-1})c_2'\left[H_1(q^{-1})u(k)\right]^2\right\}$$

with

$$k_0 = H_3(1)\left\{c_0'' + c_1''H_2(1)c_0' + c_2''\left[H_2(1)c_0'\right]^2\right\}$$

$$k_1 = k_2 = c_1'' + 2c_2''H_2(1)c_0'$$

Here $H(1)$ denotes the steady state gain of $H(q^{-1})$.

Finally, Figure 5.3.3 is the equivalent parallel representation of the cascade system. ■

Now it is easy to understand that not every parallel subsystem has its equivalent cascade system. In the case of our example, the parallel system has its cascade form only when the coefficients are the same as presented in Figure 5.3.3.

The investigation of a cascade system by its parallel form gives, however, information about its structure. The highest degree channel can be identified as a linear system by the frequency method or by step responses and the pole/zero placement – compared to that of the linear channel – gives the parameters of the pulse transfer functions of the linear dynamic terms, its location in the cascade system and the highest degree of the static polynomial between them. The details are described in Sections 5.6.

The above mentioned facts are valid not only in the discrete time case but also in the continuous time case. Then the pulse transfer functions should be replaced by transfer functions and the static gain can be calculated by replacing zero value for the Laplace operator in the transfer function.

Fig. 5.3.4 Simple Wiener–Hammerstein cascade model with polynomial
nonlinearity of arbitrary degree

The simple Wiener–Hammerstein cascade model is a special cascade model. Its structure identification is easy, even if its nonlinear element is of arbitrary degree. Its scheme (Figure 5.3.4) can be redrawn to the parallel form of Figure 5.3.5 that is a special case of Figure 5.3.6. Every channel of Figure 5.3.6 contains two linear dynamic systems and a static power between them. (In the literature Figure 5.3.6 used to be called the S_M model (Baumgartner and Rugh, 1975).)

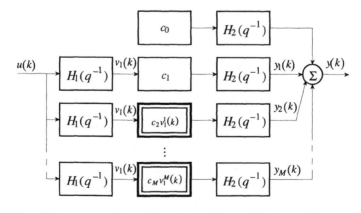

Fig. 5.3.5 Parallel representation of the simple Wiener–Hammerstein cascade model
with polynomial nonlinearity of arbitrary degree

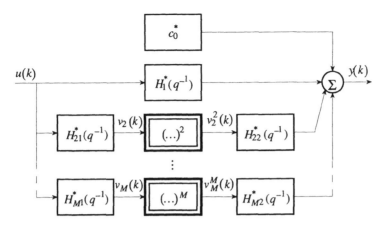

Fig. 5.3.6 Generalization of the parallel representation of the simple Wiener–Hammerstein cascade model with polynomial nonlinearity of arbitrary degree

If the linear systems before the powers are equal to each other in every channel and if the linear systems after the powers are equal to each other in every channel, then the parallel system can be transformed into a simple Wiener–Hammerstein cascade model.

Example 5.3.2   *Structure identification of the simple Hammerstein model from parallel channels of different nonlinear degree*

Example 5.3.2 can be considered as the continuation of Example 5.2.1. There the simple Hammerstein model of Figure 5.2.4 was excited by three unit steps with different amplitudes and Gardiner's method was used to separate the responses of the parallel subsystems. Easy calculations lead to the parallel scheme of Figure 5.2.6. Figure 5.2.6 is a special case of Figure 5.3.5, the parallel representation of the Wiener–Hammerstein cascade model. Since all linear systems before the nonlinear powers are unity and after them they are equal to each other, the simple Hammerstein – cascade – model of Figure 5.2.4 can be reconstructed from Figure 5.2.6.

Example 5.3.3   *Structure identification of the simple Wiener model from parallel channels of different nonlinear degree*

Example 5.3.3 can be considered as the continuation of Example 5.2.2. There the simple Wiener model of Figure 5.2.7 was excited by three unit steps of different amplitudes and Gardiner's method was used to separate the responses of the different subsystems. Easy calculations lead to the parallel scheme of Figure 5.2.9. Figure 5.2.9 is a special case of Figure 5.3.5, the parallel representation of the simple Wiener–Hammerstein cascade model. Since all linear dynamic terms are the same before the powers and there are no dynamic terms behind the powers, the simple Wiener – cascade – model of Figure 5.2.7 can be reconstructed from Figure 5.2.9.

Example 5.3.4   *Structure identification of the simple Wiener–Hammerstein model from parallel channels of different nonlinear degree*

Example 5.3.4 can be considered as the continuation of Example 5.2.3. There the simple Wiener–Hammerstein cascade model of Figure 5.2.10 was excited by three unit steps with different amplitudes and Gardiner's method was used to separate the responses of the different subsystems. Easy calculations lead to the parallel scheme of Figure 5.2.12. Figure 5.2.12 is a special case of Figure 5.3.5, the parallel representation of the simple Wiener–Hammerstein cascade model. Since all linear dynamic terms before the powers and after the powers are the same, respectively, the simple Wiener–Hammerstein cascade model of Figure 5.2.10 can be reconstructed from Figure 5.2.12.

5.4. STRUCTURE IDENTIFICATION OF SYSTEMS CONTAINING ONLY ONE NONLINEAR STATIC ELEMENT BY THE FREQUENCY METHOD

The most general scheme of a system containing only one nonlinear element is seen in Figure 5.4.1. The nonlinear element can be preceded and followed by linear dynamic parts. The cascade system can be feedbacked by a linear dynamic term. Finally, a parallel feedforward path containing a linear dynamic term can be assumed.

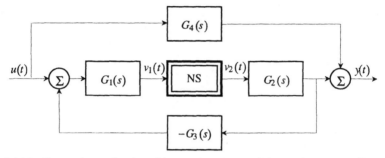

Fig. 5.4.1 Nonlinear dynamic closed loop system containing only one nonlinear static (NS) element

Singh and Subramanian (1980) recommended an excellent frequency domain method for the determination of the structure of the system. The input of the system should be excited by a sinusoidal signal and only the first harmonic has to be detected at the output. Similarly to the Nyquist plot in linear systems loci have to be drawn. The frequency has to be kept constant while the amplitude of the test signal has to be changed.

The input signal is

$$u(t) = U \sin(\omega t)$$

and the first harmonic of the output signal is

$$y_1(t) = Y_1 \sin(\omega t + \varphi)$$

The describing function is a function of both the angular frequency and the amplitude of the sinusoidal excitation

$$W(\omega, U) = \frac{Y_1}{U}(\omega, U) \exp\big(j\varphi(\omega, U)\big)$$

Now the amplitude locus is the complex gain

$$W(U) = \frac{Y_1}{U}(U) \exp\big(j\varphi(U)\big)$$

as a function of the amplitude of the input sinusoidal at a given frequency.

The considerations to be presented are exact if the nonlinear element

- is static,
- is single-valued,
- has an odd characteristic.

If these assumptions are met then the describing function of the nonlinear element is real, i.e., it causes no phase shift, and the gain depends only on the amplitude of the sinusoidal excitation at the input of the nonlinear element.

In the closed loop case we shall restrict ourselves to a low pass loop characteristic. In this case the higher order harmonics caused by the nonlinearity will be damped and at the input of the nonlinear element there will only be a first-order harmonic, thus the describing function approximation can be applied further on.

Denote the describing function of the nonlinear element by $N(v_1)$, and the frequency function of a linear term by

$$G_1(j\omega) = |G_1(\omega)|\exp(-j\varphi_1(\omega)) = |G_1|\exp(-j\varphi_1)$$

at a given frequency.

The describing function of a nonlinear static element does not cause any phase shift and does not depend on the angular frequency of the excitation.

The structure of the system can be identified as follows (Singh and Subramanian, 1980):

1. *Simple Wiener–Hammerstein cascade model*

   The complex gain at a given frequency is

   $$W(U) = |G_1|N(v_1)|G_2|\exp(-j(\varphi_1 + \varphi_2))$$

   and its amplitude depends on the amplitude of the input signal, but its phase remains constant. Therefore the amplitude locus at a given frequency is a straight line passing through the origin.

2. *Simple Wiener–Hammerstein cascade model with parallel feedforward linear term*

   The complex gain at a given frequency is

   $$W(U) = |G_4|\exp(-j\varphi_4) + |G_1|N(v_1)|G_2|\exp(-j(\varphi_1 + \varphi_2))$$

   Now, the second term gives a straight line passing through the origin and the first term is a (frequency dependent) constant. Therefore the amplitude locus is a straight line (in general) not passing the origin.

3. *Simple Wiener–Hammerstein cascade model with unity feedback*

   The complex gain at a given frequency is

   $$W(U) = \frac{|G_1|N(v_1)|G_2|\exp(-j(\varphi_1 + \varphi_2))}{1 + |G_1|N(v_1)|G_2|\exp(-j(\varphi_1 + \varphi_2))} = \frac{1}{|G_1|N(v_1)|G_2|}\exp(j(\varphi_1 + \varphi_2))$$

The amplitude locus of the denominator is a straight line passing through the point $(1,0)$. It is known from complex algebra that the inverse of a straight line is a circle. Consequently the amplitude locus of the system is a circle passing through the origin and the point $(1,0)$. The radius r of the curve is

$$r = r(\omega) = \frac{1}{2\sin(\varphi_1(\omega) + \varphi_2(\omega))}$$

and the coordinates (x_0, y_0) of the center of the circle are (Singh and Subramanian, 1980):

$$x_0 = 0.5$$

$$y_0 = y_0(\omega) = 0.5\,\mathrm{ctg}(\varphi_1(\omega) + \varphi_2(\omega))$$

4. *Simple Wiener–Hammerstein cascade model with linear dynamic feedback*
   The complex gain at a given frequency is

$$W(U) = \frac{|G_1|N(v_1)|G_2|\exp(-j(\varphi_1 + \varphi_2))}{1 + |G_1|N(v_1)|G_2||G_3|\exp(-j(\varphi_1 + \varphi_2 + \varphi_3))}$$

which at a given frequency results in an amplitude locus in the form of a circle. The circle passes through the origin (for $N(v_1) = 0$) but it does not pass through the point $(1,0)$. The parameters of the circle are (Singh and Subramanian, 1980):

$$r = r(\omega) = \frac{1}{2|G_3(\omega)|\sin(\varphi_1(\omega) + \varphi_2(\omega) + \varphi_3(\omega))}$$

$$x_0 = x_0(\omega) = \frac{\sin(\varphi_1(\omega) + \varphi_2(\omega))}{2|G_3(\omega)|\sin(\varphi_1(\omega) + \varphi_2(\omega) + \varphi_3(\omega))}$$

$$y_0 = y_0(\omega) = \frac{\cos(\varphi_1(\omega) + \varphi_2(\omega))}{2|G_3(\omega)|\sin(\varphi_1(\omega) + \varphi_2(\omega) + \varphi_3(\omega))}$$

5. *Simple Wiener–Hammerstein cascade model with linear dynamic feedback and feedforward terms*
   The complex gain at a given frequency is

$$W(U) = |G_4|\exp(-j(\varphi_4)) + \frac{|G_1|N(v_1)|G_2|\exp(-j(\varphi_1 + \varphi_2))}{1 + |G_1|N(v_1)|G_2||G_3|\exp(-j(\varphi_1 + \varphi_2 + \varphi_3))}$$

which results in an amplitude locus in the form of a circle. Because of the amplitude independent term caused by the linear feedforward term, the curve does not pass through the origin. The parameters of the curve are (Singh and Subramanian, 1980):

$$r = r(\omega) = \frac{1}{2|G_3(\omega)|\sin(\varphi_1(\omega) + \varphi_2(\omega) + \varphi_3(\omega))}$$

$$x_0 = x_0(\omega) = \frac{\sin(\varphi_1(\omega) + \varphi_2(\omega))}{2|G_3(\omega)|\sin(\varphi_1(\omega) + \varphi_2(\omega) + \varphi_3(\omega))} + |G_4(\omega)|\cos(\varphi_4(\omega))$$

$$y_0 = y_0(\omega) = \frac{\cos(\varphi_1(\omega) + \varphi_2(\omega))}{2|G_3(\omega)|\sin(\varphi_1(\omega) + \varphi_2(\omega) + \varphi_3(\omega))} - |G_4(\omega)|\sin(\varphi_4(\omega))$$

The different structures treated by the method and its characteristic amplitude loci are summarized in Table 5.4.1.

TABLE 5.4.1 Amplitude loci of different nonlinear dynamic closed loop systems containing only one nonlinear static element

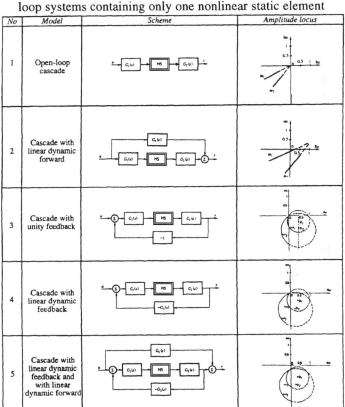

No	Model	Scheme	Amplitude locus
1	Open-loop cascade		
2	Cascade with linear dynamic forward		
3	Cascade with unity feedback		
4	Cascade with linear dynamic feedback		
5	Cascade with linear dynamic feedback and with linear dynamic forward		

Remarks:
1. The above method is not suitable for distinguishing between the simple Wiener–Hammerstein, simple Wiener, and simple Hammerstein models.
2. Furthermore, the cascade path may have more nonlinear and linear terms if the linear dynamic terms ensure – by their low pass characteristics – that the input signals of each nonlinear element have approximately only a first degree harmonic.
3. According to the experience of Singh and Subramanian (1980) the method can be applied for not strictly single valued nonlinearities (like hysteresis) and for general curves, including some even degree components. The essential point is that the

linear low pass terms should filter the higher degree harmonics and the phase shift
caused by the non-single value feature should be much less compared to the phase
lags caused by the linear terms at the frequencies investigated.
4. Vandersteen (1996) presented an alternative method for the same structure
    identification problem. The algorithm includes the following main steps:
    • estimate the multi-dimensional Fourier transform of the first, second and third
      degree Volterra kernels using a two-tone or multisines excitation,
    • estimate $G_1(j\omega)$ and $G_2(j\omega)$ from the Fourier transform of the second degree
      kernels by weighted nonlinear LS parameter estimation,
    • estimate $G_3(j\omega)$ and $G_4(j\omega)$ from the Fourier transform of the second degree
      kernels by weighted nonlinear LS parameter estimation.
5   Chen (1994) recommended also a combined structure identification and parameter
    estimation method.

5.5 STRUCTURE IDENTIFICATION OF BLOCK ORIENTED MODELS

The block oriented models consist of static nonlinear and dynamic linear terms. Such
models often occur in practice.
    Schetzen (1965b) classified the nonlinear dynamic systems according to the degree of
the steady state characteristic. He showed that several subclasses of any homogeneous
system can be distinguished. Even if we restrict ourselves to quadratic systems, a general
model can be constructed as a summated model of several (quadratic) cascade models. In
the control engineering practice of single-input single-output systems such nonlinear
models that contain only one path occur frequently. Therefore we restrict our
investigations to that class of quadratic systems which contains only one multiplier.
Such quadratic block oriented models are summarized in Section 1.3. Four different
methods will be presented for the determination of the actual structure from input–
output measurements:
    • the method based on the estimated extended Wiener–Hammerstein model;
    • the method based on the estimated Volterra kernels;
    • the frequency method;
    • the evaluation of impulse and step responses.

5.5.1Method based on the estimated extended Wiener–Hammerstein cascade model
In Chapter 1 it was shown that the different simple quadratic block oriented models are
related to each other and all are parts of the extended Wiener–Hammerstein model.
Consequently, if the process is identified in the form of the extended Wiener–
Hammerstein model then the concrete structure can be determined from the relations
between the pulse transfer functions of the linear terms in the extended Wiener–
Hammerstein model. The following rules help in the structure detection:
1. *Simple Hammerstein model structure identification*
    The two terms before the multiplier are static and the remaining blocks of the linear
    and quadratic channels have the same poles and zeros.
2. *Generalized Hammerstein model structure identification*
    The two terms before the multiplier are static.
3. *Simple Wiener model structure identification*
    The term after the multiplier is static and all remaining blocks have the same poles
    and zeros.
4. *Generalized Wiener model structure identification*

The term after the multiplier is static and all remaining blocks of the quadratic channel have the same poles and zeros.
5. *Extended Wiener model structure identification*
   The term after the multiplier is static.
6. *Simple Wiener–Hammerstein cascade model structure identification*
   The two blocks before the multiplier have the same poles and zeros. The block in the linear channel has these common poles and zeros and those of the block after the multiplier.
7. *Generalized Wiener–Hammerstein cascade model structure identification*
   The two blocks before the multiplier have the same poles and zeros.

In the sequel the above rules will be presented detailed by formulas.
   Denote the general model by

$$y(k) = c_0^* + H_1^*\left(q^{-1}\right)u(k) + H_4^*\left(q^{-1}\right)v(k)$$

$$v(k) = \left[H_2^*\left(q^{-1}\right)u(k)\right]\left[H_3^*\left(q^{-1}\right)u(k)\right]$$

where

$$H_1^*\left(q^{-1}\right) = \frac{B_1^*\left(q^{-1}\right)}{A_1^*\left(q^{-1}\right)} = \frac{b_{10}^* + b_{11}^* q^{-1} + \ldots + b_{1nb_1}^* q^{-nb_1}}{1 + a_{11}^* q^{-1} + \ldots + a_{1na_1}^* q^{-na_1}} q^{-d_1^*} = H_{d_1}^*\left(q^{-1}\right)q^{-d_1^*}$$

and $H_2^*\left(q^{-1}\right)$, $H_3^*\left(q^{-1}\right)$, and $H_4^*\left(q^{-1}\right)$ have similar forms.

The pulse transfer functions $H_2^*\left(q^{-1}\right)$, $H_3^*\left(q^{-1}\right)$, and $H_4^*\left(q^{-1}\right)$ can be determined unambiguously only if the first coefficients in the numerator polynomials of the pulse transfer functions $H_{d_2}^*\left(q^{-1}\right)$ and $H_{d_3}^*\left(q^{-1}\right)$ are equal to one:

$$b_{20}^* = q^{d_2^*} H_2^*\left(q^{-1}\right)\Big|_{q=\infty} = H_{d_2}^*(\infty) = 1$$

$$b_{30}^* = q^{d_3^*} H_3^*\left(q^{-1}\right)\Big|_{q=\infty} = H_{d_3}^*(\infty) = 1$$

(If poles and zeros were considered instead of the coefficients, a multiplicative constant, e.g., the static gain, should be considered as the normalization factor.)
   Table 5.5.1 summarizes the different block oriented models and presents the pulse transfer functions of the equivalent normalized extended Wiener–Hammerstein model.
   Looking at Table 5.5.1 one can see some relations between the pulse transfer functions of the blocks in the normalized extended Wiener–Hammerstein model. That means that if an identification has been performed using this structure the resulting pulse transfer functions point out the structure of the system. The following structure selection criteria can be used:
1. *Simple Hammerstein model structure identification*

$$H_2^*\left(q^{-1}\right) = H_3^*\left(q^{-1}\right) = 1$$

TABLE 5.5.1 The equivalent normalized extended Wiener–Hammerstein models of the block oriented models

No	Model	Scheme	c_0^*	$H_1^*(q^{-1})$	$H_2^*(q^{-1})$	$H_3^*(q^{-1})$	$H_4^*(q^{-1})$
1	Simple Hammerstein	$u \to [c_0+c_1u+c_2u^2] \to v \to [H(q^{-1})] \to y$	$c_0H(1)$	$c_1H(q^{-1})$	1	1	$c_2H(q^{-1})$
2	Generalized Hammerstein	c_0 / $H_1(q^{-1})$ / Π / $H_2(q^{-1})$, $\Sigma \to y$	c_0	$H_1(q^{-1})$	1	1	$H_2(q^{-1})$
3	Simple Wiener	$u \to [H(q^{-1})] \to v \to [c_0+c_1v+c_2v^2] \to y$	c_0	$c_1H(q^{-1})$	$\dfrac{H(q^{-1})}{H_d(\infty)}$	$\dfrac{H(q^{-1})}{H_d(\infty)}$	$c_2[H(\infty)]^2$
4	Generalized Wiener	c_0 / $H_1(q^{-1})$ / $H_2(q^{-1})$ Π, $\Sigma \to y$	c_0	$H_1(q^{-1})$	$\dfrac{H_2(q^{-1})}{H_{d_2}(\infty)}$	$\dfrac{H_2(q^{-1})}{H_{d_2}(\infty)}$	$\left[H_{d_2}(\infty)\right]^2$
5	Extended Wiener	c_0 / $H_1(q^{-1})$ / $H_2(q^{-1})$ / Π / $H_3(q^{-1})$, $\Sigma \to y$	c_0	$H_1(q^{-1})$	$\dfrac{H_2(q^{-1})}{H_{d_2}(\infty)}$	$\dfrac{H_3(q^{-1})}{H_{d_3}(\infty)}$	$H_{d_2}(\infty)\times H_{d_3}(\infty)$
6	Simple Wiener Hammerstein cascade	$\to [H_1(q^{-1})] \to [c_0+c_1v+c_2v^2] \to [H_2(q^{-1})] \to$	$c_0H_2(1)$	$\dfrac{c_1H_1(q^{-1})\times}{H_2(q^{-1})}$	$\dfrac{H_1(q^{-1})}{H_{d_1}(\infty)}$	$\dfrac{H_1(q^{-1})}{H_{d_1}(\infty)}$	$\dfrac{c_2H_2(q^{-1})\times}{[H_{d_1}(\infty)]^2}$
7	Generalized Wiener Hammerstein cascade	c_0 / $H_1(q^{-1})$ / $H_2(q^{-1})$ Π $H_3(q^{-1})$, $\Sigma \to y$	c_0	$H_1(q^{-1})$	$\dfrac{H_2(q^{-1})}{H_{d_2}(\infty)}$	$\dfrac{H_2(q^{-1})}{H_{d_2}(\infty)}$	$\dfrac{H_3(q^{-1})\times}{[H_{d_2}(\infty)]^2}$
8	Extended Wiener Hammerstein cascade	c_0 / $H_1(q^{-1})$ / $H_2(q^{-1})$ / Π $H_4(q^{-1})$ / $H_3(q^{-1})$, $\Sigma \to y$	c_0	$H_1(q^{-1})$	$\dfrac{H_2(q^{-1})}{H_{d_2}(\infty)}$	$\dfrac{H_3(q^{-1})}{H_{d_3}(\infty)}$	$\dfrac{H_4(q^{-1})\times}{H_{d_2}(\infty)\times}$ $H_{d_3}(\infty)$

$$\frac{H_1^*(q^{-1})}{H_{d_1}^*(\infty)} = \frac{H_4^*(q^{-1})}{H_{d_4}^*(\infty)}$$

2. *Generalized Hammerstein model structure identification*

$$H_2^*(q^{-1}) = H_3^*(q^{-1}) = 1$$

No	Model	Structure selection criteria	
1	Simple Hammerstein	$H_2^*(q^{-1}) = H_3^*(q^{-1}) = 1$	$\dfrac{H_4^*(q^{-1})}{H_{d_4}^*(\infty)} = \dfrac{H_1^*(q^{-1})}{H_{d_1}^*(\infty)}$
2	Generalized Hammerstein	$H_2^*(q^{-1}) = H_3^*(q^{-1}) = 1$	
3	Simple Wiener	$H_2^*(q^{-1}) = H_3^*(q^{-1}) = \dfrac{H_1^*(q^{-1})}{H_{d_1}^*(\infty)}$	$\dfrac{H_4^*(q^{-1})}{H_{d_4}^*(\infty)} = 1$
4	Generalized Wiener	$H_2^*(q^{-1}) = H_3^*(q^{-1})$	$\dfrac{H_4^*(q^{-1})}{H_{d_4}^*(\infty)} = 1$
5	Extended Wiener	$\dfrac{H_4^*(q^{-1})}{H_{d_4}^*(\infty)} = 1$	
6	Simple Wiener-Hammerstein cascade	$H_2^*(q^{-1}) = H_3^*(q^{-1})$	$\dfrac{H_1^*(q^{-1})}{H_{d_1}^*(\infty)} = H_2^*(q^{-1}) \dfrac{H_4^*(q^{-1})}{H_{d_4}^*(\infty)}$
7	Generalized Wiener-Hammerstein cascade	$H_2^*(q^{-1}) = H_3^*(q^{-1})$	
8	Extended Wiener-Hammerstein cascade		

TABLE 5.5.2 Structure selection criteria based on the identified pulse transfer functions of the normalized extended Wiener–Hammerstein models

3. *Simple Wiener model structure identification*

$$H_2^*(q^{-1}) = H_3^*(q^{-1}) = \frac{H_1^*(q^{-1})}{H_{d_1}^*(\infty)}$$

4. *Generalized Wiener model structure identification*

$$H_2^*(q^{-1}) = H_3^*(q^{-1}) = 1$$

$$\frac{H_4^*(q^{-1})}{H_{d_4}^*(\infty)} = 1$$

5. *Extended Wiener model structure identification*

$$\frac{H_4^*(q^{-1})}{H_{d_4}^*(\infty)} = 1$$

6. *Simple Wiener–Hammerstein cascade model structure identification*

$$H_2^*(q^{-1}) = H_3^*(q^{-1})$$

$$\frac{H_1^*(q^{-1})}{H_{d_1}^*(\infty)} = H_2^*(q^{-1}) \frac{H_4^*(q^{-1})}{H_{d_4}^*(\infty)}$$

7. *Generalized Wiener–Hammerstein cascade model structure identification*

$$H_2^*\left(q^{-1}\right) = H_3^*\left(q^{-1}\right)$$

8. *Extended Wiener–Hammerstein cascade model structure identification*
   No special relation can be discovered between the pulse transfer functions of the blocks.

The above structure selection criteria are summarized in Table 5.5.2.

In the case of noisy measurements the parameter estimation generally does not allow us to draw such clear relations between the coefficients of the pulse transfer functions of the blocks. It is practical to examine the poles and zeros of the pulse transfer functions, which show better the coincidence or bias of two transfer functions. Table 5.5.3 presents the structure selection criteria based on the identified poles and zeros of the pulse transfer functions of the different blocks. Of course, one has to take into account that noisy measurements lead only to approximating coincidences.

There are two ways for the estimation of the model parameters if the structure of the model is already known:

1. The special parameter estimation method can be applied for the given model type. The most important advantage is that now fewer parameters have to be estimated. (The models of Hammerstein type are linear in parameters, which fact allows their easy identification. The Hammerstein and Wiener models can be identified by an iterative identification of a linear dynamic and a static polynomial term. Further details have been treated in Chapter 3.)
2. Since the individual block oriented models are parts of the extended Wiener–Hammerstein model, their parameters can be reconstructed if the type of the model was fixed. The computations needed are very easy and are summarized in Table 5.5.4.

The reconstruction of two pulse transfer functions (or one pulse transfer function and a constant value) from a multiplication relation is not unambiguous, unless the division of the constant term is systematically done. Therefore the following assumptions were taken:

1. *Simple Hammerstein model structure identification*

   $$H_d(\infty) = 1$$

2. *Simple Wiener model structure identification*

   $$H_d(\infty) = 1$$

3. *Extended Wiener model structure identification*

   $$H_{d_2}(\infty) = 1$$

4. *Simple Wiener–Hammerstein cascade model structure identification*

   $$H_{d_1}(\infty) = 1, \qquad H_{d_2}(\infty) = 1$$

TABLE 5.5.3 Structure identification criteria based on the poles and zeros of the identified pulse transfer functions of the normalized extended Wiener–Hammerstein models (*no* denotes that neither pole or zero exists only a static gain)

No	Model	Scheme	p_{1i}^* / z_{1i}^*	p_{2i}^* / z_{2i}^*	p_{3i}^* / z_{3i}^*	p_{4i}^* / z_{4i}^*
1	Simple Hammerstein	$u \to \boxed{c_0 + c_1 u + c_2 u^2} \to v \to \boxed{H(q^{-1})} \to y$	$= p_{4i}^*$ / $= z_{4i}^*$	no / no	no / no	$= p_{1i}^*$ / $= z_{1i}^*$
2	Generalized Hammerstein				no / no	no / no
3	Simple Wiener	$u \to \boxed{H(q^{-1})} \to v \to \boxed{c_0 + qv + c_2 v^2} \to y$	$= p_{2i}^* = p_{3i}^*$ / $= z_{2i}^* = z_{3i}^*$	$= p_{1i}^* = p_{3i}^*$ / $= z_{1i}^* = z_{3i}^*$	$= p_{1i}^* = p_{2i}^*$ / $= z_{1i}^* = z_{2i}^*$	no / no
4	Generalized Wiener			$= p_{3i}^*$ / $= z_{3i}^*$	$= p_{2i}^*$ / $= z_{3i}^*$	no / no
5	Extended Wiener					no / no
6	Simple Wiener-Hammerstein cascade	$u \to \boxed{H_1(q^{-1})} \to v_1 \to \boxed{c_0 + qv_1 + c_2 v_1^2} \to v_2 \to \boxed{H_2(q^{-1})} \to y$	$= \{p_{2i}^*, p_{4i}^*\}$ / $= \{z_{2i}^*, z_{4i}^*\}$	$= p_{3i}^*$ / $= z_{3i}^*$	$= p_{2i}^*$ / $= z_{2i}^*$	
7	Generalized Wiener-Hammerstein cascade			$= p_{3i}^*$ / $= z_{3i}^*$	$= p_{2i}^*$ / $= z_{2i}^*$	
8	Extended Wiener-Hammerstein cascade					

5. *Generalized Wiener–Hammerstein cascade model structure identification*

$$H_{d_2}(\infty) = 1$$

6. *Extended Wiener–Hammerstein cascade model structure identification*

$$H_{d_2}(\infty) = 1$$
$$H_{d_3}(\infty) = 1$$

The structure identification can be performed also in the continuous Laplace domain if the discrete time pulse transfer functions of the blocks of the identified extended Wiener–Hammerstein model were transformed into their equivalent continuous transfer functions.

The method will be illustrated by some simple examples.

Example 5.5.1   *Structure and parameter estimation of the simple Hammerstein model*

The transfer function of the linear part of the process is

$$G(s) = \frac{Y(s)}{V(s)} = \frac{1}{1 + 2 \cdot 0.5 \cdot 10s + 10^2 s^2}$$

and the equation of the static polynomial is

$$v(k) = 2 + u(k) + 0.5u^2(k)$$

The scheme of the system is seen in Figure 5.5.1. Assuming $\Delta T = 4$ [s] sampling time, the equivalent pulse transfer function is

$$H(q^{-1}) = \frac{0.069412q^{-1} + 0.060714q^{-2}}{1 - 1.54q^{-1} + 0.67q^{-2}}$$

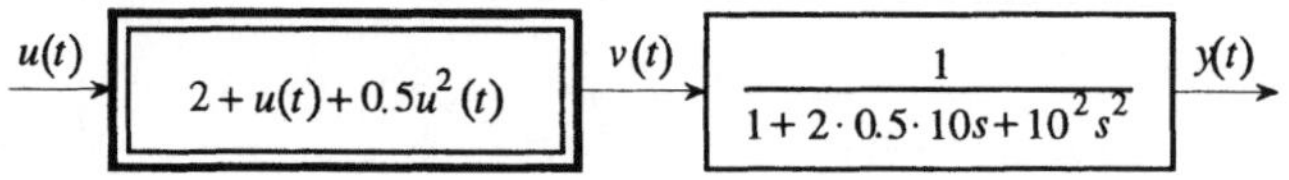

Fig. 5.5.1 Scheme of the simple Hammerstein model

The pulse transfer functions of the identified equivalent normalized extended Wiener–Hammerstein model are

$$c_0^* = c_0 H(1) = 2$$
$$H_1^*(q^{-1}) = c_1 H(q^{-1}) = \frac{0.06941 + 0.060714q^{-1}}{1 - 1.54q^{-1} + 0.67q^{-2}} q^{-1}$$
$$H_2^*(q^{-1}) = H_3^*(q^{-1}) = 1$$
$$H_4^*(q^{-1}) = c_2 H(q^{-1}) = \frac{0.0347 + 0.030357q^{-1}}{1 - 1.54q^{-1} + 0.67q^{-2}} q^{-1}$$

TABLE 5.5.4 The parameters of the block oriented models calculated
from the normalized extended Wiener–Hammerstein models

No	Model	c_0	c_1	c_2	$H(q^{-1})$	$H_1(q^{-1})$	$H_2(q^{-1})$	$H_3(q^{-1})$	$H_4(q^{-1})$	assumptions
1	Simple Hammerstein	$\frac{c_0^*}{H(1)}$	$H_{d_1}^*(\infty)$	$\frac{H_4^*(q^{-1})}{H(q^{-1})}$	$\frac{H_1^*(q^{-1})}{H_{d_1}^*(\infty)}$					$H_d(\infty)=1$
2	Generalized Hammerstein	c_0^*				$H_1^*(q^{-1})$	$H_4^*(q^{-1})$			
3	Simple Wiener	c_0^*	$H_{d_1}^*(\infty)$	$H_{d_4}^*(\infty)$	$\frac{H_1^*(q^{-1})}{H_{d_1}^*(\infty)}$					$H_d(\infty)=1$
4	Generalized Wiener	c_0^*				$H_1^*(q^{-1})$	$H_4^*(q^{-1})\times\sqrt{H_{d_4}^*(\infty)}$			
5	Extended Wiener	c_0^*				$H_1^*(q^{-1})$	$H_2^*(q^{-1})$	$H_3^*(q^{-1})\times H_{d_4}^*(\infty)$		$H_{d_4}(\infty)=1$
6	Simple Wiener-Hammerstein cascade	$\frac{c_0^*}{H_2(1)}$	$\frac{H_1^*(q^{-1})}{H_1(q^{-1})H_2(q^{-1})}$	$\frac{H_4^*(q^{-1})}{H_2(q^{-1})}$		$H_2^*(q^{-1})$	$\frac{H_1^*(q^{-1})}{H_2^*(q^{-1})H_{d_4}^*(\infty)}$			$H_{d_1}(\infty)=1$ $H_{d_4}(\infty)=1$
7	Generalized Wiener-Hammerstein cascade	c_0^*				$H_1^*(q^{-1})$	$H_2^*(q^{-1})$	$H_3^*(q^{-1})$		$H_{d_1}(\infty)=1$
8	Extended Wiener-Hammerstein cascade	c_0^*				$H_1^*(q^{-1})$	$H_2^*(q^{-1})$	$H_3^*(q^{-1})$	$H_4^*(q^{-1})$	$H_{d_2}(\infty)=1$ $H_{d_3}(\infty)=1$

The relations between the pulse transfer functions are

$$\frac{H_1^*(q^{-1})}{H_{d_1}^*(\infty)} = \frac{H_4^*(q^{-1})}{H_{d_4}^*(\infty)} = \frac{q^{-1}+0.847q^{-2}}{1-1.54q^{-1}+0.67q^{-2}}$$

and

$$H_2^*(q^{-1}) = H_3^*(q^{-1}) = 1$$

which points to the simple Hammerstein model in Table 5.5.2. Now the parameters of the model can be calculated according to Table 5.5.4:

$$c_0 = \frac{c_0^*}{H(1)} = 2$$

$$c_1 = H_{d_1}^*(\infty) = 0.06941$$

$$H(q^{-1}) = \frac{H_1^*(q^{-1})}{H_{d_1}^*(\infty)} = \frac{q^{-1}+0.8747q^{-2}}{1-1.54q^{-1}+0.67q^{-2}}$$

$$c_2 = \frac{H_4^*(q^{-1})}{H(q^{-1})} = 0.0347$$

The resulting simple Hammerstein model is equal to the initial model. The only visible difference is in the partition of some multiplicative terms between the constant terms c_1, c_2 and $H(q^{-1})$.

Example 5.5.2   *Structure and parameter estimation of the generalized Hammerstein model*

The constant term is

$$c_0 = 2$$

the transfer function of the linear part is

$$H_1(s) = \frac{1}{(1+5s)(1+10s)}$$

and that of the quadratic channel is

$$H_2(s) = \frac{0.5}{(1+10s)}$$

The scheme of the system is seen in Figure 5.5.2. The equivalent pulse transfer functions are

$$H_1(q^{-1}) = \frac{0.1087q^{-1} + 0.07286q^{-2}}{1 - 1.11965q^{-1} + 0.301195q^{-2}}$$

and

$$H_2(q^{-1}) = \frac{0.16484q^{-1}}{1 - 0.67032q^{-1}}$$

respectively, if the sampling time was $\Delta T = 4$ [s].

The pulse transfer functions of the identified equivalent normalized extended Wiener–Hammerstein model are as follows:

$$c_0^* = c_0 = 2$$

$$H_1^*(q^{-1}) = H_1(q^{-1}) = \frac{0.1087 + 0.07286q^{-1}}{1 - 1.11965q^{-1} + 0.301195q^{-2}} q^{-1}$$

$$H_2^*(q^{-1}) = 1 , \qquad H_3^*(q^{-1}) = 1$$

$$H_4^*(q^{-1}) = H_2(q^{-1}) = \frac{0.16484q^{-1}}{1 - 0.67032q^{-1}}$$

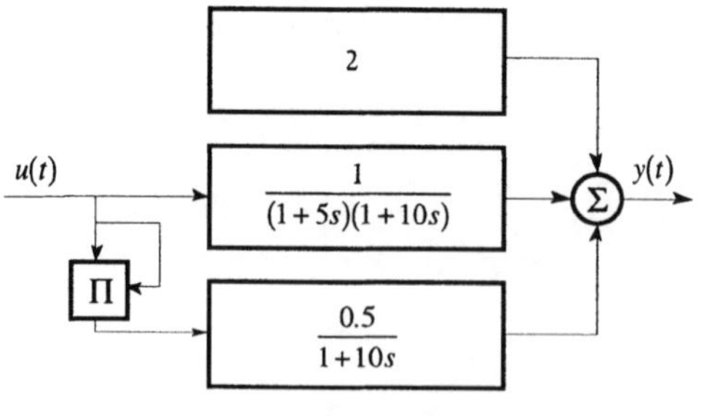

Fig. 5.5.2 Scheme of the generalized Hammerstein model

The relation between the pulse transfer functions is

$$H_2^*(q^{-1}) = H_3^*(q^{-1}) = 1$$

which assigns the generalized Hammerstein model in Table 5.5.2.
The parameters of the model can be calculated according to Table 5.5.4 as follows:

$$c_0 = c_0^* = 2$$

$$H_1(q^{-1}) = H_1^*(q^{-1}) = \frac{0.1087q^{-1} + 0.07286q^{-2}}{1 - 1.11965q^{-1} + 0.301195q^{-2}}$$

$$H_2(q^{-1}) = H_4^*(q^{-1}) = \frac{0.16484q^{-1}}{1 - 0.67032q^{-1}}$$

The reconstructed parameters are equal to those expected.

Example 5.5.3   *Structure and parameter identification of the simple Wiener model*

Let the transfer function of the linear part of the process be

$$G(s) = \frac{V(s)}{U(s)} = \frac{1}{1 + 2 \cdot 0.5 \cdot 10s + 10^2 s^2}$$

and the static polynomial be described by

$$y(k) = 2 + v(k) + 0.5v^2(k)$$

The scheme of the system is seen in Figure 5.5.3. Assuming $\Delta T = 4$ [s] sampling period, the equivalent pulse transfer function is

$$H(q^{-1}) = \frac{0.06941q^{-1} + 0.060714q^{-2}}{1 - 1.54q^{-1} + 0.67q^{-2}}$$

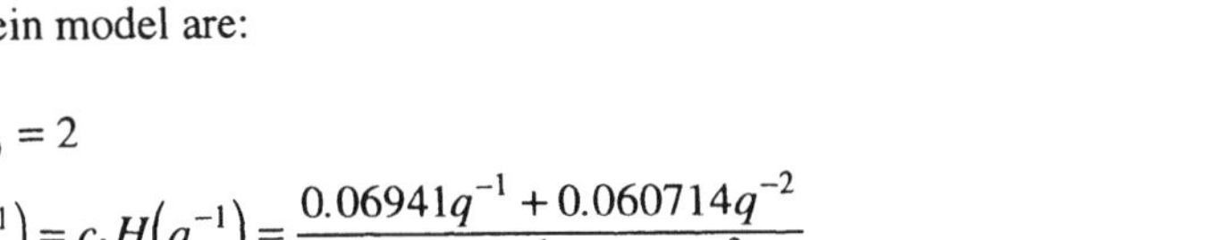

Fig. 5.5.3 Scheme of the simple Wiener model

The pulse transfer functions of the identified equivalent normalized extended Wiener–Hammerstein model are:

$$c_0^* = c_0 = 2$$

$$H_1^*(q^{-1}) = c_1 H(q^{-1}) = \frac{0.06941q^{-1} + 0.060714q^{-2}}{1 - 1.54q^{-1} + 0.67q^{-2}}$$

$$H_2^*(q^{-1}) = H_3^*(q^{-1}) = \frac{H(q^{-1})}{H_d(\infty)} = \frac{q^{-1} + 0.8747q^{-2}}{1 - 1.54q^{-1} + 0.67q^{-2}}$$

$$H_4^*(q^{-1}) = c_2[H(0)]^2 = 0.0024088$$

The relation between the pulse transfer functions is:

$$H_2^*(q^{-1}) = H_3^*(q^{-1}) = \frac{H_1^*(q^{-1})}{H_{d_1}^*(\infty)} = \frac{q^{-1} + 0.8747q^{-2}}{1 - 1.54q^{-1} + 0.67q^{-2}}$$

which assigns the simple Wiener model in Table 5.5.2.

The parameters of the model can be calculated by means of Table 5.5.4:

$$c_0 = c_0^* = 2$$

$$c_1 = H_{d_1}^*(\infty) = 0.06941$$

$$c_2 = H_{d_4}^*(\infty) = 0.0024088$$

$$H(q^{-1}) = \frac{H_1^*(q^{-1})}{H_{d_1}^*(\infty)} = \frac{q^{-1} + 0.97457q^{-2}}{1 - 1.54q^{-1} + 0.67q^{-2}}$$

The resulting simple Wiener model is equal to the original model. It can be seen better if the estimated $H(q^{-1})$ is multiplied by $H_d(\infty) = 0.06941$ (of the original model), and therefore the estimated c_1 should be divided by it and the estimated c_2 by its square.

Example 5.5.4 *Structure and parameter identification of the generalized Wiener model*

Let the constant term be

$$c_0 = 2$$

the transfer function of the linear part be

$$G_1(s) = \frac{1}{(1 + 5s)(1 + 10s)}$$

and that of the quadratic channel be

$$G_2(s) = \frac{0.5}{1 + 10s}$$

The scheme of the system is seen in Figure 5.5.4. The equivalent pulse transfer functions are

$$H_1(q^{-1}) = \frac{0.1087q^{-1} + 0.07286q^{-2}}{1 - 1.11965q^{-1} + 0.301195q^{-2}}$$

and

$$H_2(q^{-1}) = \frac{0.16484q^{-1}}{1 - 0.67032q^{-1}}$$

respectively, if the sampling period was $\Delta T = 4$ [s].

The pulse transfer functions of the identified equivalent normalized extended Wiener–Hammerstein model are as follows:

$$c_0^* = c_0 = 2$$

$$H_1^*\!\left(q^{-1}\right) = H_1\!\left(q^{-1}\right) = \frac{0.1087q^{-1} + 0.07286q^{-2}}{1 - 1.11965q^{-1} + 0.301195q^{-2}}$$

$$H_2^*\!\left(q^{-1}\right) = H_3^*\!\left(q^{-1}\right) = \frac{H_2\!\left(q^{-1}\right)}{H_{d_2}(\infty)} = \frac{q^{-1}}{1 - 0.67032q^{-1}}$$

$$H_4\!\left(q^{-1}\right) = \left[H_{d_2}(\infty)\right]^2 = 0.0271722$$

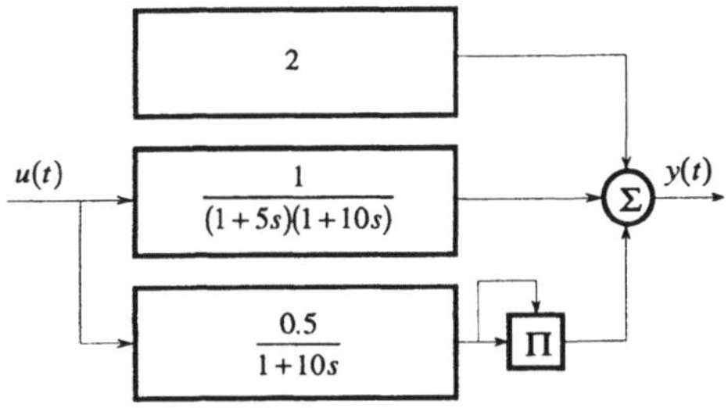

Fig. 5.5.4 Scheme of the generalized Wiener model

The relations between the pulse transfer functions are:

$$H_2\!\left(q^{-1}\right) = H_3\!\left(q^{-1}\right) = \frac{q^{-1}}{1 - 0.67032q^{-1}}$$

$$\frac{H_4\!\left(q^{-1}\right)}{H_{d_4}(\infty)} = 1$$

which assign the generalized Wiener model in Table 5.5.2.

The parameters of the model can be calculated according to Table 5.5.4 as follows:

$$c_0 = c_0^* = 2$$

$$H_1\!\left(q^{-1}\right) = H_1^*\!\left(q^{-1}\right) = \frac{0.1087q^{-1} + 0.07286q^{-2}}{1 - 1.11965q^{-1} + 0.301195q^{-2}}$$

$$H_2\!\left(q^{-1}\right) = H_2^*\!\left(q^{-1}\right)\sqrt{H_{d_4}^*(\infty)} = \frac{0.16484q^{-1}}{1 - 0.67032q^{-1}}$$

The reconstructed parameters are equal to those expected.

Example 5.5.5 *Structure and parameter estimation of the simple Wiener–Hammerstein cascade model*

Assume that the transfer function of the linear part before the static polynomial is

$$G_1(s) = \frac{V_1(s)}{U(s)} = \frac{1}{1 + 5s}$$

and that of the second linear term behind the static polynomial is

$$G_2(s) = \frac{Y(s)}{V_2(s)} = \frac{1}{1+10s}$$

and that the equation of the polynomial is

$$v_2(t) = 2 + v_1(t) + 0.5v_1^2(t)$$

Between the two linear dynamic terms there is a sampling and holding device. The scheme of the system is seen in Figure 5.5.5. The pulse transfer functions are

$$H_1(q^{-1}) = \frac{0.55067q^{-1}}{1-0.44933q^{-1}}$$

$$H_2(q^{-1}) = \frac{0.32968q^{-1}}{1-0.67032q^{-1}}$$

respectively, if the sampling period is $\Delta T = 4$ [s].
    The pulse transfer functions of the identified equivalent normalized extended Wiener–Hammerstein models are as follows

$$c_0^* = c_0 H_2(1) = 2$$

$$H_1^*(q^{-1}) = c_1 H_1(q^{-1})H_2(q^{-1}) = \frac{0.18155q^{-2}}{1-1.11965q^{-1}+0.3012q^{-2}}$$

$$H_2^*(q^{-1}) = H_3^*(q^{-1}) = \frac{H_1(q^{-1})}{H_{d_1}(\infty)} = \frac{q^{-1}}{1-0.44933q^{-1}}$$

$$H_4^*(q^{-1}) = c_2 H_2(q^{-1})\left[H_{d_1}(\infty)\right]^2 = \frac{0.05q^{-1}}{1-0.67032q^{-1}}$$

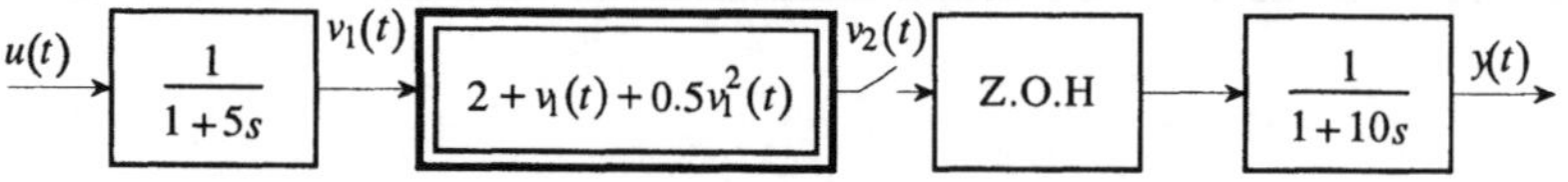

Fig. 5.5.5 Scheme of the simple Wiener–Hammerstein cascade model

The relations between the pulse transfer functions are:

$$H_2^*(q^{-1}) = H_3^*(q^{-1}) = \frac{q^{-1}}{1-0.44933q^{-1}}$$

$$\frac{H_1^*(q^{-1})}{H_{d_1}(\infty)} = H_2^*(q^{-1})\frac{H_4^*(q^{-1})}{H_{d_4}^*(\infty)} = \frac{q^{-2}}{1-1.11965q^{-1}+0.3012q^{-2}}$$

$$= \frac{q^{-2}}{\left(1-0.44933q^{-1}\right)\left(1-0.67032q^{-1}\right)}$$

The parameters of the model can be calculated according to Table 5.5.4 as follows:

$$H_1(q^{-1}) = H_2^*(q^{-1}) = \frac{q^{-1}}{1 - 0.44933q^{-1}}$$

$$H_2(q^{-1}) = \frac{H_1^*(q^{-1})}{H_2^*(q^{-1})H_{d_1}^*(\infty)} = \frac{q^{-1}}{1 - 0.67032q^{-1}}$$

$$c_0 = \frac{c_0^*}{H_2(1)} = 0.6594$$

$$c_1 = \frac{H_1^*(q^{-1})}{H_1(q^{-1})H_2(q^{-1})} = 0.18155$$

$$c_2 = \frac{H_4^*(q^{-1})}{H_2(q^{-1})} = 0.05$$

The resulting simple Wiener–Hammerstein cascade model is equivalent to the original one. It can be better seen if the estimated

- $H_2(q^{-1})$ is multiplied by $H_{d_2}(\infty)$
- $H_1(q^{-1})$ is multiplied by $H_{d_1}(\infty)$
- c_0 is divided by $H_{d_2}(\infty)$
- c_1 is divided by $H_{d_1}(\infty)H_{d_2}(\infty)$
- c_2 is divided by $\left[H_{d_1}(\infty)\right]^2 H_{d_2}(\infty)$

where $H_{d_1}(\infty)$ and $H_{d_2}(\infty)$ correspond to the original model parameters.

5.5.2 Method based on the estimated Volterra kernels

While the Volterra kernels can easily be derived from the block oriented models, the structure and the parameters of the block oriented models cannot be computed in a trivial way from the estimated Volterra kernels.

Only a few papers have dealt with the determination of the structure of the block oriented models from the estimated Volterra kernels. Marmarelis and Naka (1974) pointed out that if the two-dimensional kernel exists only at the main diagonal, then the model is of Hammerstein type, i.e., a multiplier is followed by a linear dynamic term. In the reverse configuration, i.e., if the linear dynamic term precedes the multiplier, then the two-dimensional kernel can be written as a product of two one-dimensional kernels (Marmarelis and Naka, 1974). Hung et al., (1982) showed that if the model had a multiplier and three linear dynamic terms at the two inputs and at the output of the multiplier, respectively, the time constant after the multiplier affected the kernels parallel to the main diagonal, and the time constants before multiplication dominated the time constant of the envelope of the off-diagonal curves as the curves moved away from the main diagonal.

In this section the quadratic nonlinear dynamic block oriented models will be summarized, and graphical and analytical criteria will be introduced for the structure estimation of this model type from the estimated Volterra kernels.

TABLE 5.5.5 Volterra kernels of the block oriented models

No	Model	Scheme	g_0	$g_1(\tau_1)$	$g_2^=(\tau_1,\tau_2)$
1	Simple Hammerstein	(block diagram)	$c_0 \displaystyle\int_{\zeta=0}^{\sim} g_2(\tau_1)d\tau_1$	$c_1 g(\tau_1)$	0 if $\tau_1 \neq \tau_2$ $c_2 g(\tau_1)$ if $\tau_1 = \tau_2$
2	Generalized Hammerstein	(block diagram)	c_0	$g_1(\tau_1)$	0 if $\tau_1 \neq \tau_2$ $g_2(\tau_1)$ if $\tau_1 = \tau_2$
3	Simple Wiener	(block diagram)	c_0	$c_1 g(\tau_1)$	$c_2 g(\tau_1)g(\tau_2)$
4	Generalized Wiener	(block diagram)	c_0	$g_1(\tau_1)$	$g_2(\tau_1)g_2(\tau_2)$
5	Extended Wiener	(block diagram)	c_0	$g_1(\tau_1)$	$g_2(\tau_1)g_3(\tau_2)$
6	Simple Wiener-Hammerstein cascade	(block diagram)	$c_0 \displaystyle\int_{\zeta=0}^{\sim} g_2(\tau_1)d\tau_1$	$c_1 \displaystyle\int_{\tau_1=0}^{\sim} g_2(\sigma)g_2(\tau_1-\sigma)d\sigma$	$\displaystyle\int_{\sigma=0}^{\sim} g_2(\sigma)g_1(\tau_1-\sigma)g_1(\tau_2-\sigma)d\sigma$
7	Generalized Wiener-Hammerstein cascade	(block diagram)	c_0	$g_1(\tau_1)$	$\displaystyle\int_{\sigma=0}^{\sim} g_3(\sigma)g_2(\tau_1-\sigma)g_2(\tau_2-\sigma)d\sigma$
8	Extended Wiener-Hammerstein cascade	(block diagram)	c_0	$g_1(\tau_1)$	$\displaystyle\int_{\sigma=0}^{\sim} g_4(\sigma)g_2(\tau_1-\sigma)g_3(\tau_2-\sigma)d\sigma$

The Volterra kernels of the different block oriented models are presented in Table 5.5.5. Every model in this table can be considered as a special version of the extended Wiener–Hammerstein cascade model, thus only this model is treated here in detail.

The constant term g_0 is equal to c_0 the first degree Volterra kernel $g_1(\tau_1)$ is equal to the weighting function of the linear channel. The quadratic channel can be described as

$$y_2(t) = \int_{\tau_1=0}^{\infty} \int_{\tau_2=0}^{\infty} \int_{\tau=0}^{\infty} g_4(\tau)g_2(\tau_1-\sigma)g_3(\tau_2-\sigma)u(t-\tau_1)u(t-\tau_2)d\tau d\tau_1 d\tau_2$$

thus the quadratic kernel can be calculated by the following convolution integral

$$g_2(\tau_1, \tau_2) = \int_{\tau=0}^{\infty} g_4(\tau)g_2(\tau_1 - \tau)g_3(\tau_2 - \tau)d\tau \tag{5.5.1}$$

It seems there exists an unequivocal relation between the parameters of the block oriented model and the Volterra kernels. The transformation is unequivocal only in one direction because the weighting functions $g_i(\tau_i)$ of the block oriented models cannot be reconstructed from the identified Volterra kernels.

As an example consider the case of the extended Wiener model.

Example 5.5.6   *Structure identification of the extended Wiener model*
The Volterra kernel $g_2(\tau_1, \tau_2)$ can be obtained as a product of the weighting function terms of the two linear dynamic terms in the model ($g_2(\tau_1)$ and $g_3(\tau_2)$):

$$g_2(\tau_1, \tau_2) = g_2(\tau_1)g_3(\tau_2)$$

One may think that two points of $g_2(\tau_1)$ and $g_3(\tau_2)$ can be determined by the measurement of $g_2(\tau_1, \tau_2)$ in four places on the surface (τ_1, τ_2) $\left(\left(\tau_1^0, \tau_1^0\right), \left(\tau_1^0, \tau_2^0\right),\right.$ $\left.\left(\tau_2^0, \tau_1^0\right), \left(\tau_2^0, \tau_2^0\right)\right)$. The corresponding equation system is:

$$g_2^{as}\left(\tau_1^0, \tau_1^0\right) = g_2\left(\tau_1^0\right)g_3\left(\tau_1^0\right) \tag{5.5.2a}$$

$$g_2^{as}\left(\tau_1^0, \tau_2^0\right) = g_2\left(\tau_1^0\right)g_3\left(\tau_2^0\right) \tag{5.5.2b}$$

$$g_2^{as}\left(\tau_2^0, \tau_1^0\right) = g_2\left(\tau_2^0\right)g_3\left(\tau_1^0\right) \tag{5.5.2c}$$

$$g_2^{as}\left(\tau_2^0, \tau_2^0\right) = g_2\left(\tau_2^0\right)g_3\left(\tau_2^0\right) \tag{5.5.2d}$$

Obviously we have four unknowns and four equations. The problem is that the asymmetric kernels $g_2^{as}(\tau_1, \tau_2)$ and $g_2^{as}(\tau_2, \tau_1)$ cannot be estimated separately, only their sum $\left(g_2^{as}(\tau_1, \tau_2) + g_2^{as}(\tau_2, \tau_1)\right)$. Returning to the extended Wiener model, we find that only the symmetric kernel

$$g_2^{s}(\tau_1, \tau_2) = \frac{1}{2}\left[g_2^{as}(\tau_1, \tau_2) + g_2^{as}(\tau_2, \tau_1)\right]$$

is available, and the simple algorithm (5.5.2) cannot be used.   ∎

We can generally state that instead of the so-called asymmetric kernels $g_2^{as}(\tau_1, \tau_2)$ (5.5.1) the symmetric kernels $g_2^{s}(\tau_1, \tau_2)$ are available after the parameter estimation. The reconstruction of the structure and the weighting functions of the block oriented model on their basis is not an easy task.

In the following those features will be introduced which characterize the Volterra series of the different block oriented models. The characteristic features can be presented by plotting the kernels and/or by means of analytical indices (Haber, 1985, 1989).

I. Criteria based upon the linear kernel and different sections of the quadratic kernel
1. *Simple and generalized Hammerstein model*
   The quadratic kernels of the Hammerstein models differ from zero only at the main diagonal. This feature can be seen from the plot of $h_2(\kappa_1, \kappa_2)$ and from building the following index:

$$\alpha_1 = \frac{\dfrac{2}{m(m+1)} \displaystyle\sum_{\kappa_1=0}^{m-1} \sum_{\kappa_2=\kappa_1+1}^{m} h_2^2(\kappa_1, \kappa_2)}{\dfrac{1}{m+1} \displaystyle\sum_{\kappa_1=0}^{m} h_2^2(\kappa_1, \kappa_2)} \qquad (5.5.3)$$

In (5.5.3) the mean square values of the off-diagonal elements are divided by that of the main diagonal. Here and further on any characteristic measure of the size of the element can be used instead of the square value (e.g., absolute value). The main diagonal is the straight line for which $\kappa_1 = \kappa_2$ and m is the memory of the kernels of the discretized model. The term α_1 becomes zero only at the Hammerstein model.

2. *Simple Hammerstein model*
   The linear and quadratic kernels at the main diagonal are proportional to each other in the simple Hammerstein model. This fact can be recognized from the figures of $h_1(\kappa_1)$ and $h_2(\kappa_1, \kappa_2)$ and from the following measurement number becoming zero:

$$\alpha_2 = \frac{\dfrac{1}{m+1} \displaystyle\sum_{\kappa_1=0}^{m} \left[\beta_2(\kappa_1) - \bar{\beta}_2\right]^2}{\bar{\beta}_2^2}$$

Here

$$\beta_2(\kappa_1) = \frac{h_2(\kappa_1, \kappa_2)}{h_1(\kappa_2)} \qquad (5.5.4)$$

is the ratio of the second- and first-degree kernels and β_2 is its average value,

$$\bar{\beta}_2 = \frac{1}{m+1} \sum_{\kappa_1=0}^{m} \beta_2(\kappa_1)$$

and α_2 is the normalized deviation of (5.5.4).

3. *Wiener models (simple, generalized and extended)*
   In the Wiener models where the multiplier is preceded by two equal linear terms, the sections of the quadratic kernels parallel to the axes are proportional to each other because

$$\frac{g_2(\tau_1, \tau_2 + \tau)}{g_2(\tau_1, \tau_2)} = \frac{g_2(\tau_1)g_2(\tau_2 + \tau)}{g_2(\tau_1)g_2(\tau_2)} = \text{const}(\tau_2, \tau) \qquad (5.5.5)$$

Denote the quotient (5.5.5) by $\beta_3(\kappa_1, \kappa_2, \kappa)$ and its average value parallel to the axis by $\bar{\beta}_3(\kappa_2, \kappa)$:

$$\beta_3(\kappa_1, \kappa_2, \kappa) = \frac{h_2(\kappa_1, \kappa_2 + \kappa)}{h_2(\kappa_1, \kappa_2)}$$

and

$$\bar{\beta}_3(\kappa_2, \kappa) = \frac{1}{m+1} \sum_{\kappa_1 = 0}^{m} \beta_3(\kappa_1, \kappa_2, \kappa)$$

Consider the normalized deviation of the ratio of the parallel sections at κ_2 and $\kappa_2 + \kappa$ distance

$$\alpha_3(\kappa_2, \kappa) = \frac{\dfrac{1}{m+1} \sum\limits_{\kappa_1 = 0}^{m} \left[\beta_3(\kappa_1, \kappa_2, \kappa) - \bar{\beta}_3(\kappa_2, \kappa)\right]^2}{\bar{\beta}_3^2(\kappa_2, \kappa)}$$

At the Wiener models, $\alpha_3(\kappa_2, \kappa)$ has to be zero for every (κ_2, κ), i.e., the index

$$\alpha_3 = \max_{\kappa_2, \kappa}\{\alpha_3(\kappa_2, \kappa)\} \tag{5.5.6}$$

has to be zero. It is enough to check the index (5.5.6) for $\kappa = 1$, and then it should be tested only among m measures whether there is any that differs from zero.

4. *Simple Wiener model*
   The quadratic Volterra kernels along the main diagonal are proportional to the squares of the linear kernels in the simple Wiener models. Form the ratio of them as a discrete time function

$$\beta_1(\kappa_1) = \frac{h_2(\kappa_1, \kappa_1)}{h_1^2(\kappa_1)} \tag{5.5.7}$$

and its average value

$$\bar{\beta}_4 = \frac{1}{m+1} \sum_{\kappa_1 = 0}^{m} \beta_4(\kappa_1)$$

The normalized deviation of (5.5.7)

$$\alpha_4 = \frac{\dfrac{1}{m+1} \sum\limits_{\kappa_1 = 0}^{m} \left[\beta_4(\kappa_1) - \bar{\beta}_4\right]^2}{\bar{\beta}_4^2}$$

becomes zero only at the simple Wiener model.

5. *Extended Wiener model with quadratic channel containing first-order terms*
   Haber (1985, 1989) has elaborated a new structure identification method for a special
   case of the Wiener model. Assume, e.g., that two first-order lag terms are before the
   multiplier, and the weighting functions of them are of the form

$$g_2(\tau) = K_2^0 \exp(-p_2 \tau) \tag{5.5.8}$$
$$g_3(\tau) = K_3^0 \exp(-p_3 \tau) \tag{5.5.9}$$

In this case the sections of the quadratic kernels parallel to the main diagonal are
proportional, i.e.,

$$\frac{g_2(\tau_1, \tau_1 + \tau_2 + \tau)}{g_2(\tau_1, \tau_1 + \tau_2)}$$

$$= \frac{K_2^0 \exp(-p_2 \tau_1) K_3^0 \exp(-p_3(\tau_1 + \tau_2 + \tau)) + K_2^0 \exp(-p_2(\tau_1 + \tau_2 + \tau)) K_3^0 \exp(-p_3 \tau_1)}{K_2^0 \exp(-p_2 \tau_1) K_3^0 \exp(-p_3(\tau_1 + \tau_2)) + K_2^0 \exp(-p_2(\tau_1 + \tau_2)) K_3^0 \exp(-p_2 \tau_1)}$$

$$= \frac{\exp(-p_2 \tau) + \exp(-p_3 \tau)}{2} = \text{const}(\tau)$$

Take the following ratio as an analytical index

$$\beta_5(\kappa_1, \kappa_2, \kappa) = \frac{h_2(\kappa_1, \kappa_1 + \kappa_2 + \kappa)}{h_2(\kappa_1, \kappa_1 + \kappa_2)} \tag{5.5.10}$$

Both the numerator and denominator consist of quadratic kernels laying parallel to
the main diagonal and cross the axis at discrete time units $\kappa_2 + \kappa$ and κ_2,
respectively. The average value of (5.5.10) according to the time shift is

$$\bar{\beta}_5(\kappa_2, \kappa) = \frac{1}{\kappa_1 + 1 - \kappa_2 - \kappa} \sum_{\kappa_1 = 0}^{m - \kappa_2 - \kappa} \beta_5(\kappa_1, \kappa_2, \kappa)$$

The normalized deviation

$$\alpha_5(\kappa_2, \kappa) = \frac{\frac{1}{\kappa_1 + 1 - \kappa_2 - \kappa} \sum_{\kappa_1 = 0}^{m - \kappa_2 - \kappa} \left[\beta_5(\kappa_1, \kappa_2, \kappa) - \bar{\beta}_5(\kappa_2, \kappa) \right]}{\bar{\beta}_5^2(\kappa_2, \kappa)} \tag{5.5.11}$$

shows how constant $\beta_5(\kappa_1, \kappa_2, \kappa)$ is. Finally, after having calculated (5.5.11) at
every (κ_2, κ) pairs, the index

$$\alpha_5 = \max_{\kappa_2, \kappa} \{ \beta_5(\kappa_2, \kappa) \} \tag{5.5.12}$$

becomes zero if the model is of Wiener type and has the features mentioned. It is enough to compute (5.5.12) at $\kappa_2 = 0$, which means that comparing the terms at the main and off-diagonals.

Other criteria can also be set up for the same feature.

In Wiener models where two first-order lag terms precede the multiplier, the parallel sections of the quadratic kernels perpendicular to the main diagonal are proportional to each other, because

$$\frac{g_2(\tau_1 + \tau, \tau_2 + \tau)}{g_2(\tau_1, \tau_2)}$$

$$= \frac{K_2^0 \exp(-p_2(\tau_1 + \tau)) K_3^0 \exp(-p_3(\tau_3 + \tau)) + K_2^0 \exp(-p_2(\tau_2 + \tau)) K_3^0 \exp(-p_3(\tau_1 + \tau))}{K_2^0 \exp(-p_2 \tau_1) K_3^0 \exp(-p_3 \tau_3) + K_2^0 \exp(-p_2 \tau_2) K_3^0 \exp(-p_3 \tau_1)}$$

$$= \exp(-(p_2 + p_3)\tau) = \mathrm{const}(\tau)$$

Here the notations of (5.5.8) and (5.5.9) have been applied. Take the following ratio as an analytical measure

$$\beta_6(\kappa_1, \kappa, \kappa_2) = \frac{h_2(\kappa_1 + \kappa_2, \kappa - \kappa_1 + \kappa_2)}{h_2(\kappa_1, \kappa - \kappa_1)} \tag{5.5.13}$$

Both the numerator and the denominator consist of quadratic kernels that lie on lines perpendicular to the main diagonal and cross the axis at discrete time units κ and $(\kappa_2 + \kappa)$, respectively. The average value of (5.5.13) according to the time shift is

$$\bar{\beta}_6(\kappa, \kappa_2) = \frac{1}{\kappa + 1} \sum_{\kappa_1 = 0}^{\kappa} \beta_6(\kappa_1, \kappa, \kappa_2)$$

The normalized deviation

$$\alpha_6(\kappa, \kappa_2) = \frac{\dfrac{1}{\kappa + 1} \sum_{\kappa_1 = 0}^{\kappa} \left[\beta_6(\kappa_1, \kappa, \kappa_2) - \bar{\beta}_6(\kappa, \kappa_2)\right]^2}{\bar{\beta}_6^2(\kappa, \kappa_2)} \tag{5.5.14}$$

shows how constant $\beta_6(\kappa_1, \kappa, \kappa_2)$ is. Finally, after having calculated (5.5.14) at every (κ, κ_2) pairs, the index

$$\alpha_6 = \max_{\kappa, \kappa_2} \{\beta_6(\kappa, \kappa_2)\} \tag{5.5.15}$$

becomes zero if the model is of Wiener type and has the above features. It is enough to compute (5.5.15) at $\kappa_2 = 1$ which means that α_6 has to be sought among m measures according to (5.5.12).

The different analytical criteria characterizing the identified Volterra kernels of the block oriented models are summed up in Table 5.5.6.

TABLE 5.5.6 Features of the Volterra kernels characterizing the block oriented models

No	Model	α_1	α_2	α_3	α_4	α_5, α_6	$g_2(\tau_1, \tau_2) = \text{const}$	rank of $H_2(\kappa_1, \kappa_2)$
1	Simple Hammerstein	0	0	≠0	≠0	≠0		$m+1$
2	Generalized Hammerstein	0	≠0	≠0	≠0	≠0		$m+1$
3	Simple Wiener	≠0	≠0	0	0	0 if $G(s)$ is a first-order lag term; ≠0 in other cases	G: first-order lag term / over-damped / under-damped	1
4	Generalized Wiener	≠0	≠0	0	≠0	0 if $G_2(s)$ is a first-order lag term; ≠0 in other cases	G_2: first-order lag term / over-damped / under-damped	1
5	Extended Wiener	≠0	≠0	≠0	≠0	0 if both $G_2(s)$ and $G_3(s)$ are first-order lag terms; ≠0 in other cases	G_2, G_3: over-damped / at least one of them is underdamped	2
6	Simple Wiener-Hammerstein cascade	≠0	≠0	≠0	≠0	≠0	G_1: overd. / underd. / overd. / underd.; G_2: overd. / overd. / underd. / underd.	$p = \text{grad}(G_2)$ $(m+1)$ if $p \geq m$ $(p+1)$ if $p < m$
7	Generalized Wiener-Hammerstein cascade	≠0	≠0	≠0	≠0	≠0	G_2: overd. / overd. / underd. / underd.; G_3: overd. / underd. / overd. / underd.	$p = \text{grad}(G_3)$ $(m+1)$ if $p \geq m$ $(p+1)$ if $p < m$
8	Extended Wiener-Hammerstein cascade	≠0	≠0	≠0	≠0	≠0	G_2: overd. / overd. / at least one of them is underdamped; G_3: overd. / overd. / overd. / underd.; G_4: overd. / underd. / overd. / underd.	$p = \text{grad}(G_4)$ $(m+1)$ if $p \geq m$ $\geq (2+p)$ $\leq (2p+2)$ if $p < m$

II. Conclusions drawn from the contours of the quadratic kernels

The shape of the contour of the quadratic kernel is characteristic to the structure of the block oriented models as follows:

- In Hammerstein models the points of the contours of a second degree kernel can lie only at the main diagonal because at any other place they are zero.
- In the simple and generalized Wiener models where the multiplier is preceded by two equal first-order lag terms, the contours are perpendicular to the main diagonal because (according to the notations of (5.5.8) and (5.5.9))

$$g_2(\tau_1, \tau_2) = K_2^0 \exp(-p_2 \tau_1) K_2^0 \exp(-p_2 \tau_2)$$
$$= K_2^0 \exp(-p_2(\tau_1 + \tau_2)) = \text{const}$$

if $(\tau_1 + \tau_2) = \text{const}$

- In Wiener models the contours do not close in if the linear terms before the multiplier are over-damped.
- In Wiener models the contours are closed – eventually many times, they may form *islands* – if one of the linear terms before the multiplier is an under-damped one.
- In Wiener–Hammerstein cascade models the contours close around a point of the main diagonal if all of the linear terms in the model are over-damped.
- In Wiener–Hammerstein cascade models the contours may close around more

points at the main diagonal – may build *islands* – if the linear term after the multiplier is of an under-damped type because its effect can be seen along the main diagonal.
- In Wiener–Hammerstein cascade models the contours may close in around points outside the main diagonal if at least one of the linear terms before the multiplier is of an under-damped type.

The different shapes of the contours characterizing the identified Volterra kernels of the block oriented models are summed up in Table 5.5.6.

III. Criteria based upon the rank of a matrix formed from the quadratic kernel
Consider the second-degree kernel $h_2(\kappa_1, \kappa_2)$ with the maximum memory of m and build the matrix H_2 of dimension $(m+1)(m+1)$ according to

$$H_2(\kappa_1, \kappa_2) = \left[h_2(\kappa_1, \kappa_2) \right] \tag{5.5.16}$$

The matrix H_2 can be separated into the dyadic products of ρ (a maximum of $(m+1)$) vector pairs:

$$H_2 = \sum_{i=1}^{\rho} h_{2i} h_{2i}^{T}$$

Here $\rho \leq (m+1)$ is the rank of the matrix. The dyadic vectors can be calculated by several recursive algorithms. A formal solution of the problem can be, e.g., if the left and upper non-zero vectors are always taken and normalized with the common elements. There exist many simpler methods than the dyadic separation.

Assume that the memory of all linear terms in the block oriented model amounts to at least m. This is not a strong assumption because every lag term is infinite concerning its memory. Build the vector of the weighting function series by

$$h_2 = \left[h_2(0), h_2(1), ..., h_2(m) \right]^{T}$$
$$h_3 = \left[h_3(0), h_3(1), ..., h_3(m) \right]^{T}$$

The following conclusions can be drawn:
- The rank of the simple and generalized Wiener model is

$$\rho = 1$$

because

$$g_2(\tau_1, \tau_2) = g_2(\tau_1) g_2(\tau_2)$$

and

$$H_2 = h_2 h_2^{T} \tag{5.5.17}$$

- The matrix rank of the extended Wiener model is

$$\rho = 2$$

because

$$g_2(\tau_1, \tau_2) = \frac{1}{2}\left[g_2(\tau_1)g_3(\tau_2) + g_2(\tau_2)g_3(\tau_2)\right]$$

and

$$H_2 = \frac{1}{2}\left[h_2 h_3{}^T + h_3 h_2{}^T\right] \tag{5.5.18}$$

- The matrix rank of the Hammerstein and Wiener–Hammerstein cascade models is

$$\rho = m + 1$$

The rank values characterizing the identified Volterra kernels of the block oriented models are summed up in Table 5.5.6.

On the basis of the criteria listed the structure of the block oriented models can be identified from the estimated first- and second-order Volterra kernels. The simulations consider only noise-free cases and do not contain the truncation error caused by the finite memory. These cases of neglect are acceptable as this section deals only with the structure identification. In the simulation no zero-order holding device was used, therefore the Volterra kernels of the discrete system were identified with those of the continuous system at the sampling times.

In the foregoing figures the simulated block oriented models are shown beside the Volterra kernels. They prove the rightness of the structure identification.

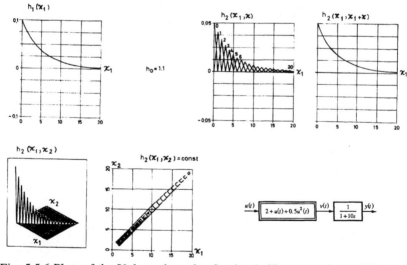

Fig. 5.5.6 Plots of the Volterra kernels of a simple Hammerstein model with first-order lag term

Example 5.5.7   *Structure determination of simple and generalized Hammerstein models (Haber, 1985, 1989)*

The Volterra kernels of Figures 5.5.6 and 5.5.7 suggest Hammerstein models because the quadratic kernels differ from zero only at the main diagonal $(\alpha_1 = 0)$. In

Figure 5.5.6 $h_1(\kappa_1)$ and $h_2(\kappa_1, \kappa_1)$ are proportional $(\alpha_2 = 0)$, which suggests a simple Hammerstein model in contradiction to Figure 5.5.7 which shows a generalized Hammerstein model.   ∎

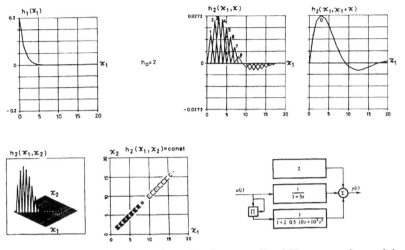

Fig. 5.5.7 Plots of the Volterra kernels of a generalized Hammerstein model with second-order (under-damped) lag term

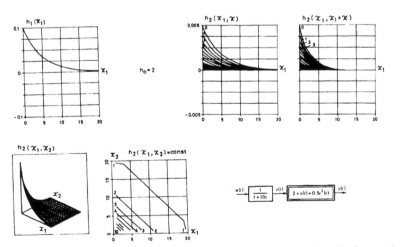

Fig. 5.5.8 Plots of the Volterra kernels of a simple Wiener model with first-order lag term

Example 5.5.8   *Structure determination of simple and generalized Wiener models (Haber, 1985, 1989)*

Figures 5.5.8 and 5.5.9 indicate simple or generalized Wiener models, because the sections $h_2(\kappa_1, \kappa)$; $\kappa = $ const are proportional to each other $(\alpha_5 = 0)$. In the first case the sections are over-damped and in the second case under-damped – which shows the type of the linear term in the quadratic channel. The aperiodic term is of first-order lag because the level lines $h_2(\kappa_1, \kappa_2) = \text{const}$ are straight. In Figure 5.5.8 $h_2(\kappa_1, \kappa_1) = h_1^2(\kappa_1)$, $(\alpha_4 = 0)$, i.e., the corresponding model is a simple Wiener model opposite to the generalized Wiener model in Figure 5.5.9.   ■

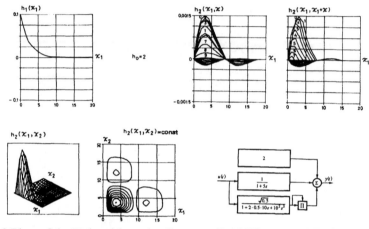

Fig. 5.5.9 Plots of the Volterra kernels of a generalized Wiener model with second-order (under-damped) lag term

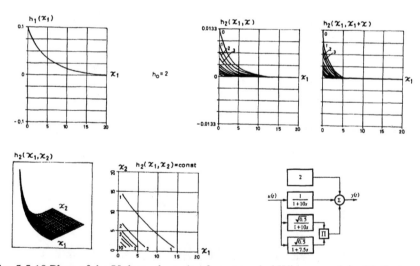

Fig. 5.5.10 Plots of the Volterra kernels of an extended Wiener model with first-order lag term

Example 5.5.9   *Structure determination of extended Wiener models with first- and second-order dynamic terms (Haber, 1985, 1989)*

In Figures 5.5.10 and 5.5.11 the quadratic kernels do exist also beside the main diagonal $(\alpha_1 \neq 0)$, i.e., the corresponding models cannot be of Hammerstein type. The sections

$h_2(\kappa_1, \kappa_2) = \text{const}$ are not proportional, therefore the assumptions of simple and generalized Wiener models are rejected. In Figure 5.5.10 the sections $h_2(\kappa_1, \kappa_1 + \kappa)$ = const are proportional to each other $(\alpha_5 = 0)$, those points at an extended Wiener model with two different first-order lag terms in the quadratic channels. The over-damped feature can also be seen at the level lines. In Figure 5.5.11 the sections $h_2(\kappa, \kappa_1 + \kappa)$, $\kappa = \text{const}$ are not proportional to each other $(\alpha_5 \neq 0)$, which fact does not exclude the model being of an extended Wiener type, but then it should have higher order lag terms in the quadratic channel. The rank of the quadratic matrix $\boldsymbol{H}_2$ formed from the values of $h_2(\kappa_1, \kappa_2)$ is 2 and that refers unequivocally to an extended Wiener model. At least one of the linear terms before the multiplier is of under-damped type as it can be seen from the sections of $h_2(\kappa_1, \kappa_2)$ and from the multiple islands in the contour plot outside the main diagonal. ∎

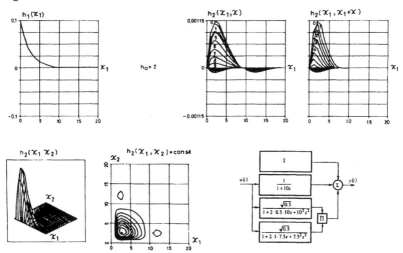

Fig. 5.5.11 Plots of the Volterra kernels of an extended Wiener model with second-order (under-damped) lag term

Example 5.5.10 *Structure identification trial of several simple Wiener–Hammerstein cascade models (Haber, 1985, 1989)*
Nothing concrete can be stated based on the criteria concerning the sections of the quadratic kernels in Figures 5.5.12 to 5.5.15 ($\alpha_1 \neq 0$, $\alpha_2 \neq 0$, $\alpha_3 \neq 0$, $\alpha_4 \neq 0$, $\alpha_5 \neq 0$, $\alpha_6 \neq 0$). Although one can guess from the contour plots that the linear dynamic terms before and after the multiplier are of the following types:

Figure 5.5.12 $G_2(s)$, $G_3(s)$, $G_4(s)$ over-damped
Figure 5.5.13 $G_2(s)$, $G_3(s)$ over-damped, $G_4(s)$ under-damped
Figure 5.5.14 at least one of $G_2(s)$ and $G_3(s)$ are under-damped, $G_4(s)$ over-damped
Figure 5.5.15 at least one of $G_2(s)$ and $G_3(s)$ under-damped, $G_4(s)$ under-damped. ∎

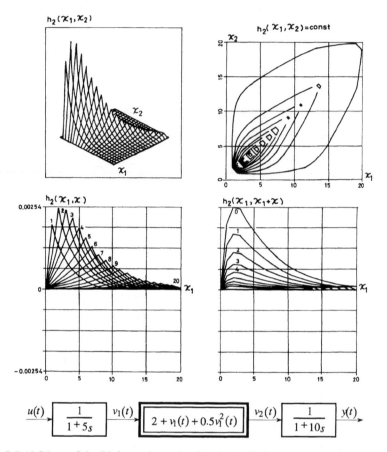

Fig. 5.5.12 Plots of the Volterra kernels of a simple Wiener–Hammerstein cascade
model (Both linear dynamic terms are over-damped)

Remarks: (Haber, 1985, 1989)

1. The widespread identification of the structure and parameter estimation of the block
   oriented models requires in the engineering practice special algorithms for each
   model type. The estimation of the Volterra kernels makes the structure identification
   in one step possible independently of the model type. One has to estimate only the
   first- and second-degree kernels up to a long enough memory and then the structure
   of the block oriented model can be determined from the plots of the first-degree –
   one-dimensional – and second-degree – two-dimensional – Volterra kernels.

2. Several criteria were listed which help in the decision about the structure. The
   analytical measures are based on the following graphic plots of the quadratic kernels:
   - axonometric plot;
   - contour plot;
   - sections parallel to an axis;
   - sections parallel to the main diagonal;
   - sections perpendicular to the main diagonal.

3. The analytical measures given in Section 5.5.2 are, of course, only statistical measures, and have to be understood as such.

4. Since most structure determination criteria rely on graphical features, it is recommended to draw the estimated kernels.

5. After having determined the right structure the parameters can be estimated either from the Volterra kernels or by means of a direct estimation of the parameters of the parametric model of known structure.

6. Emara-Shabaik *et al.*, (1995) developed also structure identification criteria for the Hammerstein and Wiener cascade models. Applying a Gaussian white noise excitation the plot of the two-dimensional spectrum of the output signal is characteristic for the cascade structure.

Keulers *et al.*, (1993) determined the structure of a fed-batch baker's yeast fermentation process from the contour plots of the estimated Volterra kernels as the (extended) Wiener–Hammerstein cascade model.

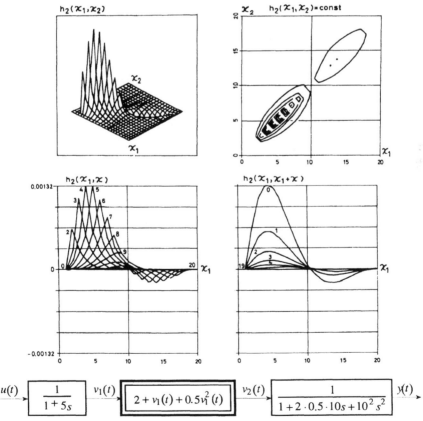

Fig. 5.5.13 Plots of the Volterra kernels of a simple Wiener–Hammerstein cascade models (The linear dynamic term before the nonlinear part is over-damped and after it is under-damped)

5.5.3 Frequency method

The structure and parameter of a quadratic block oriented model can be determined by separating the output signal at a sinusoidal excitation to the responses of the parallel constant, linear and quadratic channels. This can be done by two different methods:
- by Gardiner's method, or
- by physical considerations.

The latter method means that the second harmonic of the output signal corresponds to the quadratic channel, the first harmonic to the linear channel and only the constant term is superposed by the effect of both the constant and quadratic channels. However, it is not difficult to obtain the response of the constant channel, it is the output of the system at zero excitation.

Both the first and second harmonic responses can be identified by methods used in linear systems (e.g., by means of the Bode diagram).

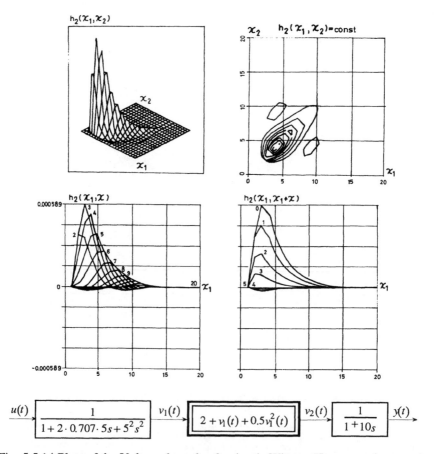

Fig. 5.5.14 Plots of the Volterra kernels of a simple Wiener–Hammerstein cascade model (The linear dynamic term before the nonlinear static part is under-damped and after it is over-damped)

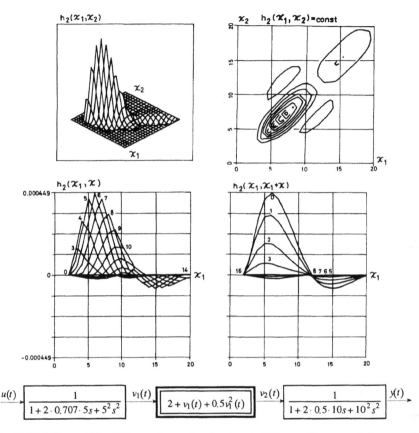

Fig. 5.5.15 Plots of the Volterra kernels of a simple Wiener–Hammerstein cascade
model (Both linear dynamic terms are under-damped)

The quadratic block oriented models with one multiplier are the special cases of the
extended Wiener–Hammerstein cascade model. Therefore it is enough to show how a
sinusoidal signal passes the quadratic channel (Figure 5.5.16) of the extended Wiener–
Hammerstein model.

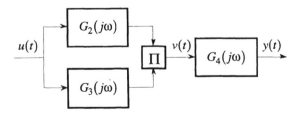

Fig. 5.5.16 The quadratic
channel of an extended Wiener–
Hammerstein cascade model

Lemma 5.5.1
A sinusoidal input signal

$$u(t) = U\cos\omega t$$

causes in a quadratic extended Wiener–Hammerstein model – also – a second harmonic response, and the equivalent frequency transfer function is

$$G(j\omega) = G_2(j\omega)G_3(j\omega)G_4(2j\omega)$$

if – theoretically – the input is assumed to be a sinusoidal signal with unity amplitude and the same – twice the actual – frequency. $G_2(s)$ and $G_3(s)$ are the transfer functions of the terms before and $G_4(s)$ after the multiplier.

Proof. The output signal after the multiplier is

$$
\begin{aligned}
v(t) &= U|G_2(\omega)|\cos(\omega t + \varphi_2(\omega))U|G_3(\omega)|\cos(\omega t + \varphi_3(\omega)) \\
&= \frac{U^2}{2}|G_2(\omega)||G_3(\omega)|\cos(2\omega t + \varphi_2(\omega) + \varphi_3(\omega)) \\
&+ \frac{U^2}{2}|G_2(\omega)||G_3(\omega)|\cos(\varphi_2(\omega) - \varphi_3(\omega))
\end{aligned}
$$

Here $|G_i(\omega)|$ and $\varphi_i(\omega)$ are the absolute and phase values at the given frequencies, respectively. If the input signal is considered – theoretically – as

$$u(t) = \cos(2\omega t)$$

i.e., its frequency is twice that used and its amplitude is unity, then the second harmonic response can be considered as the output of the linear frequency transfer function

$$G(j\omega) = \frac{U^2}{2}G_2(j\omega)G_3(j\omega)$$

Now, this signal is filtered by $G_4(j\omega)$, which results in the frequency transfer function

$$G(j\omega) = \frac{U^2}{2}G_2(j\omega)G_3(j\omega)G_4(2j\omega)$$

The reason for considering $G_4(j\omega)$ at double the frequency is that the harmonic is double. ∎

Based on the foregoing, the harmonics caused by the corresponding own channels are tabulated in Table 5.5.7. There are relations between the pole/zero configurations of the linear and second harmonic networks that are typical and select the structure of the system (Haber and Unbehauen, 1990):

1. *Simple Hammerstein model*
   The poles and zeros of the quadratic harmonics are half those of the linear channel.

2. *Simple Wiener model*
   The poles and zeros of the linear channel occur in the quadratic harmonics with multiplicity two.

TABLE 5.5.7 The frequency functions of the highest degree harmonics of the channels in the quadratic block oriented models at $u(t) = U\cos(\omega t)$ excitation

No	Model	Scheme	Equivalent frequency function of the different channels			Second-degree harmonic			First-degree harmonic		
			$i=0$	$i=1$	$i=2$	poles	zeros	static gain	poles	zeros	static gain
1	Simple Hammerstein	$[c_0+c_1u+c_2u^2]\to G(s)$	$c_0G(0)$	$Uc_1G(j\omega)$	$\frac{U^2}{2}c_2G(2j\omega)$	$\frac{p_i}{2}$	$\frac{z_i}{2}$	$\frac{U^2}{2}c_2G(0)$	p_i	z_i	$Uc_1G(0)$
2	Generalized Hammerstein	$G_1(s)$, $G_2(s)$, c_0	c_0	$UG_1(j\omega)$	$\frac{U^2}{2}G(2j\omega)$	$\frac{p_{2i}}{2}$	$\frac{z_{2i}}{2}$	$\frac{U^2}{2}G(0)$	p_{1i}	z_{1i}	$UG_1(0)$
3	Simple Wiener	$G(s)\to[c_0+c_1v+c_2v^2]$	c_0	$Uc_1G_1(j\omega)$	$\frac{U^2}{2}c_2G^2(j\omega)$	p_i, p_i	z_i, z_i	$\frac{U^2}{2}c_2G^2(0)$	p_i	z_i	$Uc_1G(0)$
4	Generalized Wiener	$G_1(s)$, $G_2(s)$, c_0	c_0	$UG_1(j\omega)$	$\frac{U^2}{2}G^2(j\omega)$	p_{2i}, p_{2i}	z_{2i}, z_{2i}	$\frac{U^2}{2}G^2(0)$	p_{1i}	z_{1i}	$UG_1(0)$
5	Extended Wiener	$G_1(s)$, $G_2(s)$, $G_3(s)$, c_0	c_0	$UG_1(j\omega)$	$\frac{U^2}{2}G_2(j\omega)G_3(j\omega)$	p_{2i}, p_{3i}	z_{2i}, z_{3i}	$\frac{U^2}{2}G_2(0)G_3(0)$	p_{1i}	z_{1i}	$UG_1(0)$
6	Simple Wiener-Hammerstein cascade	$G_1(s)\to[c_0+c_1v+c_2v^2]\to G_2(s)$	$c_0G_2(0)$	$Uc_1G_1(j\omega)\times G_2(2j\omega)$	$\frac{U^2}{2}c_2G_1^2(j\omega)\times G_2(2j\omega)$	p_{1i}, p_{1i}, $\frac{p_{2i}}{2}$	z_{1i}, z_{1i}, $\frac{z_{2i}}{2}$	$\frac{U^2}{2}c_1G_1^2(0)G_2(0)$	p_{1i}, p_{2i}	z_{1i}, z_{2i}	$Uc_1G_1(0)G_2(0)$
7	Generalized Wiener-Hammerstein cascade	$G_1(s)$, $G_2(s)$, $G_3(s)$, c_0	c_0	$UG_1(j\omega)$	$\frac{U^2}{2}G_2^2(j\omega)\times G_3(2j\omega)$	p_{2i}, p_{2i}, $\frac{p_{3i}}{2}$	z_{2i}, z_{2i}, $\frac{z_{3i}}{2}$	$\frac{U^2}{2}G_2^2(0)G_3(0)$	p_{1i}	z_{1i}	$UG_1(0)$
8	Extended Wiener-Hammerstein cascade	$G_1(s)$, $G_2(s)$, $G_3(s)$, $G_4(s)$, c_0	c_0	$UG_1(j\omega)$	$\frac{U^2}{2}G_2(j\omega)G_3(j\omega)\times G_4(2j\omega)$	p_{2i}, p_{3i}, $\frac{p_{4i}}{2}$	z_{2i}, z_{3i}, $\frac{z_{4i}}{2}$	$\frac{U^2}{2}G_2(0)G_3(0)\times G_4(0)$	p_{1i}	z_{1i}	$UG_1(0)$

3. *Generalized Wiener model*
   The poles and zeros occur with multiplicity two in the quadratic harmonics.

4. *Simple Wiener–Hammerstein cascade model*
   The poles and zeros of the first linear dynamic term occur both in the linear and quadratic harmonics, but in the latter with multiplicity two. The poles and zeros of the second linear dynamic term occur only in the linear transfer function, but its half values occur in the second harmonics.

5. *Generalized Wiener–Hammerstein cascade model*
   The poles and zeros of the linear dynamic term before the multiplier occur in the second harmonics with multiplicity two.

Some other features can be read in Table 5.5.7 that are needed for the parameter estimation of the block oriented model.

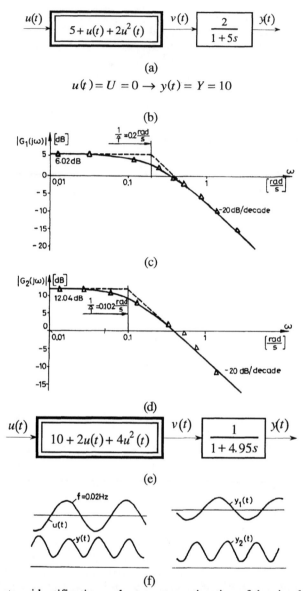

(a)

$$u(t) = U = 0 \rightarrow y(t) = Y = 10$$

(b)

(c)

(d)

(e)

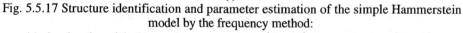

(f)

Fig. 5.5.17 Structure identification and parameter estimation of the simple Hammerstein model by the frequency method:
(a) simulated model; (b) steady state response $(Y = 10)$ to zero forcing $(U = 0)$;
(c) Bode plot of the first harmonic; (d) Bode plot of the second harmonic;
(e) identified model with estimated parameters;
(f) input signal $u(t)$, measured output signal $y(t)$ and its first-degree $y_1(t)$ and second-degree $y_2(t)$ harmonics

Example 5.5.11 *Structure identification and parameter estimation of the simple Hammerstein model (Haber et al., 1986)*

The simple Hammerstein model of Figure 5.5.17a was simulated on an analog computer MINISPACE TY865. Applying the sinusoidal test signal with unit amplitude, the output signal contained both constant (DC) value, first and second degree harmonics (Figure 5.5.17f). The different harmonics can be separated by appropriate filters. Instead of doing this, we filtered a sinusoidal signal of the same frequency and phase as the input signal by a first-order lag filter and we subtracted this filtered signal from the measured output signal. The parameters of the filter were changed until the bias signal did not contain any base harmonics (only second harmonics and DC value). Now the actual gain and time constant of the filter gave the actual amplitude and phase values of the frequency characteristics at the given frequency. The same procedure was used for filtering the second-degree harmonics; then the signal generator was working with double frequency and the filter parameters had to be changed until only first harmonics and DC value remained in the bias signal. The constant term was measured at zero input signal (Figure 5.5.17b). Figure 5.5.17c shows the Bode plot of the first harmonic and Figure 5.5.17d that of the second harmonic. Both Bode plots have the same shape, like that of a first-order term, but the cut–off frequency of the quadratic channel is half of the linear channel, i.e., the pole of the second harmonic equivalent network is the half of the pole of the linear network. According to Table 5.5.7 the above relation between the poles points at the simple Hammerstein model.

The parameters of the simple Hammerstein model can be obtained as follows. The cut–off frequency of the first order harmonics is

$$\omega_c = 0.202 \quad \left[s^{-1} \right]$$

therefore the time constant is

$$T = 1/\omega_c = 4.95 \quad [s]$$

The constant term is equal to the constant output signal at zero input signal

$$c_0 = 10$$

The static gains of the linear and quadratic channels can be read from the asymptotics of the corresponding Bode plots at zero frequencies:

$$20\log|G_1(0)| = 6.02\,\mathrm{dB} \quad \rightarrow \quad c_1 = 2$$
$$20\log|G_2(0)| = 12.04\,\mathrm{dB} \quad \rightarrow \quad c_2 = 4$$

The parameters of the identified simple Hammerstein model are almost the same as those of the simulated ones. The only difference is in replacing the gain of the linear dynamic term to the static polynomial that does not change the input–output behavior of the system.

Example 5.5.12 *Structure identification and parameter estimation of the simple Wiener model (Haber et al., 1986)*

The simple Wiener model of Figure 5.5.18a was simulated on an analog computer MINISPACES TY865. Applying sinusoidal test signal with unit amplitude, the output

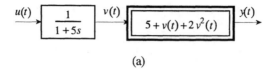

(a)

$$u(t) = U = 0 \rightarrow y(t) = Y = 5$$

(b)

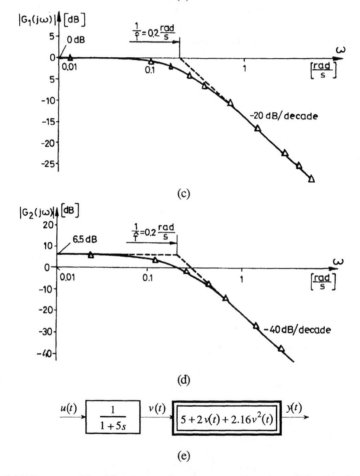

(c)

(d)

(e)

Fig. 5.5.18 Structure identification and parameter estimation of the simple Wiener
model by the frequency method:
(a) simulated model; (b) steady state response $(Y = 5)$ to zero forcing $(U = 0)$;
(c) Bode plot of the first harmonic; (d) Bode plot of the second harmonic;
(e) identified model with estimated parameters

signal consisted of a DC value, and first and second harmonics. The first- and
second-degree harmonics were separated by the procedure described in Example 5.5.11.

The Bode plot of the linear and quadratic channels (Figures 5.5.18c and 5.5.18d, respectively) had the following relation:
- the cut–off frequency was the same in both cases;
- the slope was twice in the second-degree harmonic of that in the first-order ones.

This means that the second harmonic equivalent network contains a double pole and its value is equal to the pole of the linear network. According to Table 5.5.7 this relation between the poles points at the simple Wiener model.

The parameters of the simple Wiener model could be estimated as follows. The cut–off frequency of both harmonics was

$$\omega_c = 0.2 \left[s^{-1} \right]$$

therefore the time constant is

$$T = 1/\omega_c = 5 \, [s]$$

The constant term is equal to the constant output signal at the zero input signal

$$c_0 = 5$$

The static gains of the linear and quadratic channels can be read from the asymptotics of the corresponding Bode plots at zero frequencies:

$$20\log|G_1(0)| = 0 \text{ dB} \quad \rightarrow \quad c_1 = 1$$
$$20\log|G_2(0)| = 6.5 \text{ dB} \quad \rightarrow \quad c_2 = 2.11$$

The parameters of the identified simple Wiener model are almost the same as those of the simulated one.

5.5.4 Evaluation of pulse and step responses
It is known from the frequency method that the structure can be determined from the pole/zero distribution of the equivalent transfer function to the second harmonic excitation. The step and pulse responses are also influenced by the pole/zero locations of both the linear and quadratic channels. The whole step or pulse response of the system can be considered as the response of a linear system of the same excitation and an equivalent transfer function can be determined as the ratio of the Laplace transformation of the output and input signals.

Table 5.5.8 presents the steady state relationships between the input and output signals and the equivalent transfer functions of the linear and the quadratic channels, and as well as of the whole system. The constant term was assumed to be zero, since it is very simple to estimate it by as the steady state value of the unforced system (response to a zero input signal). The structure identification of the quadratic block oriented systems can be performed based on the followings.

The equivalent transfer functions derived from the pulse responses contain the following poles (Haber *et al.*, 1986):
- the poles of the linear channels;
- the poles of the transfer functions of the terms after the multipliers in the quadratic channels;
- the sums of the poles of the transfer functions of the terms occurring before the multipliers in the quadratic channels.

TABLE 5.5.8a Steady state input–output relationships and equivalent transfer functions of the first-order quadratic block oriented models to pulse and step excitations

No	Model	Scheme	$Y = Y(U)$	$Y = Y(0)$	Equivalent transfer function ($c_0 = 0$)	
					$u(t) = U \cdot 1(t)$	$u(t) = U \cdot \delta(t)$
1	Simple Hammerstein		$(c_0 + c_1 U + c_2 U^2)K$	$c_0 K$	$\dfrac{K(c_1 + c_2 U)}{1 + sT}$	$\dfrac{K(c_1 + c_2 U)}{1 + sT}$
2	Generalized Hammerstein		$c_0 + K_1 U + K_2 U^2$	c_0	$\dfrac{K_1}{1+sT_1} + \dfrac{K_2 U}{1+sT_2} =$ $(K_1 + K_2 U)\dfrac{1 + s\frac{K_1 T_2 + K_2 U T_1}{K_1 + K_2 U}}{(1+sT_1)(1+sT_2)}$	$\dfrac{K_1}{1+sT_1} + \dfrac{K_2 U}{1+sT_2} =$ $(K_1 + K_2 U)\dfrac{1 + s\frac{K_1 T_2 + K_2 U T_1}{K_1 + K_2 U}}{(1+sT_1)(1+sT_2)}$
3	Simple Wiener		$c_0 + c_1 KU + c_2 K^2 U^2$	c_0	$\dfrac{c_1 K}{1+sT} + \dfrac{c_2 K^2 U}{(1+sT)(1+sT/2)} =$ $\dfrac{(c_1 K + c_2 K^2 U)\left(1 + sT/2\frac{c_1 K}{c_1 K + c_2 K^2 U}\right)}{(1+sT)(1+sT/2)}$	$\dfrac{c_1 K}{1+sT} + \dfrac{c_2 K^2 U/2T}{1+sT/2} =$ $K\dfrac{(c_1 + c_2 UK/2T)\left(1 + s\frac{T/2 + c_2 UK/2}{c_1 + c_2 UK/2T}\right)}{(1+sT)(1+sT/2)}$
4	Generalized Wiener		$c_0 + K_1 U + K_2 U^2$	c_0	$\dfrac{K_1}{1+sT_1} + \dfrac{K_2^2 U}{(1+sT_1)(1+sT_2/2)} =$ $(K_1 + K_2^2 U)\dfrac{1 + s\frac{3K_1 T_2/2 + K_2^2 U T_1}{K_1 + K_2^2 U} + s^2\frac{K_1 T_2^2}{2(K_1 + K_2^2 U)}}{(1+sT_1)(1+sT_1)(1+sT_2/2)}$	$\dfrac{K_1}{1+sT_1} + \dfrac{K_2^2 U/2T_2}{1+sT_2/2} =$ $(K_1 + K_2^2 U/2T_2)\dfrac{1 + s\frac{K_1 T_2/2 + K_2^2 U T_1/2T_2}{K_1 + K_2^2 U/2T_2}}{(1+sT_1)(1+sT_2/2)}$
5	Extended Wiener		$c_0 + K_1 U + K_2 K_3 U^2$	c_0	$\dfrac{K_1}{1+sT_1} + \dfrac{K_2 K_3 U\left(1 + s\frac{2T_2 T_3}{T_2 + T_3}\right)}{(1+sT_2)(1+sT_3)\left(1 + s\frac{T_2 T_3}{T_2 + T_3}\right)}$	$\dfrac{K_1}{1+sT_1} + \dfrac{K_2 K_3 U/(T_2 + T_3)}{1 + s\frac{T_2 T_3}{T_2 + T_3}} =$ $(K_1 + \frac{K_2 K_3 U}{T_2 + T_3})\dfrac{1 + s\frac{K_1 \frac{T_2 T_3}{T_2 + T_3} + K_2 K_3 U T_1}{K_1 + K_2 K_3 U}}{(1+sT_1)\left(1 + s\frac{T_2 T_3}{T_2 + T_3}\right)}$

TABLE 5.5.8b Steady state input–output relationships and equivalent transfer functions of the first-order quadratic block oriented models to pulse and step excitations

No	Model	Scheme	$Y = Y(U)$	$Y = Y(0)$	Equivalent transfer function ($c_0 = 0$)	
					$u(t) = U \cdot 1(t)$	$u(t) = U \cdot \delta(t)$
6	Simple Wiener-Hammerstein cascade		$(c_0 + c_1 K_1 U + c_2 K_1^2 U^2)K_2$	$c_0 K_2$	$\left[\dfrac{c_1 K_1}{1+sT_1} + \dfrac{c_2 K_1^2 U}{(1+sT_1)(1+sT_1/2)}\right]\dfrac{K_2}{1+sT_2} =$ $\dfrac{K_1 K_2 (c_1 + c_2 K_1 U)\left(1 + s\frac{T}{2}\frac{c_1}{c_1 + c_2 K_1 U}\right)}{(1+sT_1)(1+sT_1/2)(1+sT_2)}$	$\left(\dfrac{c_1 K_1}{1+sT_1} + \dfrac{c_2 K_1^2 U/(2T_1)}{1+sT_1/2}\right)\dfrac{K_2}{1+sT_2} =$ $\dfrac{K_1 K_2 \left(c_1 + \frac{c_2 K_1 U}{2T_1}\right)\left(1 + s\frac{c_1 T_1/2 + c_2 K_1 U/2}{c_1 + c_2 K_1 U/2T_1}\right)}{(1+sT_1)(1+sT_1/2)(1+sT_2)}$
7	Generalized Wiener-Hammerstein cascade		$c_0 + K_1 U + K_2^2 K_3 U^2$	c_0	$\dfrac{K_1}{1+sT_1} + \dfrac{K_3 K_2^2 U}{(1+sT_2)(1+sT_2/2)(1+sT_3)}$	$\dfrac{K_1}{1+sT_1} + \dfrac{K_3 K_2^2 U/(2T_2)}{(1+sT_2/2)(1+sT_3)}$
8	Extended Wiener-Hammerstein cascade		$c_0 + K_1 U + K_2 K_3 K_4 U^2$	c_0	$\dfrac{K_1}{1+sT_1} + \dfrac{K_2 K_3 K_4 U\left(1 + s\frac{2T_2 T_3}{T_2 + T_3}\right)}{(1+sT_2)(1+sT_3)\left(1 + s\frac{T_2 T_3}{T_2 + T_3}\right)(1+sT_4)}$	$\dfrac{K_1}{1+sT_1} + \dfrac{K_2 K_3 K_4 U/(T_2 + T_3)}{\left(1 + s\frac{T_2 T_3}{T_2 + T_3}\right)(1+sT_4)}$

The equivalent transfer functions derived from the step responses contain the following poles (Haber *et al.*, 1986):

- the poles of the linear channels;
- the poles of the transfer functions of the terms after the multipliers in the quadratic channels;
- the sums of the poles of the transfer functions of the terms occurring before the multipliers in the quadratic channels;
- the poles of the transfer functions of the terms before the multipliers in the quadratic channels.

The most important special cases, the simple Hammerstein and Wiener models will be treated in Section 5.6.3, where simulation runs are also given.

5.6   STRUCTURE IDENTIFICATION OF SIMPLE WIENER–HAMMERSTEIN
      CASCADE MODEL

In practice the simple cascade models appear most often. They can be of:
- Hammerstein,
- Wiener or
- Wiener–Hammerstein

type. In this section we shall assume a polynomial nonlinearity of arbitrary degree

$$v_2(k) = \sum_{i=0}^{M} c_i v_1^i(k)$$

where $v_1(k)$ is the input and $v_2(k)$ the output of the static nonlinearity. The linear filters are $H_1(q^{-1})$ and $H_2(q^{-1})$ before and after the static term, respectively. The model is seen in Figure 5.6.1. The equation of the model is:

$$y(k) = \sum_{\kappa_2=0}^{k} h_2(\kappa_2) v_2(k-\kappa_2) = \sum_{\kappa_2=0}^{k} h_2(\kappa_2) \sum_{i=0}^{M} c_i v_1^i(k-\kappa_2)$$

$$= \sum_{i=0}^{M} c_i \sum_{\kappa_2=0}^{k} h_2(\kappa_2) v_1^i(k-\kappa_2)$$

$$= \sum_{i=0}^{M} c_i \sum_{\kappa_2=0}^{k} h_2(\kappa_2) \left[\sum_{\kappa_1=0}^{k} h_1(\kappa_1) u(k-\kappa_2-\kappa_1) \right]^i$$

In the sequel three structure determination methods will be presented:
- stochastic excitation and evaluation of correlation functions;
- sinusoidal excitation;
- pulse and stepwise testing.

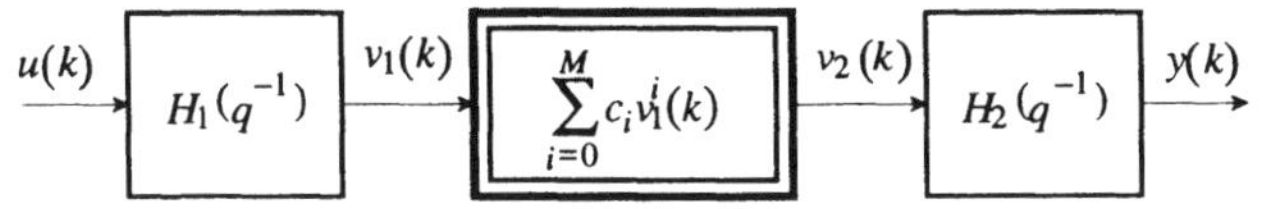

Fig. 5.6.1 Simple Wiener–Hammerstein cascade model with polynomial nonlinearity

5.6.1 Correlation analysis

Definition 5.6.1   An $x(t)$ signal is called a separable process if the function

$$w(x_2, \tau) = \int\limits_{x_1=-\infty}^{\infty} x_1 p(x_1, x_2, \tau) \, dx_1$$

can be separated into a product of two functions, where the first function depends only on x_2 and the second one on τ

$$w(x_2, \tau) = w_1(x_2) w_2(\tau)$$

The $p(x_1, x_2, \tau)$ denotes the second-order probability density function. ∎

A process is separable if

$$w_1(x_2) = x_2 p(x_2)$$

and $w_2(\tau)$ is its auto-correlation function (Billings and Fakhouri, 1978b).

Examples for separable signals are (Billings and Fakhouri, 1978b):
- sinusoidal;
- Gaussian white noise.

Lemma 5.6.1
(Billings and Fakhouri, 1978a, 1978b, 1982; Korenberg, 1985; Marmarelis and Marmarelis, 1978). Define $\Delta y(k)$ as

$$\Delta y(k) = y(k) - \mathrm{E}\{y(k)\}$$

Then for separable input signals the following relation exists between the cross-correlation functions and the weighting function series of the linear terms of the simple Wiener–Hammerstein model:

$$r_{x_1 \Delta y}(\kappa) = C_1 \sum_{\kappa_1=0}^{\kappa} h_1(\kappa_1) h_2(\kappa - \kappa_1) \tag{5.6.1}$$

if

$$x_1(k) = u(k) - \mathrm{E}\{u(k)\} \tag{5.6.2}$$

$$r_{x_2 \Delta y}(\kappa) = C_2 \sum_{\kappa_1=0}^{\kappa} h_1^2(\kappa_1) h_2(\kappa - \kappa_1) \tag{5.6.3}$$

if

$$x_2(k) = \left[u(k) - \mathrm{E}\{u(k)\} \right]^2 \tag{5.6.4}$$

Proof.   For the continuous time case see Billings and Fakhouri (1978b).

For the special case of a white Gaussian excitation the existence of Lemma 5.6.1 is shown and the coefficients C_1 and C_2 in (5.6.1) and (5.6.3) are given below.

Lemma 5.6.2
(Marmarelis and Marmarelis, 1978; Korenberg, 1985) If the test signal is a Gaussian white noise with zero mean value then the coefficients C_1 and C_2 in (5.6.1) and (5.6.3) are

$$C_1 = \sum_{i=1}^{[M/2]} \frac{(2i)!}{i!\,2^i} \sigma_u^{2i} c_{2i-1} \left[\sum_{\kappa_1=0}^{\infty} h_1^2(\kappa_1) \right]^{i-1}$$

$$= \sigma_u^2 c_1 + 3\sigma_u^4 c_3 \left[\sum_{\kappa_1=0}^{\infty} h_1^2(\kappa_1) \right] + \dots$$

(5.6.5)

$$C_2 = \sum_{i=1}^{[M/2]} \frac{i(2i)!}{i!\,2^{i-1}} \sigma_u^{2(i+1)} c_{2i} \left[\sum_{\kappa_1=0}^{\infty} h_1^2(\kappa_1) \right]^{i-1}$$

$$= 2\sigma_u^4 c_2 + 12\sigma_u^6 c_4 \left[\sum_{\kappa_1=0}^{\infty} h_1^2(\kappa_1) \right] + \dots$$

(5.6.6)

In (5.6.5) and (5.6.6) $[M/2]$ denotes the integer part of $M/2$.

Proof.   (for a quadratic system $[M = 2]$)
The output signal can be separated to subsystems attributable to the parallel channels of different degrees of nonlinearity M

$$y(k) = \sum_{\ell=0}^{M} y_\ell(k)$$

Now the cross-correlation functions are the sums of the partial cross-correlation functions belonging to the different subsystems

$$r_{x_i \Delta y}(\kappa) = \sum_{\ell=0}^{M} r_{x_i \Delta y_\ell}(\kappa) \qquad\qquad i = 1, 2$$

Here Δ denotes the subtraction of the mean value

$$\Delta y_\ell(k) = y_\ell(k) - \mathrm{E}\{y_\ell(k)\}$$

The output signals of the subsystems are

$$y_0(k) = c_0$$

$$y_1(k) = c_1 \sum_{\kappa_2=0}^{k} \sum_{\kappa_1=0}^{k} h_2(\kappa_2)h_1(\kappa_1)u(k-\kappa_2-\kappa_1)$$

$$y_2(k) = c_2 \sum_{\kappa_3=0}^{k} \sum_{\kappa_2=0}^{k} \sum_{\kappa_1=0}^{k} h_2(\kappa_3)h_1(\kappa_1)h_1(\kappa_2)u(k-\kappa_3-\kappa_1)u(k-\kappa_3-\kappa_2)$$

etc.

The cross-correlation with the degree-0 channel – constant term – is always zero because the multipliers $x_i(k)$, $i = 1, 2$, have zero mean value

$$r_{x_i \Delta y_0}(\kappa) = 0 \qquad i = 1, 2$$

The cross-correlation function between the linear multiplier $x_1(k)$ and the centered output signal of the linear subsystem $\Delta y_1(k)$ can be calculated as follows

$$r_{x_1 \Delta y_1}(\kappa) = r_{x_1 y_1}(\kappa) - \mathrm{E}\{u(k)\}\mathrm{E}\{y_1(k)\} = r_{x_1 y_1}(\kappa)$$

$$= c_1 \sum_{\kappa_2=0}^{K} \sum_{\kappa_1=0}^{K} h_2(\kappa_2)h_1(\kappa_1)\mathrm{E}\{u(k-\kappa)u(k-\kappa_2-\kappa_1)\}$$

$$= c_1 \sum_{\kappa_2=0}^{K} \sum_{\kappa_1=0}^{K} h_2(\kappa_2)h_1(\kappa_1)r_{uu}(\kappa-\kappa_2-\kappa_1)$$

$$= c_1 \sigma_u^2 \sum_{\kappa_1=0}^{K} h_1(\kappa_1)h_2(\kappa-\kappa_1)$$

The cross-correlation function between the linear multiplier $x_1(k)$ and the centered output signal of the quadratic subsystem $\Delta y_2(k)$ is zero, because

$$r_{x_1 \Delta y_2}(\kappa) = r_{x_1 y_2}(\kappa) - \mathrm{E}\{u(k)\}\mathrm{E}\{y_2(k)\} = r_{x_1 y_2}(\kappa)$$

$$= c_2 \sum_{\kappa_3=0}^{K} \sum_{\kappa_2=0}^{K} \sum_{\kappa_1=0}^{K} h_2(\kappa_3)h_1(\kappa_1)h_1(\kappa_2)$$

$$\times \mathrm{E}\{u(k-\kappa)u(k-\kappa_3-\kappa_1)u(k-\kappa_3-\kappa_2)\} = 0$$

while the average of the product of odd Gaussian white noise signals is zero.
Thus the coefficient C_1 in (5.6.5) is

$$C_1 = c_1 \sigma_u^2$$

which is equivalent to the substitution of $M = 2$ into (5.6.5).

The cross-correlation function between the nonlinear multiplier $x_2(k)$ and the centered output signal of the quadratic subsystem $\Delta y_1(k)$ is zero, because

$$r_{x_2\Delta y_1}(\kappa) = r_{x_2 y_1}(\kappa) - E\{u^2(k)\}E\{y_1(k)\} = r_{x_2 y_1}(\kappa)$$

$$= c_1 \sum_{\kappa_2=0}^{K} \sum_{\kappa_1=0}^{K} h_2(\kappa_2)h_1(\kappa_1)E\{u^2(k-\kappa_1)u(k-\kappa_2-\kappa_1)\} = 0$$

while, first, the mean value of the linear subsystem to an excitation having zero mean is zero, and, second, the average of the product of odd Gaussian white noise signals is zero.

The cross-correlation function between the nonlinear multiplier $x_2(k)$ and the centered output signal of the quadratic subsystem $\Delta y_2(k)$ can be calculated as follows:

$$r_{x_2\Delta y_2}(\kappa) = r_{x_2 y_2}(\kappa) - E\{u^2(k)\}E\{y_2(k)\}$$

The second term is the product of the variance of the input signal

$$E\{u^2(k)\} = r_{uu}(0) = \sigma_u^2$$

and the mean value of the output signal of the quadratic subsystem is:

$$E\{y_2(k)\} = c_2 \sum_{\kappa_3=0}^{K} \sum_{\kappa_2=0}^{K} \sum_{\kappa_1=0}^{K} h_2(\kappa_3)h_1(\kappa_1)h_1(\kappa_2)E\{u(k-\kappa_3-\kappa_1)u(k-\kappa_3-\kappa_2)\}$$

$$= c_2 \sum_{\kappa_3=0}^{K} \sum_{\kappa_2=0}^{K} \sum_{\kappa_1=0}^{K} h_2(\kappa_3)h_1(\kappa_1)h_1(\kappa_2)r_{uu}(\kappa_1-\kappa_2)$$

$$= c_2\sigma_u^2 \sum_{\kappa_3=0}^{K} \sum_{\kappa_1=0}^{K} h_1^2(\kappa_1)h_2(\kappa_3)$$

Consequently

$$E\{u^2(k)\}E\{y_2(k)\} = c_2\sigma_u^4 \sum_{\kappa_3=0}^{K} \sum_{\kappa_1=0}^{K} h_1^2(\kappa_1)h_2(\kappa_3)$$

The cross-correlation function between the nonlinear multiplier $x_2(k)$ and the output signal of the quadratic subsystem $y_2(k)$ is

$$r_{x_2 y_2}(\kappa) = c_2 \sum_{\kappa_3=0}^{K} \sum_{\kappa_2=0}^{K} \sum_{\kappa_1=0}^{K} h_2(\kappa_3)h_1(\kappa_1)h_1(\kappa_2)$$

$$\times E\{u^2(k-\kappa)u(k-\kappa_3-\kappa_1)u(k-\kappa_3-\kappa_2)\}$$

where

$$E\{u^2(k-\kappa)u(k-\kappa_3-\kappa_1)u(k-\kappa_3-\kappa_2)\}$$

$$= r_{uu}(0)r_{uu}(\kappa_1-\kappa_2)+2r_{uu}(\kappa-\kappa_3-\kappa_1)r_{uu}(\kappa-\kappa_3-\kappa_2)$$

Therefore

$$r_{x_2y_2}(\kappa) = c_2\sigma_u^4 \sum_{\kappa_3=0}^{K} \sum_{\kappa_1=0}^{K} h_2(\kappa_3)h_1^2(\kappa_1) + 2c_2\sigma_u^4 \sum_{\kappa_1=0}^{K} h_1^2(\kappa_1)h_2(\kappa-\kappa_1)$$

Finally, the cross-correlation function under consideration is

$$r_{x_2\Delta y_2}(\kappa) = 2c_2\sigma_u^4 \sum_{\kappa_1=0}^{K} h_1^2(\kappa_1)h_2(\kappa-\kappa_1)$$

Thus the coefficient C_2 in (5.6.6) is

$$C_2 = 2c_2\sigma_u^4$$

which is equivalent to the substitution of $M = 2$ into (5.6.6).

The cross-correlation functions with higher degree subsystems can be derived in a similar way.   ∎

So far it has been assumed that the test signal has mean value zero. The coefficients in (5.6.5) and (5.6.6) can be derived for nonzero mean value, as well.

Lemma 5.6.3
(Billings and Fakhouri, 1978a) If the test signal is a Gaussian white noise with mean value u_0, then the coefficients C_1 and C_2 in (5.6.1) and (5.6.3) are as follows

$$C_1 = \sigma_u^2 c_1 + 2u_0\sigma_u^2 c_2 \left[\sum_{\kappa_1=0}^{\infty} h_1(\kappa_1)\right]$$

$$+3\sigma_u^4 c_3 \left[\sum_{\kappa_1=0}^{\infty} h_1^2(\kappa_1)\right] + 3u_0^2\sigma_u^2 c_3 \left[\sum_{\kappa_1=0}^{\infty} h_1(\kappa_1)\right]^2 + \dots \qquad (5.6.7)$$

and

$$C_2 = 2\sigma_u^4 c_2 + 6u_0\sigma_u^4 c_3 \left[\sum_{\kappa_1=0}^{\infty} h_1(\kappa_1)\right] + \dots \qquad (5.6.8)$$

In (5.6.7) and (5.6.8) σ_u^2 is again the variance of the test signal

$$\sigma_u^2 = E\{[u(k)-u_0]^2\}$$

and the mean value of the signal is

$$u_0 = E\{u(k)\}$$

Proof.   (For a static system)
The proof is presented here for a static system, i.e., the weighting function series of the

two dynamic subsystems before and after the nonlinear static polynomial are Dirac delta functions:

$$h_1(\kappa) = \delta(\kappa)$$
$$h_2(\kappa) = \delta(\kappa)$$

Further on we shall use the notation

$$\Delta u(k) = u(k) - u_0 \qquad (5.6.9)$$

The cross-correlation function between the linear multiplier $x_1(k)$ and the centered output signal $\Delta y(k)$ is

$$r_{x_1 \Delta y}(0) = E\left\{ \sum_{i=0}^{M} c_i \left[\left[u_0 + \Delta u(k) \right]^i - E\left\{ \left[u_0 + \Delta u(k) \right]^i \right\} \right] \Delta u(k) \right\} \qquad (5.6.10)$$

The components of (5.6.10) for constant, linear, quadratic, and cubic homogeneous terms are derived below:

$$r_{x_1 \Delta y_0}(0) = E\left\{ c_0 [1-1] \Delta u(k) \right\} = 0$$
$$r_{x_1 \Delta y_1}(0) = E\left\{ c_1 \left[\left[u_0 + \Delta u(k) \right] - E\left\{ u_0 + \Delta u(k) \right\} \right] \Delta u(k) \right\} = c_1 \sigma_u^2$$
$$r_{x_1 \Delta y_2}(0) = E\left\{ c_2 \left[\left[u_0 + \Delta u(k) \right]^2 - E\left\{ \left[u_0 + \Delta u(k) \right]^2 \right\} \right] \Delta u(k) \right\} = 2u_0 c_2 \sigma_u^2$$
$$r_{x_1 \Delta y_3}(0) = E\left\{ c_3 \left[\left[u_0 + \Delta u(k) \right]^3 - E\left\{ \left[u_0 + \Delta u(k) \right]^3 \right\} \right] \Delta u(k) \right\} = 3\sigma_u^4 c_3 + 3u_0^2 \sigma_u^2 c_3$$

The derived terms are equal to those ones seen in (5.6.7), with the assumption that the system has no dynamic terms.

The cross-correlation function between the quadratic multiplier $x_2(k)$ and the centered output signal $\Delta y(k)$ is

$$r_{x_2 \Delta y}(0) = E\left\{ \sum_{i=0}^{M} c_i \left[\left[u_0 + \Delta u(k) \right]^i - E\left\{ \left[u_0 + \Delta u(k) \right]^i \right\} \right] \Delta u^2(k) \right\} \qquad (5.6.11)$$

The components of (5.6.11) for constant, linear, quadratic and cubic homogeneous terms are derived below:

$$r_{x_2 \Delta y_0}(0) = E\left\{ c_0 [1-1] \Delta u^2(k) \right\} = 0$$
$$r_{x_2 \Delta y_1}(0) = E\left\{ c_1 \left[\left[u_0 + \Delta u(k) \right] - E\left\{ u_0 + \Delta u(k) \right\} \right] \Delta u^2(k) \right\} = c_1 E\left\{ \Delta u^3(k) \right\} = 0$$
$$r_{x_2 \Delta y_2}(0) = E\left\{ c_2 \left[\left[u_0 + \Delta u(k) \right]^2 - E\left\{ \left[u_0 + \Delta u(k) \right]^2 \right\} \right] \Delta u^2(k) \right\} = 2\sigma_u^4 c_2$$
$$r_{x_2 \Delta y_3}(0) = E\left\{ c_3 \left[\left[u_0 + \Delta u(k) \right]^3 - E\left\{ \left[u_0 + \Delta u(k) \right]^3 \right\} \right] \Delta u^2(k) \right\} = 6u_0 \sigma_u^4 c_3$$

The derived terms are equal to those seen in (5.6.8) with the assumption that the system has no dynamic terms.

Remarks:
1. If the test signal has zero mean value then the coefficients C_1 and C_2 in (5.6.7) and (5.6.8) coincide with the formulas of (5.6.5) and (5.6.6).
2. If the nonlinear static term is preceded by a linear dynamic term with the pulse transfer function $H_1(q^{-1})$, then the input signal of the nonlinear static term $v_1(k)$ in Figure 5.6.1 is

$$v_1(k) = \sum_{\kappa_1=0}^{k} h_1(\kappa_1)u(k-\kappa_1)$$

The mean value of $v_1(k)$ is

$$E\{v_1(k)\} = u_0 \sum_{\kappa_1=0}^{\infty} h_1(\kappa_1) \tag{5.6.12}$$

The coefficients that include u_0 in (5.6.7) and (5.6.8) can be obtained from the coefficients derived for the static case by substituting u_0 for (5.6.12). Among the listed terms in (5.6.7) and (5.6.8) only the dynamic terms in the third element in (5.6.7) besides $3\sigma_u^4 c_3$ cannot be derived from the static case.
3. The linear and quadratic subsystems of a process can be obtained by repeated excitations using Gardiner's method. If the cross-correlation functions are calculated for the outputs of the above subsystems, then the coefficients C_1 and C_2 have a simple form which is shown in Lemma 5.6.4.

Lemma 5.6.4
Excite the process by a Gaussian test signal $u(k)$ with mean value u_0 and standard deviation σ_u. The cross-correlation functions of the multipliers (5.6.2) and (5.6.4) with the reduced output signals of the linear and quadratic subsystems

$$\Delta y_i(k) = y_i(k) - E\{y_i(k)\} \qquad i = 1, 2$$

are

$$r_{x_1 \Delta y_1}(\kappa) = C_1 \sum_{\kappa_1=0}^{\kappa} h_1(\kappa_1)h_2(\kappa-\kappa_1) \tag{5.6.13}$$

$$r_{x_2 \Delta y_2}(\kappa) = C_2 \sum_{\kappa_1=0}^{\kappa} h_1^2(\kappa)h_2(\kappa-\kappa_1) \tag{5.6.14}$$

with

$$C_1 = \sigma_u^2 c_1 \tag{5.6.15}$$
$$C_2 = 2\sigma_u^4 c_2 \tag{5.6.16}$$

Proof.   (5.6.15) results from Lemma 5.6.3 by substituting $c_i = 0$, $i \neq 1$ into (5.6.7). Similarly, (5.6.16) follows from (5.6.8) by substituting $c_i = 0$, $i \neq 2$ into (5.6.8).   ■

Billings and Fakhouri (1980) extended the above results to the case of a pseudo-random test signal. The test signal is the superposition of two uncorrelated sequences

$$u(k) = u_1(k) + u_2(k) \qquad E\{u_1(k)\} = 0 \qquad E\{u_2(k)\} = 0$$

and the multipliers are

$$x_1(k) = u_1(k)$$
$$x_2(k) = u_1(k)u_2(k)$$

The constants C_1 and C_2 of the cross-correlation functions (5.6.13) and (5.6.14) are given in Billings and Fakhouri (1980) for maximum length PRBS and PRTS signals (see also Lemma 3.11.3).

Furthermore, we use $r_{x_1 \Delta y}(\kappa)$ and $r_{x_2 \Delta y}(\kappa)$ for structure identification. As is seen, this is right only for Gaussian excitation. The cross-correlation functions $r_{x_1 \Delta y_1}(\kappa)$ and $r_{x_2 \Delta y_2}(\kappa)$ can be used both for Gaussian and pseudo-random input signals.

Although the constants C_1 and C_2 depend on the characteristics of the test signal and the process, it is generally true that $r_{x_1 \Delta y}(\kappa)$ is proportional to the weighting function series of the linear subsystem of the process (see (5.6.1)) and $r_{x_2 \Delta y}(\kappa)$ is proportional to the weighting function series of an other linear cascade system of two linear elements (see (5.6.3)). In the latter case the second linear part is equal to that of the process behind the static nonlinearity and the weighting function series of the first term are the squares of those in the original cascade system. The two cross-correlation functions can detect the structure of the system without any parameter estimation.

1. *Simple Hammerstein model*
    Since

$$H_1\left(q^{-1}\right) = 1$$

the two cross-correlation functions

$$r_{x_1 \Delta y}(\kappa) = C_1 h_2(\kappa)$$
$$r_{x_2 \Delta y}(\kappa) = C_2 h_2(\kappa)$$

are proportional to the weighting function series of the linear dynamic part of the model and consequently they are proportional to each other

$$r_{x_2 \Delta y}(\kappa) = \frac{C_2}{C_1} r_{x_1 \Delta y}(\kappa)$$

Define

$$\rho_1(\kappa) = \frac{r_{x_2 \Delta y}(\kappa)}{r_{x_1 \Delta y}(\kappa)}$$

and its average value over memory m is

$$\bar{\rho}_1 = \frac{1}{m+1} \sum_{\kappa=0}^{m} \rho_1(\kappa)$$

Now, the following measurement number

$$\zeta_1 = \frac{\frac{1}{m+1} \sum_{\kappa=0}^{m} \left[\rho_1(\kappa) - \bar{\rho}_1\right]^2}{\bar{\rho}_1^2}$$

becomes zero if the model is of simple Hammerstein type.

2. *Simple Wiener model*
   Since

$$H_2\left(q^{-1}\right) = 1$$

the two cross-correlation functions become

$$r_{x_1 \Delta y}(\kappa) = C_1 h_1(\kappa)$$
$$r_{x_2 \Delta y}(\kappa) = C_2 h_1^2(\kappa)$$

i.e., the second, nonlinear cross-correlation function is proportional to the square of the first, linear function

$$r_{x_2 \Delta y}(\kappa) = \frac{C_2}{C_1^2} r_{x_1 \Delta y}^2(\kappa)$$

Define the ratio

$$\rho_2(\kappa) = \frac{r_{x_2 \Delta y}(\kappa)}{r_{x_1 \Delta y}^2(\kappa)}$$

and its average value over memory m

$$\bar{\rho}_2 = \frac{1}{m+1} \sum_{\kappa=0}^{m} \rho_2(\kappa)$$

The measurement number

$$\zeta_2 = \frac{\frac{1}{m+1} \sum_{\kappa=0}^{m} \left[\rho_2(\kappa) - \bar{\rho}_2\right]^2}{\bar{\rho}_2^2}$$

becomes zero if the model is of simple Wiener type.

3. *Linear model*
    If

    $$r_{x_2 \Delta y}(\kappa) = 0 \qquad \forall \kappa$$

    then the model is linear. This agrees with the nonlinear cross-correlation nonlinearity test of Section 4.9.

4. *Static model*
    Since

    $$H_1(q^{-1}) = 1$$

    and

    $$H_2(q^{-1}) = 1$$

    both cross-correlation functions are zero, except for $\kappa = 0$,

    $$r_{x_1 \Delta y}(\kappa) = 0 \qquad \kappa \geq 1$$
    $$r_{x_2 \Delta y}(\kappa) = 0 \qquad \kappa \geq 1$$

    If the process is not of higher nonlinear degree than quadratic then
    *   the calculated linear cross-correlation function is proportional to the first-degree Volterra kernels of the system

    $$h_1(\kappa) = c_1 \sum_{\kappa_1=0}^{\kappa} h_1(\kappa_1) h_2(\kappa - \kappa_1) \tag{5.6.17}$$

    *   the calculated quadratic cross-correlation function is proportional to the second-degree Volterra kernel of the system at the main diagonal

    $$h_2(\kappa, \kappa) = c_2 \sum_{\kappa_1=0}^{\kappa} h_1^2(\kappa_1) h_2(\kappa - \kappa_1) \tag{5.6.18}$$

Consequently all structure identification methods elaborated for quadratic block oriented models using the estimated Volterra kernels (see Section 5.5.2) can be applied if they do not require the off-diagonal values of the quadratic Volterra kernel. It is easy to check that the tests for the simple Wiener and simple Hammerstein models in Section 5.5.2 use only the main diagonal elements of the second order Volterra kernels, and in the case of an *a priori* assumption that the model is of simple Wiener–Hammerstein cascade type the estimation of the off-diagonal elements is superfluous.

Example 5.6.1 *Structure determination of a simple Hammerstein model using correlation method (Haber and Zierfuss, 1987)*

The simple Hammerstein model with static polynomial

$$v(k) = 2 + u(k) + 0.5u^2(k)$$

and with the differential equation of the linear term

$$10\dot{y}(t) + y(t) = v(t)$$

was excited by a pseudo-random ternary test signal (PRTS) with maximum length 728, amplitude ± 2 and mean value 1. The sampling time was equal to the minimum switching time of the PRTS that was $\Delta T = 2$ [s]. The linear $r_{x_1 \Delta y}(\kappa)$ and nonlinear $r_{x_2 \Delta y}(\kappa)$ cross-correlation functions are plotted on Figure 5.6.2 for the shifting time domain $(-20 \le \kappa \le 80)$. The plots indicate a simple Hammerstein model because the linear and nonlinear cross-correlation functions are proportional to each other in the domain $(-20 \le \kappa \le 80)$.   ∎

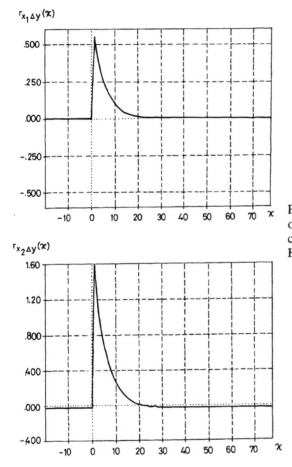

Fig. 5.6.2 Linear (a) and second-order (nonlinear) (b) cross-correlation function of the simple Hammerstein model

Example 5.6.2   *Structure determination of a simple Wiener model using correlation method (Haber and Zierfuss, 1987)*

The simple Wiener model with the differential equation of the linear part

$$10\dot{v}(t) + v(t) = u(t)$$

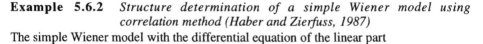

and with the static polynomial

$$y(k) = 2 + v(k) + 0.5v^2(k)$$

was excited by the same test signal as used in Example 5.6.1. Figure 5.6.3 shows the linear $r_{x_1 \Delta y}(\kappa)$ and nonlinear $r_{x_2 \Delta y}(\kappa)$ cross-correlation functions in the shifting time domain $(-20 \le \kappa \le 80)$. The plots point to a simple Wiener model because the nonlinear cross-correlation function is proportional to the square of the linear cross-correlation function in the domain $(0 \le \kappa \le 10)$. (The above relation is not valid outside the region because of known anomalies which occur if pseudo-random test signals are applied.) ∎

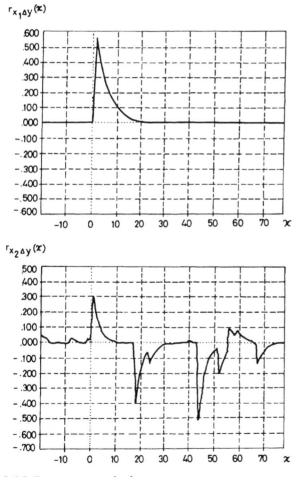

Fig. 5.6.3 Linear (a) and second-order (nonlinear) (b) cross-correlation function of the simple Wiener model

5.6.2 Frequency method

If a cascade model contains only polynomial nonlinearities – or it can be well approximated by them – then sinusoidal test signals can show the structure of the system, i.e., it is possible

- to determine the structure of the system, i.e., the allocation of the linear dynamic terms between the static polynomial elements;
- to determine the poles and zeros of the transfer functions of the linear dynamic terms;
- to determine the highest powers of the nonlinear polynomial elements.

To get an insight into the method, consider an example.

Example 5.6.3 *Structure identification of a cascade system having three linear dynamic terms and a cubic and a quadratic power static nonlinearity between them*

The scheme of the system can be seen in Figure 5.6.4.

The input signal is a sinusoidal function with amplitude U

$$u(t) = U \cos\omega(t)$$

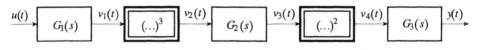

Fig. 5.6.4 Cascade system consisting of three linear dynamic terms and a cubic and a quadratic power static nonlinearity

The output of the first linear term is

$$v_1(t) = U|G_1(\omega)| \cos(\omega t + \varphi_1(\omega))$$

where $|G_i(\omega)|$ and $\varphi_i(\omega)$ will denote the absolute values and phases at the given angular frequency. Before calculating the cubic power of $v_1(t)$, notice that the multiplication of an nth harmonic signal by the base harmonic results in $(n+1)$-th and $(n-1)$-th harmonics, because

$$u_1(t) = U_1 \cos(\omega t + \varphi_1)$$
$$u_n(t) = U_n \cos(n\omega t + \varphi_n)$$

and

$$u_n(t)u_1(t) = \frac{U_n U_1}{2} \cos((n+1)\omega t + \varphi_n + \varphi_1) + \frac{U_n U_1}{2} \cos((n-1)\omega t + \varphi_n - \varphi_1)$$

Therefore the output signal of the cubic power term is:

$$v_2(t) = \frac{U^3}{2^2} |G_1(\omega)|^3 \cos(3\omega t + 3\varphi_1(\omega)) + \dots$$

where the remaining terms are subharmonics. The second linear dynamic term filters $v_2(t)$ by $G_2(3j\omega)$, since the frequency is three times as $G_2(j\omega)$ was defined. The output signal of the linear dynamic term characterized by its (linear) frequency function $G_2(j\omega)$ is

$$v_3(t) = \frac{U^3}{2^2} |G_1(\omega)|^3 |G_2(3\omega)| \cos(3\omega t + 3\varphi_1(\omega) + \varphi_2(3\omega)) + \ldots$$

The signal $v_3(t)$ is squared by the next nonlinear element. The resulting signal has six harmonic and subharmonics

$$v_4(t) = \frac{U^6}{2^5} |G_1(\omega)|^6 |G_2(3\omega)|^2 \cos(6\omega t + 6\varphi_1(\omega) + 2\varphi_2(3\omega)) + \ldots$$

Finally, the last linear filter works at the sixth harmonic, therefore the output signal of the whole cascade system is

$$y(t) = \frac{U^6}{2^5} |G_1(j\omega)|^6 |G_2(3j\omega)|^2 |G_3(6j\omega)|$$
$$\times \cos(6\omega t + 6\varphi_1(\omega) + 2\varphi_2(3\omega) + \varphi_3(6\omega)) + \ldots$$

If, theoretically, we assume that the input signal were a sinusoidal with a sextuple frequency and unit amplitude of that in reality, then the sixth harmonics of the output signal could be considered as the output of an equivalent linear network with frequency function

$$G^*(j\omega) = \frac{U^6}{2^5} G_1^6(j\omega) G_2^2(3j\omega) G_3(6j\omega)$$

or in transfer function form:

$$G^*(s) = \frac{U^6}{2^5} G_1^6(s) G_2^2(3s) G_3(6s) \qquad \blacksquare$$

If a degree-n nonlinear system is excited by a sinusoidal test signal then the (measured) output signal has harmonics up to nth degree. The equivalent linear frequency of function of the ith $(0 < i \leq n)$ subharmonic is defined as follows. If a sinusoidal input signal with amplitude one and frequency i-times that of the test signal is put to a linear system with the above frequency function, then the output signal is equal to the ith harmonic of the (measured) output signal.

It can be generally stated that the equivalent frequency function of the highest (nth) harmonic consists of the product of the following terms:

- the nth degree power of the amplitude of the sinusoidal input signal divided by 2^{n-1};
- the nth power of the frequency function of the linear dynamic term before the static nonlinearity;
- a frequency function which is equal to the frequency function of the linear dynamic term after the static nonlinearity at n-times the value of the actual frequency.

From the above it can be concluded that the poles and zeros of the transfer function of the linear dynamic terms occur in the transfer function of the linear equivalent frequency function of the highest harmonic in the following forms (Gardiner, 1973a,b):

- The zeros and poles of the linear dynamic term succeeding the static
  nonlinearity occur with their values divided by the degree of the polynomial;
- The zeros and poles of the linear dynamic term preceding the static nonlinearity
  occur with multiplicity of the degree of the polynomial.

Of course, a multiplicity in the poles and zeros does not show a nonlinear power
unambiguously because it can be in a multiplicative form in the cascade system itself.
Therefore one has to identify the linear and the highest power equivalent networks and to
compare the two pole/zero configurations. Here the linear network should consist of the
effects of only the linear terms, i.e., the subsystems attributable to different nonlinear
degrees have to be separated by Gardiner's method. (The special case where the linear
dynamic terms contain multiplicative poles and zeros, and the poles and zeros of
different transfer functions coincide, are not treated by this method (Baumgartner and
Rugh, 1975).)

The simple Hammerstein and Wiener models are both cascade systems and quadratic
block oriented models. Their structure identification and parameter estimation by the
frequency method was presented in Section 5.5.3.

5.6.3 Evaluation of pulse and step responses

Both the simple Hammerstein and Wiener models contain one linear dynamic term. The
basic idea of the structure determination is that all step or pulse responses of the linear
element at different input magnitudes have the same form. In the Wiener model the
input signal of the linear element is the input excitation itself and in the Hammerstein
model its scaled values through the static nonlinear transformation.

The nonlinear static curve can be determined in both cases from several step
responses. The relation between the steady state values of the input and output signals
of the process give the static nonlinear curve and allow a static regression for its
parameters. It is usual to approximate the static nonlinearity by a polynomial equation,
but any other single-valued function can be used.

The input and output signals of the linear dynamic part can be calculated as follows
(Haber *et al.*, 1986):

1. *Simple Hammerstein model*
   - the input signal of the linear dynamic part is the nonlinear transformation of
     the real input signal of the process;
   - the output signal of the linear dynamic part is the same as that of the whole
     process.
2. *Simple Wiener model*
   - the input signal of the linear dynamic part is the same as that of the process;
   - the output signal of the linear dynamic part is the input signal of the nonlinear
     term which can be obtained by the transformation of the output signal of the
     whole process through the inverse characteristic of the static nonlinearity.

Now the calculated input–output signals are step – or pulse – responses of the linear
dynamic parts. If they are normalized to the steady state values then they should
coincide.

The normalization can be carried out in such a way that the initial value of the
response is considered to be zero and the final value one.

The structure checking should be done in different working points and by different
amplitudes of the input signal.

The simple Wiener model should have a single valued characteristic – in the
investigated input range – to allow an inverse transformation. Such a limitation is not

known in the case of the Hammerstein model.

In the case of the simple Hammerstein model any step or pulse response of the system is proportional to the step or pulse response (weighting function) of the linear dynamic part of the system that makes its parameter estimation easy.

Example 5.6.4   *Structure identification of the simple Hammerstein model from evaluation of three step responses (Haber and Unbehauen, 1990)*

The equation of the quadratic static part is

$$V = 2 + U + 0.5U^2$$

and the transfer function of the linear dynamic part is

$$\frac{Y(s)}{V(s)} = \frac{1}{1 + 10s}$$

The process was excited by three unit steps with the amplitudes 1, 2, and 4, respectively:

$$u_1(t) = 1 \cdot 1(t) \qquad u_2(t) = 2 \cdot 1(t) \qquad u_3(t) = 4 \cdot 1(t)$$

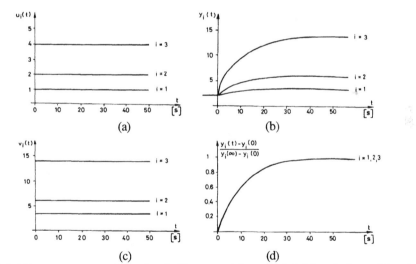

Fig. 5.6.5 Step responses of the simple Hammerstein model: (a) input steps; (b) output step responses; (c) inner variable between the linear and nonlinear terms; (d) normalized output step responses

The three input steps are seen in Figure 5.6.5a and the output responses in Figure 5.6.5b. The normalized output signals are seen in Figure 5.6.5d, and since they coincide the system can be modeled by the simple Hammerstein model. The parameters of the model can be estimated as follows.

The steady state characteristic of the process – which is equal to that of the nonlinear element of the model – could be calculated from the three data pairs

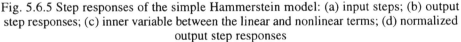

and is seen in Figure 5.6.6.

The inner variable $v(t)$ behind the nonlinear element can be calculated by transforming the input signals through the nonlinear static curve of Figure 5.6.6. They are seen for the three excitations in Figure 5.6.5c. The transfer function of the linear dynamic term can be estimated from any sampled data sequence pairs $\{v_i(k)\}, \{y_i(k)\}, \ i = 1, 2, 3$.

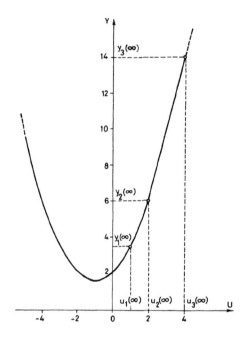

Fig. 5.6.6 Estimated steady state characteristic

Example 5.6.5   *Structure identification of the simple Wiener model from evaluation of three step responses (Haber and Unbehauen, 1990)*

The equation of the quadratic static part is

$$Y = 2 + V + 0.5V^2$$

and the transfer function of the linear dynamic part is

$$\frac{V(s)}{U(s)} = \frac{1}{1 + 10s}$$

The process was excited by three unit steps with the amplitudes 1, 2 and 4, respectively.

$$u_1(t) = 1 \cdot 1(t) \qquad u_2(t) = 2 \cdot 1(t) \qquad u_3(t) = 4 \cdot 1(t)$$

The three input steps are seen in Figure 5.6.7a and the output responses in Figure 5.6.7b. The steady state characteristic of the process – which is equal to that of the nonlinear element of the model – can be calculated from the three data pairs

$$\left(u_1(\infty), y_1(\infty)\right) \qquad \left(u_2(\infty), y_2(\infty)\right) \qquad \left(u_3(\infty), y_3(\infty)\right)$$

which can be seen in Figure 5.6.6. The inner variable $v(t)$ before the nonlinear element can be calculated by transforming the output signals through the inverse of the nonlinear static curve of Figure 5.6.6. They are seen for the three excitations in Figure 5.6.7c. The normalized values of the step responses of the inner variable are seen in Figure 5.6.7d. Since they all coincide the process can be modeled by the simple Wiener model. The transfer function of the linear dynamic part can be estimated from any sampled data pairs $\{u_i(k)\}, \{v_i(k)\}$, $i = 1, 2, 3$.

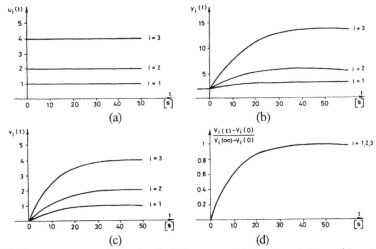

Fig. 5.6.7 Step responses of the simple Wiener model: (a) input steps; (b) output step responses; (c) inner variable between the linear and nonlinear terms; (d) normalized step responses of the inner variable

5.7 STRUCTURE IDENTIFICATION AND PARAMETER ESTIMATION OF QUASI-LINEAR MODELS HAVING SIGNAL DEPENDENT PARAMETERS

A nonlinear process can be described by a quasi-linear system with signal dependent parameters in a domain of the input signal:
- if the system can be linearized for small excitations around all possible working points;
- if the parameters are functions of any measurable or computable signal.

A quasi-linear model with signal dependent parameters consists of a linear differential equation, where each parameter (gain factor, time constant, damping factor, zero, pole and/or coefficient) depends on a measurable (computable) signal or state variable. A detailed introduction of the quasi-linear models was given in Section 1.6.

Using delta or bilinear transformation for the discretization of the differential equation, nonlinear difference equations linear in the parameters can be obtained (Haber and Keviczky, 1985). Instead of assuming that the parameters of the differential equation are signal dependent, one can consider that the parameters of the equivalent difference equation are signal dependent. The advantage of the proposed method is that it leads to a difference equation that is linear-in-parameters. It is disadvantageous, however, that any simple (e.g., linear) dependence of the parameters of the difference equation cannot be transformed so easily into a simple (e.g., linear) functional dependence of the parameters of the continuous time differential equation.

5.7.1 Method using normal operating data

First we treat the case when the identification experiment cannot be planned, i.e., no special test signal can be applied. The discretization of the continuous time model leads to difference equations linear-in-parameters if the Euler or bilinear transformation is used. The model components are tabulated in Tables 1.6.5 to 1.6.10 for the usual models. The selection of the best structure can be performed by repeated parameter estimation and model validity tests. However, there is a difference from the usual selection rules given in Section 5.9. Here only a few models which have a physical meaning have to be checked. For example, if the model is of first-order and only linear and inverse linear dependence of the parameters is assumed, then only 9 different cases exist (Haber and Keviczky, 1985) and the parameter estimation of all possible (9) structures is an easy task. The number of models is limited by the *a priori* knowledge on the process.

There is an important point that has to be underlined. One has to know which type of model can be identified and which not from the input–output records. Therefore the models have to be checked for whether the components in it are linearly dependent or not. If there is a model component that depends linearly on the others then the corresponding model cannot be identified from the data. One has to know whether a model has to be dropped because of whether the data are not persistently exciting or that the assumed model is not adequate.

Example 5.7.1 *Structure identification and parameter estimation of a first-order linear system with output signal dependent time constant (Haber et al., 1986; Haber and Unbehauen, 1990)*

The process is of a first-order quasi-linear system

$$y(t) + T\dot{y}(t) = Ku(t)$$

with constant gain and output signal dependent time constant

$$T(t) = T_0 + T_1 y(t) = 10 - 2.5y(t) \quad [s]$$

The parameter dependence of the system is seen in Figure 5.7.1. The input and output signals used for the identification are plotted in Figure 5.7.2. The input signal consists of four constant periods, each being 60 [s] long. The value of the input signal was

$$u(t): \begin{cases} 1.0 & \text{if} & t < 60[s] \\ 0.0 & \text{if} & 60\,[s] \le t < 120[s] \\ -0.5 & \text{if} & 120\,[s] \le t < 180[s] \\ 0.0 & \text{if} & 180\,[s] \le t < 240[s] \end{cases}$$

The entirely of records were sampled by $\Delta T = 2$ [s] and the LS method was applied for the identification. Different first-order models, all being linear-in-parameters, were fitted to the input–output records. The component $y(k-1)$ was always included in the model to get a parametric model with few parameters. The following models were considered as possible model components:

1. $u(k)$
2. $u^2(k)$
3. $u(k)y(k)$
4. $y^2(k)$

These model components were chosen *a priori* because:
- they gave a good model matching for several systems;
- similar components occur in the difference equations of the first-order quasi-linear models with signal dependent parameters.

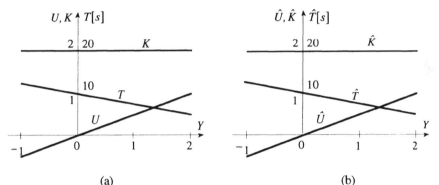

(a)   (b)

Fig. 5.7.1 Gain and time constant against the steady state value of the output signal:
(a) true parameters, (b) estimated parameters

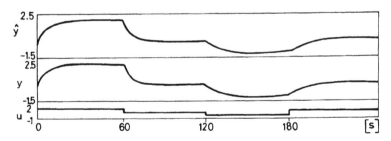

Fig. 5.7.2 Measured input u and output y signals and the computed output signal $\hat{y}$
based on the estimated model of the first-order quasi-linear model with output signal
dependent time constant

Consequently the structures fitted to the input–output data were not selected according to physical meaning, i.e., those 9 structures were not chosen which correspond to the first-order quasi-linear systems with linear or inverse linear dependence of the parameters.

Usually the model components are considered by their delayed values in the difference equation models because the continuous–discrete transformation assuming a zero order holding device does not lead to terms of the same discrete time as the output signal. However, quasi-linear systems with signal dependent parameters are more easily discretized by the bilinear transformation that leads to difference equations where both the actual and the delayed values of the model components occur. Therefore both cases were assumed during the regressions.

TABLE 5.7.1 Estimated parameters, their standard deviations, and standard deviations of the residuals, at the fitting of all possible first-order quadratic models linear-in-parameters

No	$-y(k-1)$	1	$u(k)$	$u(k-1)$	$u^2(k)$	$u^2(k-1)$	$u(k)y(k)$	$u(k-1)y(k-1)$	$y^2(k)$	$y^2(k-1)$	σ_ε
1	-0.79203 ±0.00538		0.50211 ±0.01662	-0.083664 ±0.019463							0.02003
2	-0.7810 ±0.01576			0.4211 ±0.02929							0.05874
3	-0.8200 ±0.0013		0.3560 ±0.0046	0.00401 ±0.0045					0.1152 ±0.0023	0.1153 ±0.0023	0.00420
4	-0.7834 ±0.0169			0.4211 ±0.0293						-0.001774 ±0.004648	0.05871
5	-0.8187 ±0.0031		0.3530 ±0.0176	0.00977 ±0.01588	0.03076 ±0.05134	0.01685 ±0.04474	0.03394 ±0.04024	-0.01807 ±0.02536	0.09815 ±0.04479	-0.09937 ±0.02701	0.00803
6	-0.8106 ±0.0193			0.3667 ±0.0350		0.0382 ±0.0555		0.07143 ±0.05989		-0.04810 ±0.02204	0.05696
7	-0.8174 ±0.0017	0.000458 ±0.000628	0.3533 ±0.00973	0.00928 ±0.00867	-0.0319 ±0.0278	0.01685 ±0.02449	0.03613 ±0.02235	-0.01831 ±0.01390	0.09363 ±0.02500	-0.09888 ±0.01493	0.00435
8	-0.8102 ±0.0194	0.001858 ±0.0007914		0.3655 ±0.0354		0.04407 ±0.06100		0.06671 ±0.06323		-0.04619 ±0.02347	0.05695
9	-0.8200 ±0.00135	0.0002719 ±0.0005427	0.3565 ±0.0046	0.003845 ±0.004502					0.11508 ±0.00231	-0.1153 ±0.0023	0.00426
10	-0.7838 ±0.0173	0.0007420 ±0.0074202		0.4211 ±0.0293						-0.002187 ±0.006217	0.05871

TABLE 5.7.2 Standard deviations of the residuals σ_ε, subjective measures of the model matching based on the simulated and fitted models' output signals and on the significance of the estimated parameters at the fitting of all possible first-order quadratic models

No	1	u	u^2	uy	y^2	b_0	σ_ε	Acceptance of the model	
								output matching	parameter estimation
1	0	1	0	0	0	1	0.0200	1	1
2	0	1	0	0	0	0	0.0587	0.5	1
3	0	1	0	0	1	1	0.0042	1	0.9
4	0	1	0	0	1	0	0.0587	0	0
5	0	1	1	1	1	1	0.0080	1	0
6	0	1	1	1	1	0	0.0569	0.5	0
7	1	1	1	1	1	1	0.0044	1	0
8	1	1	1	1	1	0	0.0569	0.5	0
9	1	1	0	0	1	1	0.0042	1	0.5
10	1	1	0	0	1	0	0.0587	1	0.7

Table 5.7.1 summarizes the estimated parameters, their standard deviations and the standard deviations of the residuals at the fitting of all possible first-order quadratic models linear-in-parameters. Table 5.7.2 presents the standard deviations of the residuals, the subjective measures of model matching, based on the simulated and fitted models' output signals and that of the significance of the estimated parameters at the fitting of all possible first-order quadratic models linear-in-parameters. The subjective measures were among 0 and 1, the bigger value means the better model. The column b_0 shows whether the actual and delayed values of the model components were considered (1) or only their values (0) shifted by one. The best model had the lowest residual mean square, excellent output matching, and their parameters were significant except one. (The parameter not significant belongs to the term $u(k-1)$, but since the parameter belonging to $u(k)$ is significant, the model component $u(k)$ is itself therefore a significant component of the model.)

The equation of the best estimated model is

$$\hat{y}(k) = (0.820 \pm 0.0013)\hat{y}(k-1) + (0.356 \pm 0.0046)u(k)$$
$$+(0.00401 \pm 0.0045)u(k-1) + (0.1152 \pm 0.0023)\hat{y}^2(k)$$
$$-(0.1153 \pm 0.0023)\hat{y}^2(k-1)$$

We show that the identified model is a first-order quasi-linear one with output signal dependent parameters. Observe that

$$0.1152\hat{y}^2(k) - 0.1153\hat{y}^2(k-1)$$
$$= 0.1152[\hat{y}(k) - \hat{y}(k-1)][\hat{y}(k) + \hat{y}(k-1)] - 0.0001\hat{y}^2(k-1)$$

By introducing

$$\Delta\hat{y}(k) = \hat{y}(k) - \hat{y}(k-1)$$

we obtain

$$\Delta\hat{y}(k)\{1 - 0.1152[\hat{y}(k) + \hat{y}(k-1)]\}$$
$$= -0.18y(k-1) + 0.356u(k) + 0.0041u(k-1) - 0.001\hat{y}^2(k-1)$$

and further on

$$\Delta\hat{y}(k) = \frac{0.18}{1 - 0.2304[\hat{y}(k) + \hat{y}(k-1)]/2}$$
$$\times\left[1.978u(k) + 0.0222u(k-1) - 0.000556\hat{y}^2(k-1) - \hat{y}(k-1)\right]$$

The above form has a similarity to the discretization of the first-order differential equation

$$y(t) + T\dot{y}(t) = Ku(t)$$

by the bilinear transformation

$$\Delta y(k) = \frac{2\Delta T}{\Delta T + 2T_0}\left[K_0\frac{u(k) + u(k-1)}{2} - y(k-1)\right]$$

Comparing the estimated model equation with the theoretically derived one we obtain two equations:

$$\frac{2\Delta T}{\Delta T + 2T_0} = \frac{0.18}{1 - 0.2304[\hat{y}(k) + \hat{y}(k-1)]/2}$$

and

$$K_0\frac{u(k) + u(k-1)}{2} = 1.978u(k) + 0.0222u(k-1) - 0.000556\hat{y}^2(k-1)$$

By replacing $u(k)$ and $u(k-1)$ by the steady state value of the input signal U and $y(k-1)$ by the steady state value of the output signal Y we obtain the following parameter dependencies. The static gain is

$$\hat{K} = \frac{Y}{U} = \frac{2.0002}{1 + 0.000556Y}$$

which is practically constant in the working domain investigated $(-1 \leq Y \leq 2)$, and the time constant depends linearly on the output signal as

$$\hat{T} = 10.11 - 2.56Y$$

which is almost the same equation as was simulated. The good coincidences between simulated and estimated time constants in different working points are seen in Table 5.7.3 and Figure 5.7.1.

TABLE 5.7.3 True and estimated values of the input signal U, gain K and time constant T for different steady state values of the output signal

Y	U	$\hat{U}$	K	$\hat{K}$	T	$\hat{T}$
-1.0	-0.50	-0.50	2.0	2.0	12.50	12.67
-0.5	-0.25	-0.25	2.0	2.0	11.25	11.39
0.0	0.0	0.0	2.0	2.0	10.00	10.11
0.5	0.25	0.25	2.0	2.0	8.75	8.83
1.0	0.50	0.50	2.0	2.0	7.50	7.55
1.5	0.75	0.75	2.0	2.0	6.25	6.27
2.0	1.0	1.0	2.0	2.0	5.00	4.99

5.7.2 Evaluation of step responses

Any linearizable system can be linearized around a number of operating points and then the whole operating range can be covered with linear approximations. Therefore several identification tests with small amplitudes of the test signal have to be repeated in several working points and the approximating linear models have to be estimated.

Since step responses can be easily evaluated by grapho-analytical methods, it is practical to excite the system with unit steps of small amplitudes, starting with the lowest allowed input level. One should continue this procedure until the maximum allowed value is reached and then repeat the procedure in reverse mode to discover any direction-dependent phenomenon, as well. The linear approximating models of the system can be estimated, e.g., by a grapho-analytical method at every working point. Then the estimated parameters belong to the middle of the working point of the step response, i.e., to the mean value of the initial and final values of the output response. The amplitudes of the unit steps should be chosen small enough to ensure a linear approximation of the desired step responses.

The estimated parameters should be drawn against the working point (the value of the output signal) to recover any analytical relation. The whole identification method is therefore a two-stage procedure:

- the parameters of approximating linear systems should be estimated in the neighborhood of many working points;
- the analytical relation between the estimated parameters and the output signal can be obtained by fitting a static function to the estimated parameters.

Since step-like test signals do not give enough persistent excitation it is expedient to apply more exciting test signals, if possible. For example, PRBS test signals with small amplitudes are well suited to the parameter estimation of the approximating linear models at different working points.

An example for identification of a first-order quasi-linear process with constant gain and output signal dependent time constant from step responses with small amplitudes in different working points has already been presented in Example 3.15.2.

Dunoyer *et al.,* (1996) report on the grapho-analytical identification of first- and second-order bilinear systems, where both the static gain and the time constant depend on the input signal.

5.8 TWO-STEP STRUCTURE IDENTIFICATION METHOD: BEST INPUT–OUTPUT MODEL APPROXIMATION FROM NORMAL OPERATING DATA AND EVALUATION OF ITS STEP RESPONSES

If there is no *a priori* knowledge of the process to be identified the best way towards modeling would be to register some step responses that show:
- the steady state equation,
- the working point dependence,
- the amplitude dependence,
- the direction dependence.

of the system and can result in a satisfactory model. In many applications there exists no possibility of designing and executing active experiments, the identification has to be performed from normal operating data. In this case it can be advantageous to apply a two-step structure and parameter identification method (Haber, 1987):
- to identify the best input–output model;
- to excite the identified model in several working points by unit steps with different amplitudes.

There is no restriction for the best input–output model, for a model linear-in-parameters is generally sufficient and easy to estimate. Stepwise regression and the group method of data handling are appropriate procedures for obtaining a good process approximation. The output of the model should approximate well the measured output signal, i.e., a model validity test has to be used to verify the goodness of the approximation.

The evaluation of the step responses can lead to different structures, e.g.:
- quasi-linear models with signal dependent parameters;
- block oriented models;
- cascade models with higher nonlinear power, etc.

The final verification of the model requires that the originally recorded input signal be given to the model and its computed output signal be compared to the measured ones and model validity tests be applied at the residuals.

Some simulations illustrate the method.

Example 5.8.1 *Two-step structure and parameter estimation of a first-order quasi-linear system with input signal dependent time constant (Haber, 1987)*

The process is a first-order quasi-linear system

$$y(t) + T\dot{y}(t) = Ku(t)$$

with constant gain and input signal dependent time constant

$$K(t) = K_0 = 2$$
$$T(t) = T_0 + T_1 u(t) = 10 + 5u(t) \, [\text{s}]$$

The above relations are presented in Figure 5.8.1. Figure 5.8.2 shows the used input signal and the measured output signal during the identification. The input signal consists of four constant periods, each being 60 [s] long. The values of the signal were:

$$u(t): \begin{cases} 1.0 & \text{if} & t < 60 \, [\text{s}] \\ 0.0 & \text{if} & 60 \, [\text{s}] \le t < 120 \, [\text{s}] \\ -0.5 & \text{if} & 120 \, [\text{s}] \le t < 180 \, [\text{s}] \\ 0.0 & \text{if} & 180 \, [\text{s}] \le t < 240 \, [\text{s}] \end{cases}$$

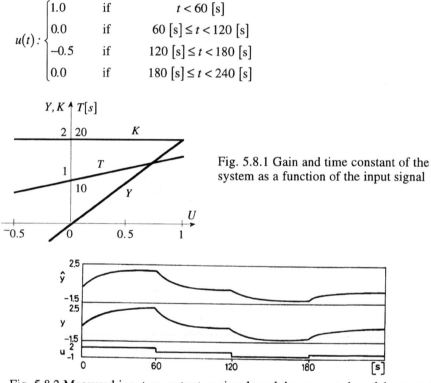

Fig. 5.8.1 Gain and time constant of the system as a function of the input signal

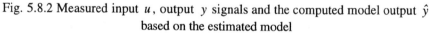

Fig. 5.8.2 Measured input u, output y signals and the computed model output $\hat{y}$ based on the estimated model

The entirely of the records were sampled by $\Delta T = 2$ [s] and an LS method was applied to identify a global valid nonlinear dynamic model. Different models, all being linear in the parameters, were fitted to the input–output records. The possible model components were 1, $u(k)$, $u^2(k)$, $u(k)y(k)$, $y^2(k)$. Stepwise regression led to the best model

$$\hat{y}(k) = (0.838 \pm 0.0034)\hat{y}(k-1) + (0.4075 \pm 0.0156)u(k)$$
$$-(0.08592 \pm 0.161)u(k-1) - (0.05016 \pm 0.0092)u(k)\hat{y}(k)$$
$$+(0.04422 \pm 0.0091)u(k-1)\hat{y}(k-1)$$

Since the test signal was in the range of

$$-0.5 \le u(t) \le 1 \qquad \forall t$$

the model has to be valid also in this range. The estimated model was excited by four unit steps with very small amplitudes in order to obtain the local valid linearized model of the system.

The initial and final values of the unit steps were as follows:

$$u_1(t): -0.5 \to -0.4 \quad u_2(t): 0.0 \to 0.1 \quad u_3(t): 0.5 \to 0.6 \quad u_4(t): 1.0 \to 1.1$$

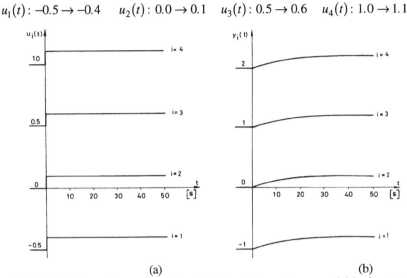

(a)   (b)

Fig. 5.8.3 Excitation of the estimated globally valid nonlinear model in four working points: (a) unit steps; (b) step responses

The size of the steps was always 0.1, that is, only 6.7 percent of the whole amplitude range of the input signal used at the identification. The input steps and the corresponding step responses for $i = 1, 2, 3, 4$ are drawn in Figure 5.8.3. Figure 5.8.4 shows each step response in a normalized form. The initial – working point – values were subtracted and the change in the output signal was normalized first to the change in the input signal (Δu) (Figure 5.8.4a) and second to the steady state change in the output signal (Figure 5.8.4b). The steady state value of

$$y'(t) = \frac{y_i(t) - y_i(0)}{\Delta u}$$

is the estimated static gain and can be easily read in Figure 5.8.4a. All normalized step responses

$$y''(t) = \frac{y_i(t) - y_i(0)}{y_i(\infty) - y_i(0)}$$

led to 1 and show an exponential behavior, therefore the time constants can be read at their 63 percent value. Finally, the calculated gains and time constants are shown in Figures 5.8.4c and 5.8.4d, where the continuous lines are the true values. The fact that the gain values coincide with the true ones – but some biases are seen between the true and estimated time constants – can be explained by the insufficiently excited form of the test signal used for the identification.   ■

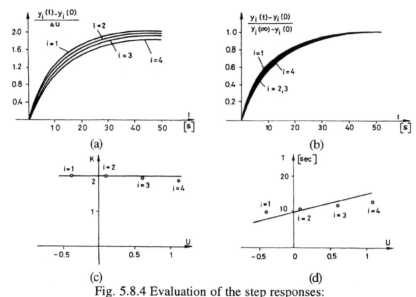

Fig. 5.8.4 Evaluation of the step responses:
(a) normalized responses $y_i'(t) = [y_i(t) - y_i(0)]/\Delta u$; (b) normalized responses
$y_i''(t) = [y_i'(t) - y_i'(0)]/[y_i'(\infty) - y_i'(0)]$; (c) true (—) and estimated (o) static gains; (d) true
(—) and estimated (o) time constants

Example 5.8.2   *Two-step structure and parameter estimation of a first-order*
quasi-linear system with output signal dependent time constant
(Haber, 1987)

The process is a first-order quasi-linear system

$$y(t) + T\dot{y}(t) = Ku(t)$$

with constant gain and output signal dependent time constant

$$K(t) = K_0 = 2$$
$$T(t) = T_0 + T_1 y(t) = 10 - 2.5y(t) \text{ [s]}$$

The above relations are presented in Figure 5.8.5. Figure 5.8.6 shows the input signal
used and the measured output signal during the identification. The input signal consists
of four constant periods, each being 60 [s] long. The values of the signal were

$$u(t): \begin{cases} 1.0 & \text{if} & t < 60 \text{ [s]} \\ 0.0 & \text{if} & 60 \text{ [s]} \le t < 120 \text{ [s]} \\ -0.5 & \text{if} & 120 \text{ [s]} \le t < 180 \text{ [s]} \\ 0.0 & \text{if} & 180 \text{ [s]} \le t < 240 \text{ [s]} \end{cases}$$

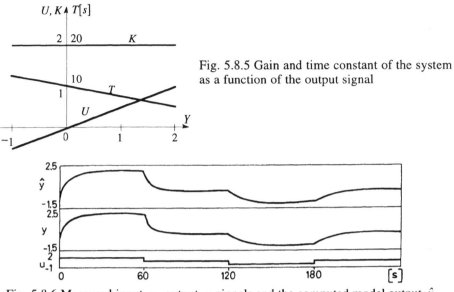

Fig. 5.8.5 Gain and time constant of the system as a function of the output signal

Fig. 5.8.6 Measured input u, output y signals and the computed model output $\hat{y}$ based on the estimated model

The entirely of the records were sampled by $\Delta T = 2$ [s] and an LS method was applied to identify a global valid nonlinear dynamic model. Different models, all being linear in the parameters, were fitted to the input–output records. The possible model components were 1, $u(k)$, $u^2(k)$, $u(k)y(k)$, $y^2(k)$. Stepwise regression lead to the best model:

$$\hat{y}(k) = (0.820 \pm 0.0013)\hat{y}(k-1) + (0.356 \pm 0.0046)u(k)$$
$$+ (0.00401 \pm 0.0045)u(k-1) + (0.1152 \pm 0.0023)\hat{y}^2(k)$$
$$- (0.1153 \pm 0.0023)\hat{y}^2(k-1)$$

Since the test signal was in the range of

$$-0.5 \le u(t) \le 1 \qquad \forall t$$

the model also has to be valid in this range. The estimated model was excited by four unit steps with very small amplitudes in order to obtain the local valid linearized model of the system.

The initial and final values of the unit steps were as follows:

$u_1(t):$ $-0.5 \rightarrow -0.4$

$u_2(t):$ $0.0 \rightarrow 0.1$

$u_3(t):$ $0.5 \rightarrow 0.6$

$u_4(t):$ $1.0 \rightarrow 1.1$

The size of the steps was always 0.1, that is, only 6.7 percent of the whole amplitude

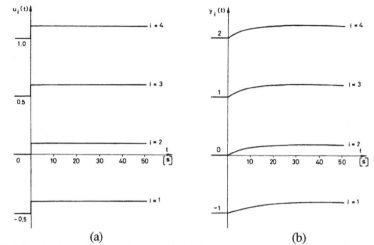

(a)   (b)

Fig. 5.8.7 Excitation of the estimated globally valid nonlinear model in four working
points: (a) unit steps; (b) step responses

range of the input signal used at the identification. The input steps and the
corresponding step responses for $i = 1, 2, 3, 4$ are drawn in Figure 5.8.7. Figure 5.8.8
shows each step response in a normalized form. The initial – working point – values
were subtracted and the change in the output signal was normalized first to the change in
the input signal (Δu) (Figure 5.8.8a), and second to the steady state change in the
output signal (Figure 5.8.8b). The steady state value of

$$y'(t) = \frac{y_i(t) - y_i(0)}{\Delta u}$$

is the estimated static gain and can be easily read in Figure 5.8.8a. All normalized step
responses

$$y''(t) = \frac{y_i(t) - y_i(0)}{y_i(\infty) - y_i(0)}$$

led to 1 and show an exponential behavior, therefore the time constants can be read at
their 63 percent value. Finally, the calculated gains and time constants are shown in
Figures 5.8.8c and 5.8.8d, where the continuous lines are the true values. The fact that
the gain values coincide with the true ones, but some biases are seen between the true
and estimated time constants, can be explained by the insufficiently excited form of the
test signal used for the identification.   ∎

Example 5.8.3   *Two-step structure and parameter estimation of a simple Wiener*
                    model (Haber, 1987)

The simple Wiener model under investigation had the linear dynamic transfer function

$$\frac{V(s)}{Y(s)} = \frac{1}{1 + 10s}$$

and the quadratic static part was

$$y(k) = 2 + v(k) + 0.5v^2(k)$$

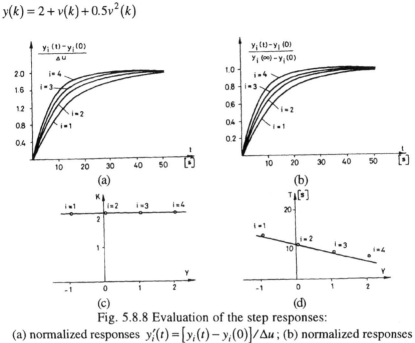

Fig. 5.8.8 Evaluation of the step responses:
(a) normalized responses $y_i'(t) = [y_i(t) - y_i(0)]/\Delta u$; (b) normalized responses
$y_i''(t) = [y_i'(t) - y_i'(0)]/[y_i'(\infty) - y_i'(0)]$; (c) true (—) and estimated (o) static gains;
(d) true (—) and estimated (o) time constants

The test signal was a PRTS with maximum length 26, amplitude ± 2, and mean value
1. The sampling time was $\Delta T = 2$ [s] and the minimum switching time of the PRTS
signal was 5-times more. $N = 26 \cdot 5 = 130$ data pairs were used for the identification.

Different models, all being linear in the parameters, were fitted to the input–output
records. The possible model components were 1, $u(k)$, $u^2(k)$, $u(k)y(k)$, $y^2(k)$.
Stepwise regression led to the best model:

$$\hat{y}(k) = (0.8975 \pm 0.1155)\hat{y}(k-1) + (0.2818 \pm 0.0215)u(k)$$

$$+(0.03114 \pm 0.007489)u(k-1) + (0.01645 \pm 0.002637)u^2(k-1)$$

$$+(0.07652 \pm 0.002105)u(k-1)\hat{y}(k-1) - (0.01959 \pm 0.001396)\hat{y}^2(k-1)$$

Figure 5.8.9 shows the measured input and output signals and the computed output
signal based on the estimated model.
Since the test signal was in the range of

$$-1 \le u(t) \le 3 \qquad \forall t$$

the model has to be valid in this range also. The estimated model was excited by eight
unit steps, all started by zero to check whether the estimated steady state characteristic

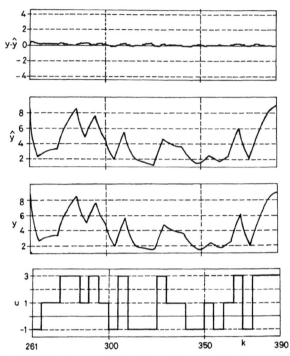

Fig. 5.8.9 Measured input, output signals and the computed output based on the
estimated globally valid nonlinear model

coincides with the true curve. The amplitudes of the unit steps were -1, -0.5, 0, 0.5, 1,
1.5, 2, 2.5 and 3. The steady state output values are drawn in Figure 5.8.10 by circles.
The continuous curve shows the true characteristic. As is seen, the coincidence is very
good.

Four step tests were selected for structure identification. The input unit steps were:

$u_1(t):$ $0.5 \cdot 1(t)$

$u_2(t):$ $1.0 \cdot 1(t)$

$u_3(t):$ $1.5 \cdot 1(t)$

$u_4(t):$ $2.0 \cdot 1(t)$

Figure 5.8.11a and 5.8.11b show the unit steps and the corresponding step responses.

The process is of simple Wiener model type if the inner variables $v_i(t)$ between the
linear dynamic and static nonlinear parts are proportional to each other for $i = 1, 2, 3$ and
4, as explained in Section 5.6.3. Therefore the step responses have to be transformed
through the inverse steady state characteristic of the whole model. As is seen in
Figure 5.8.10, the nonlinear curve is single-valued in the region investigated.

The steady state characteristic of the model can be calculated by replacing all delayed
values $u(k-1), y(k-1)$ by their steady state values U, Y

$$0.01959\hat{Y}^2 + 0.1025\hat{Y} - 0.07652U\hat{Y} - 0.03114U - 0.01645U^2 - 0.2818 = 0$$

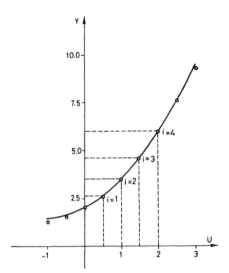

Fig. 5.8.10 Steady state characteristic of the
Wiener model
(true (——) and estimated (o) values)

The inverse relation can be calculated from

$$\hat{U}^2 + \hat{U}(1.893 + 4.6517Y) + 17.1307 - 6.231Y - 1.19089Y^2 = 0$$

Since the relation between $v(k)$ and $y(k)$ is static, the above equation can be used to compute $v(k)$ from $y(k)$ at every time point.

The four step responses in the inner variable $v_i(t)$ are drawn in Figure 5.8.11c. In

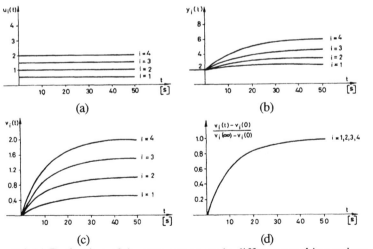

Fig. 5.8.11 Evaluation of the step responses in different working points:
(a) unit steps $u_i(t)$; (b) step responses $y_i(t)$; (c) computed inner variables $v_i(t)$;
(d) normalized inner variables $v_i'(t) = [v_i(t) - v_i(0)] / [v_i(\infty) - v_i(0)]$

Figure 5.8.11d they are normalized to their steady state values

$$v_i'(t) = \frac{v_i(t) - v_i(0)}{v_i(\infty) - v_i(0)}$$

All four normalized step responses in the inner variable coincide, which shows that the *a priori* assumption is proved by an *a posteriori* analysis of the data. Finally, the time constant of the linear dynamic part can be read at the 63 percent value of its stationary value and it is equal to 10 [s], as in the original model.   ∎

5.9 SELECTION OF THE MOST SIGNIFICANT MODEL COMPONENTS OF MODELS LINEAR-IN-PARAMETERS

The nonlinear model is assumed to be a difference equation linear in the parameters. The possible model components can be linear and nonlinear – usually polynomial – functions of the input and output signals. Such models were proposed for system identification by Haber and Keviczky (1974, 1976). Leontaritis and Billings (1985b) extended this idea for stochastic cases, and suggested building the model components from the input, output and – iteratively computed – residuals. The task of the structure identification is to select the significant components among all possible ones.

Denote the possible components by $\phi_i(k)$, then a model linear-in-parameters would be

$$y(k) = \boldsymbol{\phi}^T(k)\boldsymbol{\theta}$$

where the memory vector

$$\boldsymbol{\phi}(k) = \left[\phi_1(k), \dots, \phi_{n_\theta}(k)\right]^T$$

contains all possible components.

5.9.1 All possible regressions

The trivial way of searching for the best structure is to perform a parameter estimation for all possible structures and to choose the basis of comparing certain performance indices.

Below we list several performance indices that are used for the decision of which model can be expected and which not. Before doing so, some known mathematical definitions are summarized. (Denote the residuals by $\varepsilon(k)$, the number of data pairs by N, and the number of model components (and parameters) by n_θ. The predicted value is marked by $\hat{(...)}$ and the mean value by $\overline{(...)}$.)

1. Residual sum of squares (RSS):

$$\text{RSS} = \sum_{k=1}^{N} \left[\hat{y}(k) - y\right]^2 = \sum_{k=1}^{N} \varepsilon^2(k)$$

2. Total sum of squares (TSS) of the residuals:

$$TSS = \sum_{k=1}^{N} [y(k) - \bar{y}]^2$$

3. Sum of squares owed to regression (SS) of the residuals:

$$SS = \sum_{k=1}^{N} [\hat{y}(k) - \bar{y}]^2$$

4. normalized residual sum of squares (NRSS):

$$NRSS = \frac{RSS}{N}$$

The following performance indices are characteristic for the model fit and can be used when the best fitting model is sought (e.g., Draper and Smith, 1966):

1. Residual sum of squares (RSS)
   The model that has a smaller RSS value (RSS>0) is better

2. Multiple correlation coefficient (R^2)

$$R^2 = \left(\frac{SS}{TSS}\right) = \left(1 - \frac{RSS}{TSS}\right)$$

The model that has a greater R^2 value $(0 < R^2 < 1)$ is better.

The problem of using these criteria is that they point out the best model with too many – often with all possible – model components. To overcome this problem selection rules that take the complexity of the model, i.e., the number of parameters, into account (e.g., Draper and Smith, 1966) have been introduced:

1. Adjusted multiple correlation coefficient (R_a^2)

$$R_a^2 = 1 - (1 - R^2)\frac{N-1}{N-n_\theta}$$

A model with higher R_a^2 is better than one with smaller value;

2. Mallow's C_p statistic

$$C_p = \frac{RSS}{\hat{\sigma}_\varepsilon^2} - (N - 2n_\theta)$$

where $\hat{\sigma}_\varepsilon^2$ is the variance of the source noise that can be approximated by the square of the least value of all possible normalized residual sum of squares.
If the estimated model is adequate the expected value of the residual sum of squares is:

$$E\{RSS\} = (N - n_\theta)\sigma_\varepsilon^2$$

and thus

$$E\{C_p\} = n_\theta$$

is equal to the number of parameters. The C_p statistics can be calculated for all possible models, and that model is selected such that:
- the C_p value is closest to the number of parameters;
- if there are more models having the same deviation between C_p and the number of parameters, the model which has the fewest parameters must be chosen.

The C_p plot contains the C_p statistics of the different models plotted as a function of the number of parameters.

3. Overall F-test (OVF)

$$OVF = \frac{SS}{RSS}\frac{N-n_\theta}{n_\theta-1}\left(=\frac{TSS-RSS}{RSS}\frac{N-n_\theta}{n_\theta-1}\right)$$

A model with greater OVF value shows a better model fit. (As is seen from the term in the bracket, the overall F-test value is an F-test that compares a model under investigation with a model with a constant term only.)

The performance indices are closely related to each other, which is not discussed here.

As is seen in Mallow's statistic, the performance index is the weighted sum of the residual sum of squares and a penalty term increasing with the number of parameters. By means of this performance index a model that has small residuals and a few termed parameters simultaneously is optimal. Mallow's statistic uses the variance of the source noise that is known only if all possible regressions have been performed.

The disadvantage of the decision based on the residual sum of squares is that the optimal model has too many parameters. It is straightforward also to include the number of parameters into the criterion that is used for determining the optimal structure. The aim is quite natural, an engineer would like to have a model of the process that is as simple as possible (the principle of parsimony). Kortmann and Unbehauen (1987, 1988a, b) suggest four different performance indices for using them for the structure determination:

1. Final prediction error technique (FPE) (Akaike, 1970)

$$FPE = N\ln(NRSS) + N\ln[(N+n_\theta)/(N-n_\theta)]$$

2. Akaike's information criterion (AIC) (Akaike, 1972; Bhansali and Downham, 1977)

$$AIC(\rho) = N\ln(NRSS) + \rho n_\theta \qquad \rho > 0$$

3. Khinchin's law of iterated logarithm criterion (LILC) (Hannan and Quin, 1979)

$$LILC(\rho) = N\ln(NRSS) + 2n_\theta \rho \ln(\ln(N))$$

4. Bayesian information criterion (BIC) (Kashyap, 1977)

$$BIC = N\ln(\text{NRSS}) + n_\theta \ln(N)$$

Common in the above performance indices is that they add to the N-times logarithm of the residual sum of squares a second penalty function that increases with the number of parameters. For large data pairs FPE is equal to AIC with $\rho = 2$ (Sriniwas *et al.*, 1995).

In practice it is not usual to estimate all possible models because other effective methods exist. Most of the performance indices above are used in those techniques.

A more economical way of building up a model in a successive way is to apply forward, backward, or stepwise regression (e.g., Draper and Smith, 1966).

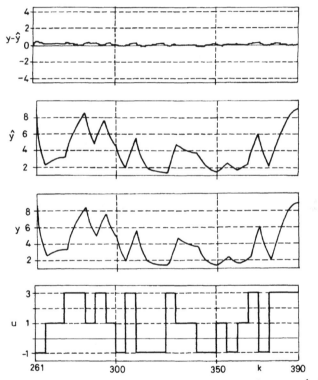

Fig. 5.9.1 Measured input u, output y signals, the computed output $\hat{y}$ based on the estimated nonlinear model and the output error $(y - \hat{y})$

Example 5.9.1 *Structure identification and parameter estimation of the quadratic first-order simple Wiener model by all possible regressions.*

The simple Wiener model under investigations had the linear dynamic function

$$\frac{V(s)}{Y(s)} = \frac{1}{1+10s}$$

and the quadratic static part was

$$y(k) = 2 + v(k) + 0.5v^2(k)$$

The test signal was a PRTS with maximum length 26, amplitude ± 2, and mean value 1. The sampling time was $\Delta T = 2$ [s] and the minimum switching time of the PRTS signal was 5 times more. $N = 26 \cdot 5 = 130$ data pairs were used for the identification. The plot of the simulated input and output signals is seen in Figure 5.9.1.

Since the Wiener model is not linear in the parameters, and therefore the estimation of the parameters would need too many computations, nonlinear models linear in the parameters, were chosen for approximating the input–output behavior of the system.

The possible model components were chosen as follows:

1. 1;
2. $u(k)$;
3. $u^2(k)$;
4. $u(k)y(k)$;
5. $y^2(k)$.

First-order models were assumed only, i.e., all model included the term $y(k-1)$ and the other model components were taken into account by their shifted (by one) values. The *a priori* assumption that the term $y(k-1)$ was always in the model reduced the number of regressions significantly (from $\left(2^6 - 1 = 63\right)$ to $\left(2^5 - 1 = 31\right)$).

Table 5.9.1 presents the estimated parameters, their standard deviations, and the standard deviations of the residuals at fitting all possible first-order quadratic models linear in the parameters. Table 5.9.2 summarizes the variances of the residuals, the overall F-values, the multiple correlation coefficients, and the estimated parameter/standard deviation ratios of the above models. The standard deviations of the residuals, the variance of the residuals (the normalized loss functions), and the multiple correlation coefficients are drawn against the number of parameters in Figures 5.9.2 to 5.9.4, respectively, for the best models (a) and for the average values of all models (b) having the same number of parameters. All features improve when the number of parameters is increased. This fact can be best seen in the plots of the average values (b). Even the plots of the performance indices of the best models, having the lowest mean square residuals by the given number of parameters, show a little improvement by increasing the number of components. However, there is a break in the curves at $n_\theta = 2$, which fact shows that the model improvement is much less if further model components are taken into account. A model has to have significant estimated parameters besides that it has to show as good as possible model matching. Therefore the estimated parameter/standard deviation ratios have to be at least about 3 to 4. According to Table 5.9.2 only some models meet this requirement. That model should finally be accepted which has the least residual mean square and whose parameters are significant. As is seen, the best model includes all five model components. (Of course, if more components were considered the optimal model would not include all possible components.)

Perhaps the C_p plot shows most clearly that the model with five parameters fits the simple Wiener model much better than the models with fewer parameters. This fact can be seen by the fact that all other models have so big C_p values that they cannot be seen in the C_p plot of Figure 5.9.5.

TABLE 5.9.1a Estimated parameters, their standard deviations, and the standard deviations of the residuals at fitting all possible first-order quadratic models linear in the parameters

No	No.of comp.	1	2	3	4	5	Components	σ_ε	1	2	3	4	5	$-y(k-1)$
									Estimated parameters and standard deviations					
1	1	x					1	0.5014	0.08747 ±0.09228					-0.9908 ±0.02150
2	1		x				2	0.2512		0.3027 ±0.01584				-0.9171 ±0.006963
3	1			x			3	0.2336			0.1337 ±0.006216			-0.8791 ±0.007702
4	1				x		4	0.2993				0.06457 ±0.004239		-0.8882 ±0.01001
5	1					x	5	0.5013					-0.005066 ±0.005208	-1.038 ±0.03215
6	2	x	x				1,2	0.2056	0.3099 ±0.03887	0.3264 ±0.01301				-0.8466 ±0.01053
7	2	x		x			1,3	0.2335	0.04551 ±0.04302		0.1334 ±0.00622			-0.8700 ±0.01149
8	2	x			x		1,4	0.1149	0.6585 ±0.02428			0.08946 ±0.001868		-0.7070 ±0.007706
9	2	x				x	1,5	0.5032	0.03609 ±0.2062				-0.003246 ±0.01164	-1.020 ±0.1079
10	2		x	x			2,3	0.1919		0.1540 ±0.01953	0.08073 ±0.008433			-0.8838 ±0.006356
11	2		x		x		2,4	0.2422		0.2272 ±0.002758		0.02064 ±0.006341		-0.9014 ±0.008263
12	2		x			x	2,5	0.2274		0.3139 ±0.01417			-0.01286 ±0.002388	-0.9889 ±0.01475
13	2			x	x		3,4	0.1970			0.09573 ±0.007408	0.02857 ±0.003943		-0.8625 ±0.006886
14	2			x		x	3,5	0.2340			0.1334 ±0.006239		-0.001855 ±0.002435	-0.8902 ±0.01653
15	2				x	x	4,5	0.08504				0.0914 ±0.001396	-0.03894 ±0.0010243	-1.066 ±0.005470

TABLE 5.9.1b Estimated parameters, their standard deviations, and the standard deviations of the residuals at fitting all possible first-order quadratic models linear in the parameters

No	No.of comp.	1	2	3	4	5	Components	σ_ε	1	2	3	4	5	$-y(k-1)$
									Estimated parameters and standard deviations					
16	3	x	x	x			1,2,3	0.1702	0.2068 ±0.03486	0.2039 ±0.01925	0.06221 ±0.008104			-0.8444 ±0.008717
17	3	x	x		x		1,2,4	0.08754	0.5827 ±0.02011	0.1040 ±0.01084		0.06650 ±0.002785		-0.7339 ±0.006506
18	3	x	x			x	1,2,5	0.2009	0.5094 ±0.08432	0.3308 ±0.01282			0.01241 ±0.004685	-0.7318 ±0.04451
19	3	x		x	x		1,3,4	0.09517	0.5313 ±0.002608		0.03554 ±0.004640	0.07129 ±0.002833		-0.7325 ±0.007196
20	3	x		x		x	1,3,5	0.2343	0.08007 ±0.09601		0.1335 ±0.006248		0.002185 ±0.005423	-0.8501 ±0.05087
21	3	x			x	x	1,4,5	0.06151	0.2727 ±0.02533			0.09252 ±0.001015	-0.0256 ±0.001443	-0.9300 ±0.01323
22	3		x	x	x		2,3,4	0.1855		0.1031 ±0.02482	0.07780 ±0.008203	0.01538 ±0.004887		-0.8733 ±0.006986
23	3		x	x		x	2,3,5	0.1820		0.1787 ±0.01958	0.07100 ±0.008378		-0.007790 ±0.002002	-0.9313 ±0.01362
24	3		x		x	x	2,4,5	0.08502		0.01249 ±0.01205		0.08845 ±0.003714	±0.001273	-1.063 ±0.006123
25	3			x	x	x	3,4,5	0.07523			0.02332 ±0.003885	0.07911 ±0.002392	-0.03328 ±0.001244	-1.036 ±0.006910
26	4	x	x	x	x		1,2,3,4	0.07871	0.5169 ±0.02165	0.08112 ±0.01058	0.02307 ±0.004168	0.05975 ±0.002785		-0.7445 ±0.006156
27	4	x	x	x		x	1,2,3,5	0.1672	0.3581 ±0.07302	0.2112 ±0.01917	0.06016 ±0.008009		0.009201 ±0.003923	-0.7594 ±0.03723
28	4	x	x		x	x	1,2,4,5	0.05598	0.3086 ±0.02407	0.04296 ±0.008280		0.08250 ±0.002141	-0.02115 ±0.001569	-0.9023 ±0.001317
29	4	x		x	x	x	1,3,4,5	0.05211	0.2529 ±0.02164		0.01922 ±0.002714	0.08231 ±0.001679	-0.02235 ±0.001306	-0.9154 ±0.01139
30	4		x	x	x	x	2,3,4,5	0.07554		-0.000517 ±0.01093	0.02336 ±0.003984	0.07921 ±0.003230	-0.03385 ±0.001349	-1.036 ±0.007093
31	5	x	x	x	x	x	1,2,3,4,5	0.04900	0.2818 ±0.02150	0.03114 ±0.007489	0.01645 ±0.002637	0.07652 ±0.002105	-0.01959 ±0.001396	-0.8975 ±0.01155

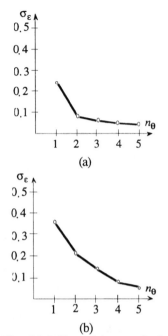

(a)

(b)

Fig. 5.9.2 Plot of the standard deviations of the residuals against the number of parameters at fitting of first-order quadratic models linear in the parameters: (a) best models; (b) average value of all models

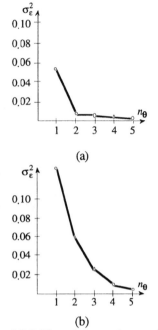

(a)

(b)

Fig. 5.9.3 Plot of the variance of the residuals against the number of parameters at fitting of first-order quadratic models linear in the parameters: (a) best models; (b) average value of all models

The estimated parameters (with their standard deviations) of the best model are given by the difference equation

$$\hat{y}(k) = (0.8975 \pm 0.01155)\hat{y}(k-1) + (0.2818 \pm 0.0215)u(k)$$

$$+ (0.03114 \pm 0.007489)u(k-1) + (0.01645 \pm 0.002637)u^2(k-1)$$

$$+ (0.07652 \pm 0.002105)u(k-1)\hat{y}(k-1) - (0.01959 \pm 0.001396)\hat{y}^2(k-1)$$

The output signal of the simulated model, the computed output signal based on the estimated model and the biases between them are seen in Figure 5.9.1. Although the approximating model is linear in the parameters it approximates the simple Wiener model very well.   ∎

5.9.2 Forward and backward regression
The philosophy of the two methods is as follows:
- *forward regression:*
  The model is built up by starting with one component and in turn adding new terms to the already existing model components. The model building is completed if the model approximates the process sufficiently;

- *backward regression:*
  The initial model has to include all possible components. Terms not significant are omitted in turn, thus already neglected components cannot be included any more. The model building is completed if the model approximates the process sufficiently.

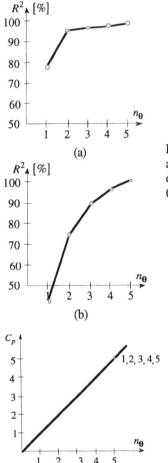

(a)

(b)

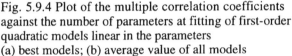

Fig. 5.9.4 Plot of the multiple correlation coefficients against the number of parameters at fitting of first-order quadratic models linear in the parameters
(a) best models; (b) average value of all models

Fig. 5.9.5 C_p plot at fitting of first-order quadratic models linear in the parameters

Both regressions usually apply the partial F-test to decide whether to include or exclude a model component. The calculation of the partial F-value is the following. The model containing n_θ parameters has the total sum of squares $TSS(n_\theta)$, the sum of squares owed to regression $SS(n_\theta)$, and the residual sum of squares $RSS(n_\theta)$. Now repeat the parameter estimation by $(n_\theta - 1)$ parameters, i.e., by eliminating the ith model component. The total sum of squares remains the same, the sum of squares owed to regression becomes $SS_i(n_\theta - 1)$, and the residual sum of squares $RSS_i(n_\theta - 1)$. The partial F-values are

$$F = \frac{SS(n_\theta) - SS_i(n_\theta - 1)}{RSS(n_\theta)/(N - n_\theta)} = \frac{RSS_i(n_\theta - 1) - RSS(n_\theta)}{RSS(n_\theta)/(N - n_\theta)}$$

which must be compared with the F-value $F(1, N - n_\theta, \alpha)$, where $100(1 - \alpha)$ is the significance level in percent. The usual values of α are $\alpha = 0.05$ or $\alpha = 0.10$, and if N is large then

$$F(1, \infty, 0.05) \approx 4$$

Using the partial F-test the techniques of the two regressions are the following:
- *Forward regression*
  Only new model components can be checked for partial F-tests (F to enter) but no earlier accepted components can be removed. The algorithm starts with a model having no components;
- *Backward regression*
  Only existing model components can be checked for partial F-tests (F to remove), and no new model component can be inserted. The algorithm has to start by the most complicated model, i.e., a model having all possible model components.

Remarks:
1. Neither of the two methods can be recommended for practical use. Forward regression does not allow removing an old component if the later accepted components cover the information given by the earlier accepted component. Backward regression has the disadvantage that the starting model has too many parameters, and the solution of the least squares estimation may cause numerical problems.
2. The efficiency of both forward and the backward regression depends on the postulated ordering of the model components, which is not the case with stepwise regression and with forward regression using orthogonal model components.
3. Leontaritis and Billings (1987) used forward and backward regression and found that use of only one information criterion AIC(4) was sufficient and optimal.
4. Deergha and Reddy (1992) and Thomson *et al.*, (1995) used the forward regression in combination with model validity test.
5. Backward regression was used by Simeu (1995) for the identification of an eddy current brake.

5.9.3 Stepwise regression
Stepwise regression starts by considering only the most significant model component, then extending the model by the second most significant one, etc. Since the model components can be correlated to each other, it is possible that a component that was chosen at an earlier stage may be superfluous at a later stage. Therefore the stepwise regression algorithm consists of the following basic steps:
1. Select the most significant model component for possible entry into the model;
2. Include the selected component in the model if it passes a selection stage, the so-called inclusion threshold;
3. Calculate a test for emission for every model component;
4. Eliminate those components from the model which do not pass the emission threshold

There is a possibility of using higher emission threshold values (e.g., F to remove) for the earlier selected components than for the inclusion threshold values (e.g., F to enter) for the newly selected component. This means that the elimination of an already accepted model component is more difficult than entering a new component to the model. (The forward and backward regressions are special cases with infinite emission and inclusion threshold values, respectively.)

There are several procedures for selecting the most significant model component for possible entry into the model. From the available set of possible components, the variable which:

- has the greatest partial correlation with the output signal, taking the already chosen model components into account;
- produces the greatest increase in the multiple correlation;
- causes the greatest decrease in the residual sum of squares;
- which makes the greatest increase in the sum of squares due to regression,

has to be chosen.

To obtain a satisfactory model with as few as parameters possible the performance indices FPE, AIC, LILC and BIC should be used instead of the RSS, both when checking a new term for possible entry into the model (c.f. F-test to enter) and when the in-terms are checked for possible removement from the model (c.f. F-test to remove). The rules are as follows:

- add a new component to the model if the performance indices of the enlarged model are bigger than that of the old model;
- remove any component from the model if the reduced model has a bigger performance index as the original one.

We can use either only one performance index or more. The best procedure is to use all indices, and to add the new component if all indices are bigger and remove a component if any criterion is bigger than that of the original one. (Bigger means bigger in absolute value, since the new performance indices are negative.)

Stepwise regression has been applied by some authors for the identification of nonlinear dynamic systems. The experiences are summarized in the next remarks.

Remarks:

1. Bard and Lapidus (1970) evaluated the cross-correlation of the possible model components with the residuals in the inclusion test and removed a component if the parameter/standard deviation relation was less than a prescribed value. Since they applied a Volterra series approximation it was important to select only the significant components of the series up to a predefined memory. The basic steps of their algorithm were as follows:

   1. Divide the possible terms into two parts, terms in the model, and terms not in it;
   2. Estimate the parameters of the model having only the chosen components. Calculate the residuals;
   3a Eliminate those components of the model that has less parameter/standard deviation than an omission threshold;
   3b Include those components that have greater cross-correlation value with the residuals than an inclusion threshold;
   4. Replace the components to be removed by the new components as suggested in Step 3;
   5. Continue the algorithm at Step 2 while the new model differs from the previous one.

Clearly there is a close relationship between the algorithms presented by Draper and Smith (1966) and that of Bard and Lapidus (1970). The omission threshold corresponds to the F to remove value and the inclusion threshold of the F to enter value.

2. Haber and Zierfuss (1987) applied only the F-test for checking the components for including or removing from the model.
3. Kortmann and Unbehauen (1988a) used the overall F-test, the multiple correlation coefficient and the information criteria FPE, AIC(2), LILC(1), BIC for checking whether a model component can be included, and they compared all the information criteria FPE, AIC(2), LILC(1), BIC and the partial F-test while checking a component for possible omission from the model. They concluded that the best procedure is to add the new component if all performance criteria support it, and to remove an already existing component if any criterion indicates so.
4. Billings and Voon (1986) and Leontaritis and Billings (1987) used not only the functions of the measured input and output signals but built terms also from the – iteratively computed – residuals.
5. Billings and Voon (1986) used the F-ratio test, which is based on the partial correlation coefficients, for possible entry into the model and the t-test for checking the components for removal from the model.
6. Leontaritis and Billings (1987) used the information criterion AIC(4) for the selection of the components.
7. Thouverez and Jezequel (1996) applied a pre-selection before applying the stepwise regression in order to limit the possible components.
8. Cao and Feng (1995) applied the information criterion BIC for the selection of the components.

Example 5.9.2   *Structure identification and parameter estimation of the quadratic, first-order simple Wiener model by first-order linear-in-parameters models using the stepwise regression (Continuation of Example 5.9.1)*

The process and the test signal are the same as in Example 5.9.1.

31 regressions were needed if all possible regressions were done. The number of regressions could be drastically decreased to 21 if stepwise regression were used. Table 5.9.3 summarizes the procedure. In the first step all possible model components were considered for possible entry into the model. The least residual mean square was obtained if the 3rd component $u^2(k-1)$ was taken into account in addition to the fix component $y(k-1)$. In the second step the 2nd component $u(k-1)$ was added to the model. Since the partial F-test values are bigger than 4 the model is significant. The third step led to the inclusion of the term $u(k-1)y(k-1)$ and the last component $y^2(k-1)$ was introduced into the model at the last step. All components are significant, as shown by their partial F-tests.

The F-test does not take the complexity of the model into account. That can be investigated by the different information criteria introduced in Section 5.9.1. In Table 5.9.4 they are calculated for the best models at every stage of the computation of the stepwise regression. As is seen, the absolute values of the information criteria of the whole models increased by increasing the number of parameters, and all components were significant since the absolute values of the partial information criteria decreased if any component was omitted from the model. The reason for obtaining the same result by the F-test and the information criteria is that the residual sum of squares term in the information criteria is much more dominant if the number of parameters is very few. ∎

TABLE 5.9.2 Variances of the residuals, overall F-values, multiple correlation coefficients, and estimated parameter/standard deviation ratios at fitting all possible first-order quadratic models linear in the parameters

No.	No. of comp.	1	2	3	4	5	Compo-nents	σ^2_ε	Overall F	$R^2[\%]$	C_p	1	2	3	4	5	$\dfrac{|a_i|}{\sigma\{a_i\}}$	Model accept-able
1	1	x					1	0.2514		0	13328.68	0.95					46.08	0
2	1		x				2	0.0631	∞	74.90	3249.55		19.55				131.71	1
3	1			x			3	0.0546	∞	78.28	2794.57			21.51			144.14	1
4	1				x		4	0.0896	∞	64.36	4668.02				15.23		88.88	1
5	1					x	5	0.2513	∞	0.04	13323.33					0.97	32.29	0
6	2	x	x				1,2	0.042	638.17	83.29	2102.70	7.97	25.09				80.40	1
7	2	x		x			1,3	0.0545	462.24	78.32	2768.61	1.058		21.45			75.72	0
8	2	x			x		1,4	0.0132	2309.82	94.74	575.08	27.12			47.89		91.75	1
9	2	x				x	1,5	0.2532	-0.91	-0.72	13321.97	0.175				0.198	12.91	0
10	2		x	x			2,3	0.0368	746.43	85.36	1828.53		7.88	9.57			139.05	1
11	2		x		x		2,4	0.0587	420.20	76.65	2991.68		8.24		3.26		109.09	0
12	2		x			x	2,5	0.0517	494.42	79.44	2619.89		22.15			5.39	67.04	1
13	2			x	x		3,4	0.0388	701.36	84.57	1934.75			12.92	7.25		125.25	1
14	2			x		x	3,5	0.0548	459.21	78.20	2784.54			21.38		0.762	53.85	0
15	2				x	x	4,5	0.0072	4341.33	97.13	256.41				65.47	38.03	194.88	1
16	3	x	x	x			1,2,3	0.0290	486.98	88.46	1404.22	5.93	10.59	7.68			96.87	1
17	3	x	x		x		1,2,4	0.00766	2020.56	96.95	279.66	28.97	9.59		23.88		112.80	1
18	3	x	x			x	1,2,5	0.0404	331.65	83.93	2004.97	6.04	25.80			2.65	16.44	0
19	3	x		x	x		1,3,4	0.00905	1700.47	96.40	352.91	20.37		7.66	25.16		101.79	1
20	3	x		x		x	1,3,5	0.0549	227.28	78.16	2769.07	0.834		21.37		0.40	16.71	0
21	3	x			x	x	1,4,5	0.0038	4137.53	98.49	76.25	10.77			91.15	17.74	70.29	1
22	3		x	x	x		2,3,4	0.0344	400.57	86.32	1688.78		4.15	9.48	3.15		125.01	0
23	3		x	x		x	2,3,5	0.0331	418.79	86.83	1620.27		9.13	8.47		3.90	68.38	0
24	3		x		x	x	2,4,5	0.0072	2153.71	97.14	255.42		1.04		27.87	29.97	173.61	0
25	3			x	x	x	3,4,5	0.0057	2737.18	97.73	176.37			6.00	33.07	27.19	149.93	1
26	4	x	x	x	x		1,2,3,4	0.0062	1661.03	97.53	202.15	23.88	7.67	5.54	21.45		120.94	1
27	4	x	x	x		x	1,2,3,5	0.02796	335.64	88.87	1339.81	4.90	11.02	7.51		2.35	20.40	0
28	4	x	x		x	x	1,2,4,5	0.0031	3364.06	98.77	40.07	12.82	5.19		38.53	13.48	68.51	1
29	4	x		x	x	x	1,3,4,5	0.0027	3868.67	98.92	19.69	11.69		7.08	49.02	17.11	80.37	1
30	4		x	x	x	x	2,3,4,5	0.0057	1810.42	97.73	177.13		0.047	5.86	24.52	25.09	146.06	0
31	5	x	x	x	x	x	1,2,3,4,5	0.002401	3240.82	99.04	5.00	13.11	4.16	6.24	36.35	14.03	77.71	1

TABLE 5.9.3 Selection of the best structure from all possible first-order quadratic models linear in the parameters by means of the stepwise regression based on the F-test

No	No of comp.	1	2	3	4	5	Compo-nents	σ^2_ε	F to enter	Overall F	$R^2[\%]$	1	2	3	4	5	Best model
1	1	x					1	0.2514			0						
2	1		x				2	0.0631		∞	74.90						
3	1			x			3	0.0546		∞	78.28						3
4	1				x		4	0.0896		∞	64.36						
5	1					x	5	0.2513		∞	0.04						
6	2	x		x			1,3	0.0545	0.2349	462.24	78.32						
7	2		x	x			2,3	0.0368	61.91	746.43	85.36		61.91 (0.0368)	91.47 (0.0631)			2,3
8	2			x	x		3,4	0.0388	52.12	701.36	84.57						
9	2			x		x	3,5	0.0548	-0.46	459.21	78.20						
10	3	x	x	x			1,2,3	0.0290	34.16	486.98	88.46	34.16 (0.0368)	111.67 (0.0545)	56.93 (0.042)			1,2,3
11	3		x	x	x		2,3,4	0.0344	8.86	400.57	86.32						
12	3		x	x		x	2,3,5	0.0331	14.20	418.79	86.83						
(13)	2	x	x				1,2	0.042									
14	4	x	x	x	x		1,2,3,4	0.0062	463.35	1661.03	97.53	573.10 (0.0344)	57.92 (0.00905)	29.67 (0.00766)	463.35 (0.0290)		1,2,3,4
15	4	x	x	x		x	1,2,3,5	0.02796	4.69	335.64	88.87						
(16)	3	x		x	x		1,3,4	0.00905									
(17)	3	x	x		x		1,2,4	0.00766									
18	5	x	x	x	x	x	1,2,3,4,5	0.00240	197.78	3240.82	99.04	171.75 (0.0057)	15.56 (0.0027)	36.39 (0.0031)	1332.73 (0.02796)	197.78 (0.0062)	1,2,3,4,5
(19)	4		x	x	x	x	2,3,4,5	0.0057									
(20)	4	x		x	x	x	1,3,4,5	0.0027									
(21)	4	x	x		x	x	1,2,4,5	0.0031									

TABLE 5.9.4 Selection of the best structure from all possible first-order quadratic models linear in the parameters by means of the stepwise regression based on different information criteria

No	No of comp.	Components 1 2 3 4 5	Compo-nents	σ_ε^2	FPE	AIC	LLIC	BIC	Partial information criteria				
									Comp	FPE	AIC	LLIC	BIC
1	1	x	3	0.0546	-376.00	-376.00	-374.83	-373.13					
2	2	x x	2,3	0.0368	-425.29	-425.29	-426.12	-419.55	1	-357.20	-375.20	-356.03	-354.33
									2	-376.00	-376.00	-374.83	-373.13
3	3	x x x	1,2,3	0.0290	-454.26	-454.26	-450.76	-445.66	1	-425.29	-425.29	-426.12	-419.55
									2	-374.24	-374.24	-371.91	-368.50
									3	-408.11	-408.11	-405.71	-402.37
4	4	x x x x	1,2,3,4	0.0062	-652.82	-652.82	-648.16	-641.35	1	-432.06	-432.06	-428.56	-423.46
									2	-306.31	-306.31	-302.81	-297.71
									3	-627.33	-627.33	-623.83	-618.72
									4	-454.26	-454.26	-450.76	-445.66
5	5	x x x x x	1,2,3,4,5	0.0024	-774.19	-774.19	-768.36	-759.85	1	-663.75	-663.75	-659.09	-652.28
									2	-760.88	-760.88	-756.22	-749.41
									3	-742.93	-742.93	-738.26	-731.45
									4	-457.00	-457.00	-452.34	-445.53
									5	-652.82	-652.82	-648.16	-641.35

Example 5.9.3 *Structure identification and parameter estimation of the quadratic, first-order simple Wiener model by second-order models linear-in-parameters using the stepwise regression (Continuation of Example 5.9.2, Zierfuss and Haber, 1987)*

The process and the test signal are the same as in Examples 5.9.1 and 5.9.2. The possible models are second-order, quadratic models linear-in-parameters. (In Example 5.9.2 only first-order models were assumed.)

The possible model components were chosen as follows:

1:	1	2:	$u(k-1)$
3:	$u(k-2)$	4:	$y(k-1)$
5:	$y(k-2)$	6:	$u^2(k-1)$
7:	$u^2(k-2)$	8:	$u(k-1)y(k-1)$
9:	$u(k-1)y(k-2)$	10:	$u(k-2)y(k-1)$
11:	$u(k-2)y(k-2)$	12:	$y^2(k-1)$
13:	$y^2(k-2)$		

Table 5.9.5 shows the steps of the stepwise regression. The F threshold value chosen was 4. The first component that is included is component 4, $y(k-1)$. Up to seven model components, none of the already existing terms were removed, because the partial F to remove values became bigger than 4. Only after the inclusion of term 3: $u(k-2)$, component 2: $u(k-1)$ is removed. Component 13: $y^2(k-2)$ is included in step 8 and component 5: $y(k-2)$ in step 9. Term 3: $u(k-2)$ will be removed in the same step. The stages of the stepwise regression are also illustrated graphically. Both the F-values (Figure 5.9.6) and the normalized loss functions (the standard deviations of the residuals) (Figure 5.9.7) are plotted against the actual number of parameters.

The equation of the resulting model is

TABLE 5.9.5a Stepwise regression of the simple Wiener model by fitting a quadratic, second-order model liner-in-parameters

Step	No	Number of compon. (n)	Components	$\sigma^2_{\varepsilon_n}$	F to enter	Best model F to remove and $\sigma^2_{\varepsilon_{n-1}}$
1	1	1	1	4.414		
	2	1	2	8.614		
	3	1	3	7.728		
	4	1	4	0.251		best model:
	5	1	5	0.856		4
	6	1	6	5.607		
	7	1	7	4.683		
	8	1	8	5.593		
	9	1	9	5.939		
	10	1	10	4.836		
	11	1	11	5.185		
	12	1	12	2.298		
	13	1	13	3.129		
2	14	2	4,1	0.2514	-0.20	best model:
	15	2	4,2	0.0631	381.16	4, 6
	16	2	4,3	0.1770	53.51	4: 13 016.60
	17	2	4,5	0.1356	108.93	(5.607)
	18	2	4,6	0.0546	406.42	
	19	2	4,7	0.1585	74.70	6: 460.42
	20	2	4,8	0.0896	230.57	(0.251)
	21	2	4,9	0.1093	165.94	no remove
	22	2	4,10	0.2036	29.80	
	23	2	4,11	0.2154	21.16	
	24	2	4,12	0.2513	-0.15	
	25	2	4,13	0.2110	24.27	

$$\hat{y}(k) = (1.117 \pm 0.0249)\hat{y}(k-1) + (0.018 \pm 0.002165)u^2(k-1)$$
$$+(0.28 \pm 0.01751) + (0.07919 \pm 0.0014)u(k-1)\hat{y}(k-1)$$
$$+(-0.03689 \pm 0.00239)\hat{y}^2(k-1) + (0.01738 \pm 0.00248)\hat{y}^2(k-2)$$
$$+(-0.2177 \pm 0.0253)\hat{y}(k-2)$$

The computed model output signal and the output error are also plotted in Figure 5.9.8. The model, being linear in the parameters, approximates the input–output behavior of the simple Wiener model, being nonlinear in its parameters, very well.

TABLE 5.9.5b Stepwise regression of the simple Wiener model by
fitting a quadratic, second-order model liner-in-parameters

Step	No	Number of compon. (n)	Components	$\sigma^2_{\varepsilon_n}$	F to enter	Best model F to remove and $\sigma^2_{\varepsilon_{n-1}}$
	26	3	4,6,1	0.0545	0.23	best model:
	27	3	4,6,2	0.0366	61.43	4, 6, 2
	28	3	4,6,3	0.0516	7.38	4: 19 354.39
	29	3	4,6,5	0.048	17.46	(5.645)
	30	3	4.6,7	0.0550	-0.92	6: 90.76
3	31	3	4,6,8	0.0388	51.72	(0.0631)
	32	3	4,6,9	0.0400	46.36	2: 61.43
	33	3	4,6,10	0.0542	0.94	(0.0546)
	34	3	4,6,11	0.0537	2.13	no remove
	35	3	4,6,12	0.0548	-0.46	
	36	3	4,6,13	0.0532	3.34	
	37	2	6,2	5.645		
	38	4	4,6,2,1	0.0290	33.89	best model:
	39	4	4,6,2,3	0.0361	2.44	4, 6, 2, 1
	40	4	4,6,2,5	0.0355	4.61	4: 9363.10
	41	4	4,6,2,7	0.0368	0	(2.184)
	42	4	4,6,2,8	0.0344	8.79	6: 57.79(0.04)
4	43	4	4,6,2,9	0.0351	6.10	2: 110.79
	44	4	4,6,2,10	0.0363	1.74	(0.0545)
	45	4	4,6,2,11	0.0364	1.38	1: 33.89
	46	4	4,6,2,12	0.0331	14.08	(0.0368)
	47	4	4,6,2,13	0.0304	26.53	no remove
	48	3	4,2,1	0.0423		

If all possible models were to be fitted then $2^{13} - 1 = 8191$ regression must be done.
Applying stepwise regression, 125 parameter estimations were enough to estimate a
properly fitting model.

5.9.4 Regression analysis using orthogonal model components
By comparing the advantages and disadvantages of the forward, backward and stepwise
regression procedures one can draw the conclusion that both the number of regressions
and the parameter estimation algorithm itself could be reduced if orthogonal model
components had been applied. The advantages of the new procedure are as follows:
1. The estimation of the parameters of the orthogonalized model is easy and
   independent of each other. Entering a new model component does not alter the values
   of the previously estimated parameters.

TABLE 5.9.5c Stepwise regression of the simple Wiener model by fitting a quadratic, second-order model liner-in-parameters

Step	No	Number of compon. (n)	Components	$\sigma^2_{\varepsilon_n}$	F to enter	Best model F to remove and $\sigma^2_{\varepsilon_{n-1}}$
	50	5	4,6,2,1,3	0.029378	-1.61	best model:
	51	5	4,6,2,1,5	0.027126	8.64	4,6,2,1,8
	52	5	4,6,2,1,7	0.029378	-1.61	4: 14 625.30
	53	5	4,6,2,1,8	0.006195	460.15	(0.731025)
	54	5	4,6,2,1,9	0.00995	239.32	6: 29.62 (0.0076)
	55	5	4,6,2,1,10	0.027159	8.47	2: 57.75 (0.009057)
5	56	5	4,6,2,1,11	0.026471	11.94	1: 569.11 (0,034)
	57	5	4,6,2,1,12	0.02796	4.65	8: 460.15 (0.029)
	58	5	4,6,2,1,13	0.029036	-0.15	no remove
	59	4	6,2,1,8	0.731025		
	60	4	4,2,1,8	0.007663		
	61	4	4,6,1,8	0.009057		
	62	6	4,6,2,1,8,3	0.005389	18.55	best model:
	63	6	4,6,2,1,8,5	0.005927	5.60	4,6,2,1,8,12
	64	6	4.6,2,1,8,7	0.006107	1.78	4: 6032.10 (0.1192)
	65	6	4,6,3,1,8,9	0.005164	24.76	6: 37.86 (0.003134)
	66	6	4.6,2,1,8,10	0.006289	-1.84	2: 16.22 (0.002715)
6	67	6	4,6,2,1,8,11	0.006279	-1.66	1: 170.79 (0.005706)
	68	6	4.6,2,1,8,12	0.002401	195.94	8: 1320.00 (0.02796)
	69	6	4,6,2,1,8,13	0.003979	69.05	12: 195.94 (0.006195)
	70	5	6,2,1,8,12	0.1192		no remove
	71	5	4,2,1,8,12	0.003134		
	72	5	4,6,1,8,12	0.002715		
	73	5	4,6,2,8,12	0.005706		

2. The sum of squares owed to regression of a model having more components is the sum of squares owed to regression only if the individual components were included in the model.
3. As a consequence of items 1 and 2 the best model components to be considered for possible entry into the model can be computed by simple calculation of the sum of squares owed to regression for all model components not in the model, and choosing the one which results in the largest sum of squares owed to the regression value.
4. The different tests based on the residual sum of squares can be calculated without knowing the estimated parameters.
5. In consequence of the orthogonal property, between the model components there is no need for a stepwise regression, but a forward regression is sufficient. The formal application of the stepwise regression would result in a procedure that is equivalent to the forward regression.

TABLE 5.9.5d Stepwise regression of the simple Wiener model by fitting a quadratic, second-order model liner-in-parameters

Step	No	Number of compon. (n)	Components	$\sigma^2_{\varepsilon_n}$	F to enter	Best model F to remove and $\sigma^2_{\varepsilon_{n-1}}$
	74	7	4,6,2,1.8.12,3	0.002028	22.62	best model:
	75	7	4,6,2,1.8,12,5	0.0021169	16.51	4,6,2,1,8,12,13
	76	7	4.6.2.1.8.12,7	0.0021846	12.18	4: 6142.24 (0.1033)
	77	7	4,6.2.1,8,12,9	0.002403	-0.10	6: 44.76 (0.002766)
	78	7	4,6,2,1,8,12,10	0.0021828	12.30	2: 2.30 (0.002066)
	79	7	4.6.2.1.8.12,11	0.0021753	12.76	1: 227.56 (0.00578)
	80	7	4,6,2,1,8,12,13	0.0023416	3.12	8: 1593.42 (0.0283)
	81	6	6,2,1.8,12,3	0.1033		12: 203.85 (0.005389)
7	82	6	4,2,1.8,12,3	0.002766		3: 22.62 (0.002401)
	83	6	4,6,1.8,12,3	0,002066		remove component 2
						new model: 4,6,1,8,12,3
	84	6	4,6,2.8.12,3	0.00578		4: 6244.05 (0.1061)
	85	6	4,6,2.1.12,3	0.0283		6: 51.98 (0.002932)
	86	5	6,1,8,12,3	0.1061		1: 220.15 (0.005734)
	87	5	4,1,8,12,3	0.002932		8: 2676.86 (0.04666)
	88	5	4,6,8,12,3	0.005734		12: 247.16 (0.006184)
	89	5	4,6,1,12,3	0.04666		3: 38.95 (0.002715)
	90	5	4,6,1,8,3	0.006184		no remove
	91	7	4,6,1,8,12,3,5	0.002073	-0.42	best model:
	92	7	4,6,1,8,12,3,7	0.0020557	0.61	4,6,1,8,12,3,13
	93	7	4,6,1,8,12,3,9	0.0020223	2.66	4: 4975.28 (0.0820823)
	94	7	4,6,1,8,12,3,10	0.002079	3.56	6: 57.52 (0.0029063)
	95	7	4,6,1,8,12,3,11	0.0020331	1.99	1: 220.44 (0.0055294)
	96	7	4,6,1,8,12,3,13	0.0019803	5.32	8: 2788.32 (0.0468723)
8	97	6	6,1,8,12,3,13	0.0820823		12: 201.05 (0.0052172)
	98	6	4,1,8,12,3,13	0.0029063		3: 44.24 (0.0026926)
	99	6	4,6,8,12,3,13	0.0055294		13: 5.32 (0.002066)
	100	6	4,6,1,12,3,13	0.0468723		no remove
	101	6	4,6,1,8,3,11	0.0052172		
	102	6	4,6,1.8,12,13	0.0026926		

To obtain the orthogonal model's components, all components not in the model have to be orthogonalized to the components already in the model, at every step. Thus the components already in the model are orthogonal to each other, as well. Only the first step is an exception, the possible components for entering the model can be the original components, since there is no component in the model yet.

The procedure was originally suggested by Desrochers and Saridis (1978, 1980) and Desrochers (1981) and will be presented in a slightly modified form here.

Algorithm 5.9.1 The forward regression algorithm using orthogonal model components consists of the following steps (Desrochers and Saridis (1978, 1980) and

TABLE 5.9.5e Stepwise regression of the simple Wiener model by fitting a quadratic, second-order model liner-in-parameters

Step	No	Number of compon. (n)	Components	$\sigma^2_{\varepsilon_n}$	F to enter	Best model F to remove and $\sigma^2_{\varepsilon_{n-1}}$
	103	8	4,6,1,8,12,3,13,2	0.00199	-0.59	best model:
	104	8	4,6,1.8,12,3,13,5	0.0017007	20.06	4,6,1,8,12,3,13,5
	105	8	4,6,1.8,12,3,13,7	0.0019963	-0.90	4: 378.27 (0.0060739)
	106	8	4,6,1.8,12,3,13,9	0.0019643	0.99	6: 77.98 (0.0027878)
	107	8	4,6,1.8,12,3,13,10	0.0019625	1.11	1: 114.51 (0.003297)
	108	8	4,6,1.8,12,3,13,11	0.0019123	4.39	8: 3187.79 (0.0031013)
	109	7	6,1,8,12,3,13,5	0.0069739		12: 100.47 (0.0031013)
	110	7	4,1,8,12,3,13,5	0.0027878		3: -1.06 (0.0016859)
9	111	7	4,6,8,12,3,13,5	0.003297		13: 26.70 (0.0020729)
	112	7	4,6,1,12,3,13,5	0.046139		5: 20.06 (0.0019803)
	113	7	4,6,1,8,3,13,5	0.0031013		remove component 3
	114	7	4,6,1,8,12,13,5	0.0016859		new model:
	115	7	4,6,1,8,12,3,5	0.0020729		4,6,1,8,12,13,5
	116	6	6,1,8,12,13,5	0.02948		4: 2027.80 (0.02948)
	117	6	4,1,8,12,13,5	0.002628		6: 68.73 (0.002628)
	118	6	4,6,8,12,13,5	0.005209		1: 257.04 (0.005209)
	119	6	4,6,1,12,13,5	0.04580		8: 3218.48 (0.04580)
	120	6	4,6,1,8,12,5	0.0023513		12: 166.72 (0.003971) 13: 48.55 (0.0023513) 5: 73.45 (0.0026926) no remove
	121	8	4,6,1,8,12,13,5,2	0.0016966	-0.769	no more inclusion
	122	8	4,6,1,8,12,13,5,7	0.0016966	-0.769	best model: 4,6,1,8,12,13,5
10	123	8	4,6,1,8,12,13,5,11	0.0016966	-0.769	
	124	8	4,6,1,8,12,13,5,9	0.001681	0.35	
	125	8	4,6,1,8,12,13,5,10	0.0016711	1.0656	

Desrochers (1981)):

1. Denote the orthonormal components by $\phi^o_i(k)$ where i corresponds to the order of getting into the model. Orthonormalize all possible components not in the model to the model's components already there. (See Chapter 3 for details in orthogonalization.)

$$\phi^o_j(k) = \phi_j(k) - \sum_{i=1}^{n_1} \frac{\sum_{i=1}^{N} \phi_j(k)\phi_i(k)}{\sum_{i=1}^{N} \phi^2_i(k)} \phi_i(k), \qquad j = 1, \ldots, n_2, \qquad k = 1, \ldots, N$$

Here n_1 is the number of components already in the model and n_2 is the number of components considered for possible entry into the model. If n_0 is the total number of all possible components, then

$$n_\theta = n_1 + n_2$$

2. Calculate the sum of squares owed to regression for every orthogonal component not yet in the model

$$\Delta SS_j(1) = \frac{\left[\sum\limits_{k=1}^{N} y(k)\phi_j^o(k)\right]^2}{\sum\limits_{k=1}^{N}\left[\phi_j^o(k)\right]^2} \qquad j = 1,\ldots, n_2 \qquad (5.9.1)$$

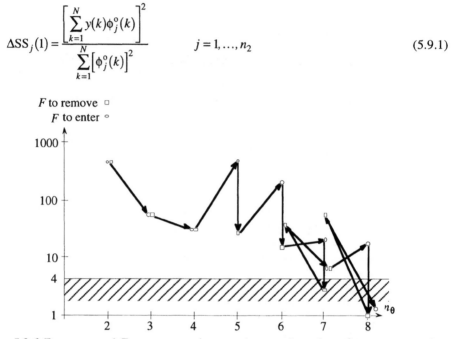

Fig. 5.9.6 *F to enter* and *F to remove* values vs. the actual number of parameters at the identification of the simple Wiener model with the stepwise regression. The arrow shows how a new component was added to the model (arrow points to the right) or how a component was eliminated from the model (arrow points to the left). The vertical position of the arrow shows only from the *F to enter* value to the smallest *F to remove* value of the same model

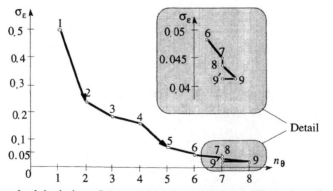

Fig. 5.9.7 Standard deviation of the residuals against the actual number of parameters at the identification of the simple Wiener model with the stepwise regression

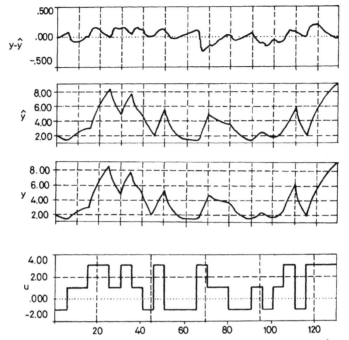

Fig. 5.9.8 Simulated input u, output y signals, the computed output $\hat{y}$ based on the estimated model, and the output error $(y - \hat{y})$ at the identification of the simple Wiener model with the stepwise regression

3. Select that component for possible entry to the model whose sum of squares owed to regression is the greatest value.
4. Enter the new component to the model if its F-value

$$F = \frac{SS(n_1 + 1) - SS(n_1)}{RSS(n_1 + 1)/(N - n_1 - 1)} = \frac{\Delta SS_{n_1 + 1}(1)}{\left[TSS - \sum_{i=1}^{n_1 + 1} \Delta SS_i(1)\right]/(N - n_1 - 1)}$$

is greater than the given threshold. Otherwise stop the algorithm.

5. If the new component has entered the model set

$$n_1 \quad \rightarrow \quad n_1 + 1$$
$$n_2 \quad \rightarrow \quad n_2 - 1$$

and repeat the algorithm from Step 1 until all model components have been checked for possible entry. ∎

Remarks:
1. The structure selection did not need parameter estimation.

2. The equation of the model whose structure has been chosen optimal is

$$y(k) = \phi^{oT}\theta^o = \phi^{*T}\theta^*$$

where the memory vector ϕ^o contains the orthogonal components

$$\phi^o(k) = \left[\phi_1^o(k), \ldots, \phi_{n_1}^o(k)\right]^T$$

and the memory vector ϕ^* contains the corresponding non-orthogonal components

$$\phi^*(k) = \left[\phi_1^*(k), \ldots, \phi_{n_1}^*(k)\right]^T$$

There are two ways of computing the parameters of the model:
- LS parameter estimation of the parameter vector θ^*;
- independent, orthogonal estimation of the parameters $\theta_j^o, j = 1, \ldots, n_1$;

and then the transformation of them to the parameter vector θ^* which has physical meaning. (See Section 3.3 for details.)

3. In the paper of Desrochers (1981) there are some tricks presented which make the computations easier.

Desrochers and Mosheni (1984) recommended a new algorithm for structure identification that seems to be different from the one described. However, the steps are the same, except one, Step 2, that has to be replaced by three new steps.

Algorithms 5.9.2 (Desrochers and Mosheni, 1984) The (forward) regression algorithm (by orthogonalization) has the following steps:
1. As 1 in Algorithm 5.9.1
2. Calculate the projection matrix P_i

$$P_i = \frac{\phi_i^o\phi_i^{oT}}{\phi_i^{oT}\phi_i^o} \qquad i = 1, \ldots, n_1$$

for the orthogonal components already in the model. Form the transformation matrix M_{n_1},

$$M_{n_1} = I - \sum_{i=1}^{n_1} P_i = I - \sum_{i=1}^{n_1} \frac{\phi_i^o\phi_i^{oT}}{\phi_i^{oT}\phi_i^o}$$

Transform the vector of output measurements by the matrix M_{n_1},

$$y^o = M_{n_1}y = \left[I - \sum_{i=1}^{n_1} \frac{\phi_i^o\phi_i^{oT}}{\phi_i^{oT}\phi_i^o}\right]y$$

The last transformation is the orthogonalization of the measurement record to all components being already in the model.

3. Project the orthogonalized output vector to all possible orthogonal components being no more in the model. The orthogonality means that the possible components are orthogonal to those being already in the model:

$$y^{*p} = P_j y^o = \frac{\phi_j \phi_j^T}{\phi_j^T \phi_j} \qquad j = 1, \ldots, n_2$$

4. Compute the scalar measure of the projected orthogonalized output vectors,

$$\Delta SS_j(1) = y^{opT} y^{op} \qquad j = 1, \ldots, n_2 \tag{5.9.2}$$

5-7. Steps 3-5 in Algorithm 5.9.1.   ∎

As it is not by chance that the scalar measure (5.9.2) is denoted $\Delta SS_j(1)$, we shall now prove that it is actually equivalent to the sum of squares owed to regression (5.9.1) for every orthogonal component not yet in the model.

Lemma 5.9.1
The scalar measure (5.9.2) is the sum of squares owed to regression (5.9.1).

Proof.

$$y^{opT} y^{op} = y^{oT} P_j^T P_j y^o = y^{oT} P_j y^o$$

because

$$P_j^T P_j = \frac{\phi_j \phi_j^T}{\phi_j^T \phi_j} \frac{\phi_j \phi_j^T}{\phi_j^T \phi_j} = \frac{\phi_j \phi_j^T}{\phi_j^T \phi_j} = P_j$$

Furthermore

$$y^{oT} P_j y^o = y^T \left[I - \sum_{i=1}^{n_1} \frac{\phi_i \phi_i^T}{\phi_i^T \phi_i} \right] \frac{\phi_j \phi_j^T}{\phi_j^T \phi_j} \left[I - \sum_{i=1}^{n_1} \frac{\phi_i \phi_i^T}{\phi_i^T \phi_i} \right] y$$

$$= y^T \frac{\phi_j \phi_j^T}{\phi_j^T \phi_j} y = \frac{\left[\sum_{k=1}^{N} y(k)\phi_j(k) \right]^2}{\sum_{k=1}^{N} \left[\phi_j(k) \right]^2} = \Delta SS_j(1)$$

which has the form of (5.9.1).   ∎

Thus it is proven that the Algorithm 5.9.1 of Desrochers (1981) and Algorithm 5.9.1 of Desrochers and Mosheni (1984) are identical, in spite of the different formalisms and computations.

Procedures were suggested and simulations were carried out by Desrochers and Saridis (1978, 1980), Desrochers (1981), Desrochers and Mohseni (1984), Janiszowski (1986),

Kortmann and Unbehauen (1987, 1988a), Zierfuss and Haber (1987), Korenberg *et al.*, (1988) and Keulers *et al.*, (1993).

Step	Component with the greatest ΔSS	ΔSS	F to enter
1	$y(k-1)$	533.150	
2	$u^2(k-1)$	25.220	290.56
3	$u(k-1)$	2.290	32.97
4	1	1.933	35.36
5	$u(k-1)y(k-1)$	3.892	162.27
6	$u(k-2)$	0.241	10.84
7	$y(k-2)$	0.156	7.38
8	$u^2(k-2)$	0.025	1.18

TABLE 5.9.6 Decrease of the residual sum of squares and F to enter values by regression analysis using orthogonal model components

Example 5.9.4 *Structure identification and parameter estimation of the quadratic, first-order simple Wiener model by second-order models linear-in-parameters using regression analysis and orthogonal model components (Continuation of Example 5.9.1, Zierfuss and Haber, 1987)*

The same input–output data pairs that were used in Examples 5.9.1 to 5.9.3 were identified. The possible model components were chosen as in Example 5.9.3. Table 5.9.6 show the model components in the sequence as they were involved in the model. (ΔSS means the decrease of the sum of squares of the residuals.) The resulting model contains seven components because the F to enter value became less than the chosen threshold (4) when the eighth component was included in the model. The equation of the resulting model with the orthogonal model components – denoted by $[\ ]^\circ$ – is as follows:

$$\hat{y}(k) = (1.009 \pm 0.0029)[\hat{y}(k-1)]^\circ + (0.1336 \pm 0.0037)[u^2(k-1)]^\circ$$

$$+(0.01540 \pm 0.0143)[u(k-1)]^\circ + (0.2790 \pm 0.0281)[1]^\circ$$

$$+(0.0678 \pm 0.0048)[u(k-1)\hat{y}(k-1)]^\circ + (0.0550 \pm 0.01568)[u(k-2)]^\circ$$

$$+(0.14130 \pm 0.0502)[\hat{y}(k-2)]^\circ$$

Instead of transforming the equation of the orthogonal model components into the resulting model with the original components, we estimated the parameters of the model with the selected but not orthogonal model components:

$$\hat{y}(k) = (-0.6377 \pm 0.0454)\hat{y}(k-1) + (0.02375 \pm 0.0039)u^2(k-1)$$

$$+(0.03597 \pm 0.01405)u(k-1) + (0.05939 \pm 0.02838)$$

$$+(0.06568 \pm 0.00318)u(k-1)\hat{y}(k-1) + (0.06622 \pm 0.01622)u(k-2)$$

$$+(0.07298 \pm 0.0357)\hat{y}(k-2)$$

The computed model output signal and the output error are also plotted in Figure 5.9.9. The model, being linear in the parameters, approximates the input–output behavior of the simple Wiener model nonlinear in the parameters, as well.

Comparing the estimated models of Examples 5.9.3 and 5.9.4 we can observe that both models have seven parameters but the model components selected by the different automatic search procedures were different.

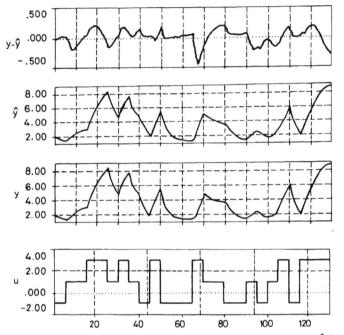

Fig. 5.9.9 Simulated input u, output y signals, the computed output $\hat{y}$ based on the estimated model, and the output error $(y - \hat{y})$ at the identification of the simple Wiener model with the regression analysis using orthogonal components

5.9.5 Term clustering
Aguirre and Billings (1995) called the sum of the estimated parameters of the same model components delayed differently as cluster coefficient. For example two different cluster components, one belonging to the linear and the second one to the nonlinear components, are given below:
(a)   linear cluster coefficient

$$\left(\hat{\theta}_1 + \hat{\theta}_2\right) \text{ results from } \hat{\theta}_1 u(k-1) + \hat{\theta}_2 u(k-2)$$

(b)   nonlinear cluster coefficient

$$\left(\hat{\theta}_3 + \hat{\theta}_4\right) \text{ results from } \hat{\theta}_3 u^2(k-1)y(k-1) + \hat{\theta}_4 u^2(k-2)y(k-3)$$

Those components are significant whose cluster coefficient is significantly larger than that of the others.

5.9.6 Genetic algorithm

Genetic algorithm is a multi-dimensional search with randomly generated initial selection and with a systematical modification of the chosen components. Genetic algorithms were applied for the selection of the components in the following models linear-in-parameters:
- polynomial difference equation model (e.g., Li and Jeon, 1993);
- Volterra series model (Yao, 1996).

5.10 GROUP METHOD OF DATA HANDLING (GMDH)

Ivakhnenko (1970, 1971) introduced the Group Method of Data Handling (GMDH) which is an automatic, so-called self-organization method. GMDH can be used to build up a hierarchical multi-level model that can be transformed into a usual input–output equation description.

The model has a cascade-like structure with one input and one output signal, but more parallel paths between them. The difference from the usual cascade structure is that the submodels may have more input signals and that they are the output signals of the previous submodels in the parallel paths. Because of this cascade-like, but not cascade, structure the model is called hierarchical. The hierarchy of the model is shown by the fact that each subsequent submodel can be estimated only in the knowledge of the output signals of the previous submodels. The submodels estimated from the same set of possible input signals (components) build a layer. The complexity of the model is characterized by the maximal number of submodels in the *longest* path. The serial number of the submodels in this path is called the serial number of the layers. Each layer consists not only of the corresponding submodel in the *longest* path, but, in addition to it, all submodels having the same set of input signals as the submodel are considered in that layer.

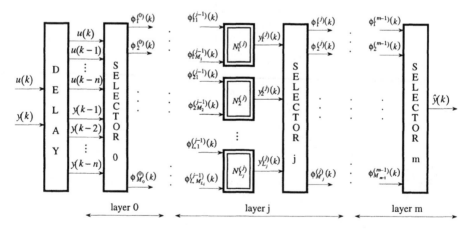

Fig. 5.10.1 Scheme of the group method of data handling (GMDH)

In each layer submodel with simple structures are assumed. The task of the lowest (zero-th) selector is to choose the possible model components of the first layer. The model components will be linear and nonlinear functions of the delayed input and output signals,

$$\phi^{(0)}(k) = \phi^{(0)}(k)\big(u(k), y(k)\big)$$

$$= \phi^{(0)}\big(u(k), \ldots, u(k-n), y(k-1), \ldots, y(k-n)\big)$$

At the first layer several models will be estimated which fit approximately to the measured output signal. The output signals of the models are functions of the model components chosen earlier,

$$y_i^{(1)}(k) = \phi_i^{(1)}\Big(\phi_1^{(0)}(k), \ldots, \phi_{M_0}^{(0)}(k)\Big)$$

The components of the submodels of the other layers are functions of the computed output signals of the estimated submodels of the previous layers. They are denoted by $\phi_i^{(j-1)}(k)$, which means the ith model component of the jth layer. The last layer is denoted by m and the number of components of the jth layer by M_j. Thus $0 \le j \le m$ and $1 \le i \le M_j$. The hierarchical structure of the model is seen in Figure 5.10.1.

To understand the philosophy of the hierarchical model consider an example.

Example 5.10.1   *Hierarchical description of nonparametric linear, generalized Hammerstein and Volterra models with memory 5.*
If the memory is 5, the input–output model has the form

$$y(k) = \phi\big(u(k)\big) = \phi\big(u(k-1), \ldots, u(k-5)\big)$$

The linear model

$$y(k) = h_0 + \sum_{i=1}^{5} h_1(i)u(k-i)$$

has 6 unknown parameters. The generalized Hammerstein model

$$y(k) = h_0 + \sum_{i=1}^{5} h_1(i)u(k-i) + \sum_{i=1}^{5} h_2(i,i)u^2(k-i)$$

has 11 unknown parameters, and the Volterra series

$$y(k) = h_0 + \sum_{i=1}^{5} h_1(i)u(k-i) + \sum_{i=1}^{5}\sum_{j=i}^{5-i} h_2(i,j)u(k-i)u(k-j)$$

has 21 unknown parameters.

Assume that the part models must not contain more than 6 parameters. Then the nonlinear models have to be reconstructed in a hierarchical structure.

Make two partitions in the Hammerstein model: the linear part including the constant term

$$y_1^{(1)}(k) = h_0 + \sum_{i=1}^{5} h_1(i)u(k-i)$$

and the quadratic part containing the quadratic terms

$$y_2^{(1)}(k) = \sum_{i=1}^{5} h_2(i, i) u^2(k - i)$$

The final model results in the second level as follows:

$$y(k) = y_1^{(2)}(k) = x_1^{(1)}(k) + x_2^{(1)}(k)$$

where the $x_i^{(1)}(k)$ input signals have to be chosen from the output signals of the first level,

$$x_1^{(1)}(k) = y_1^{(1)}(k)$$
$$x_2^{(1)}(k) = y_2^{(1)}(k)$$

The Volterra series has $4 + 3 + 2 + 1 = 10$ parameters, more than the Hammerstein model. Two further submodels have to be set up at the first level to cover these components:

$$y_3^{(1)}(k) = \sum_{i=1}^{4} h_2(i, i+1) u(k-i) u(k-i-1)$$

and

$$y_4^{(1)}(k) = \sum_{j=2}^{4} \sum_{i=1}^{5-j} h_2(i, i+j) u(k-i) u(k-i-j)$$

The measured output signal can be achieved by a four-variable model at the second level

$$y(k) = y_1^{(2)}(k) = \sum_{i=1}^{4} x_i^{(1)}(k)$$

where

$$x_i^{(1)}(k) = y_i^{(1)}(k) \quad i = 1, 2, 3, 4 \qquad \blacksquare$$

The example shows the hierarchical structure of complex, nonlinear dynamic processes. The submodels are described by only a few parameters. The GMDH gives an automatic, so-called self-organization method of building up a hierarchical model from the input–output measurement. Both the hierarchical structure and the estimation of the submodels will be presented simultaneously.

The parameter estimation of the submodels is performed by fitting their output signals to the measured output signal. Only those model outputs are considered for possible input signals for the next layer which fit the measured output signal to a certain extent. The output signal with the best fit of the submodels in the last layer is chosen as the computed model output signal. There are several criteria known for the determination of the number of layers. Experts in GMDH state that it is principally important to use different criteria for sorting out the submodels and performing the parameter estimation.

In the sequel the method is introduced in detail.

Algorithm   5.10.1   (Ivakhnenko (1970, 1971)) The GMDH contains the following steps:

1. The SISO model is built up from the delayed input

$$u(k), u(k-1), \ldots, u(k-n)$$

and output signals

$$y(k-1), y(k-2), \ldots, y(k-n)$$

The choice of the maximal memory n is similar to that of determining the maximal order of a linear system.

On the basis of the delayed input and output signals, the possible input signals $\phi_i^{(0)}(k)$ of the models at the first level can be determined. If the task is to fit a linear overall model, then they are equal to the delayed input and output signals

$$\phi_i^{(0)}(k) = \{1, u(k-\ell), y(k-j)\} \qquad 0 \le \ell \le n \qquad 1 \le j \le n$$

otherwise nonlinear terms (e.g., square of a signal, cross-product terms, etc.) can occur. The choice of the outputs of the zero-th selector and the structures of the submodels of the different levels are closely related to the structure of the resulting model. If we restrict ourselves to a model quadratic in the input–output signals, then $\phi_i^{(0)}$ should contain all linear and quadratic terms possible to derive them from the shifted input and output signals

$$\phi_i^{(0)}(k) = \{1, u(k-\ell), y(k-j), u(k-\ell)u(k-j), y(k-\ell)y(k-j), u(k-\ell)y(k-j)\}$$
$$0 \le \ell \le n \qquad 1 \le j \le n$$

2. Let us set the model components after layer zero $\left(\phi_i^{(0)}(k); \ i = 1, \ldots, M_0\right)$ equal to the input components of layer j $\left(\phi_i^{(j-1)}(k); \ i = 1, \ldots, M_{j-1}\right)$,

$$\phi_i^{(j-1)}(k) = \phi_i^{(0)}(k) \qquad i = 1, \ldots, M_0 = M_{j-1}$$

3. Estimate L_j submodels with few model components

$$\hat{y}_i^{(j)}(k) = \sum_{\ell=1}^{M_i^{(j)}} \theta_{i_\ell}^{(j)} \phi_{i_\ell}^{(j)}(k)$$

by fitting the models to the measured output signal

$$\sum_{k=1}^{N} \left[y(k) - \hat{y}_i^{(j)}(k)\right]^2 \Rightarrow \min_{\theta_i^{(j)}}$$

where the parameter vector contains the unknown coefficients

$$\theta_i^{(j)} = \left[\theta_{i_1}^{(j)}, \theta_{i_2}^{(j)}, \ldots, \theta_{i_{M_i}}^{(j)}\right]$$

If M_{j-1} signals passed the $(j-1)$th selector and each model may contain n_{θ_i} parameters (input signals), then

$$L_j = \binom{M_j}{n_{\theta_i}}$$

models can be fitted. The output signals

$$\hat{y}_i^{(j)}(k) \qquad i = 1, \ldots, L_j$$

approximate the measured output signal to a certain extent.

4. If the best calculated model outputs are accurate enough, consider the best one, or the mean value of the best ones, as the model output. Stop the algorithm.

5. Consider the best output signals of the best models as the output components of the selector j. They are denoted by

$$\phi_i^{(j)}(k) \qquad i = 1, \ldots, M_j$$

6. Set the input components of the layer $(j+1)$ equal to the output components of selector j.

7. Repeat Steps 3 to 5 until the stop condition in Step 4 is satisfied or other terminating conditions become active.   ∎

There are several ways of finding out which layer is the last one, i.e., when the algorithm has to be terminated:
- The number of layers can be predetermined ($m = 3 \div 4$ are usual values).
- The output signals of submodels of the m-th layer $\hat{y}_i^{(m)}(k)$ are almost the same, i.e., they are highly correlated to each other and to the measured output signal;
- Although the submodels of the m-th layer have more input signals, only one of them behaves significantly, i.e., the resulting computed output signal based on the estimated model does not differ from an output signal in the previous layer;
- The performance index of the parameter estimation of the normalized loss function does not decrease from the increase of the level of the hierarchy.

The main advantages of the method are as follows:
- The submodels have much fewer parameters than a complex model would need. The estimation of a few parameters can be performed more easily since the parameter estimation of too many unknown parameters causes numerical problems.
- Generally, more input–output data pairs are needed for the identification than the number of unknown parameters. Since only submodels with only a few unknown parameters are estimated, fewer measured data are enough.

The algorithm presented is the basic algorithm of the GMDH. There are several papers in the literature that suggest different modifications to improve the classical GMDH. Most of them are just summarized in the form of the following remarks (Haber and Unbehauen, 1990):

1. Ivakhnenko (1971) suggested the use of quadratic models with two input signals at every layer.
2. Ivakhnenko and Müller (1984) suggested the use of different submodels such as:
    *   linear ones;
    *   quadratic ones having only cross-product terms of the input signals;
    *   rational transfer functions, etc.
3. Ikeda *et al.,* (1976) showed that the convergence of the iteration procedure could be improved by considering also the measured input signal and its delayed values as possible input signals of the submodels in each layers.
4. Yoshimura *et al.,* (1982a), Yoshimura *et al.,* (1985) suggested computing the correlation coefficients between the possible input variables of the model and the output signal, and selecting only those limited number of signals for submodel inputs which have the largest correlation coefficients.
5. Mital (1984), Duffy and Franklin (1975), Yoshimura *et al.,* (1982a) suggested estimating submodels at higher levels, not by structures fixed in advance, but by a structure identification method (e.g., stepwise regression) to obtain a significant subset of the allowed structure.
6. Shimizu and Nishikawa (1984) applied the ridge estimation (e.g., Draper and Smith, 1966) to stabilize the GMDH algorithm.
7. Yoshimura *et al.,* (1982b) compared the different performance indices of parameter estimation. Experience showed that usage of the criteria AIC and PRESS (prediction sum of squares) resulted in a smaller number of layers than by using RSS. The model was more accurate, however, when applying RSS. In practice, one aims to obtain a model that is accurate and has relatively few parameters; in this respect the criterion AIC proved to be the best.
8. Mehra (1977) recommended the use of the BIC criterion.
9. Haber and Perényi (1987) proposed choosing only the most significant model components from the possible linear and nonlinear components as the components of the submodels at each layer. They estimated all possible one-parameter models and ranked them according to the residual sum of squares. In addition, they reduced the computations by building models only from the neighboring components. This suboptimal strategy led to acceptable results in several simulated cases.
10. Kortmann and Unbehauen (1988b) proposed preselecting the possible linear and nonlinear components of the submodels such that only those that were least linearly dependent (correlated to each other) were used for estimating the submodels. They did this by building orthogonal model components, because the selection of the most significant model components was then an easy task.
11. The three selection criteria (regularity, unbiased and combined) suggested in Farlow (1984) are compared in Ravindra *et al.,* (1994). Aksenova (1995) compared six different quadratic loss functions for model selection.
12. Ivakhnenko and Müller (1994) applied linear programming for selecting the components in the GMDH structure.

In most of the papers on GMDH the submodels have two input signals and constant, linear and quadratic terms. This fact leads to nonlinear models of very high degree if more layers are used for an accurate fitting. Sometimes we know that the model is quadratic, then it would be better to use linear submodels at least from the second layer. The best method seems to consider all possible model components at the zero-th layer and then to use only linear MISO submodels at every further layer. Then the model can contain only those components that were allowed at the zero-th layer. This procedure was shown in Example 5.10.1 and will be shown later in the simulation. There is a big disadvantage in the method if at every layer all possible submodels with the given structure are estimated. Table 5.10.1 shows how many linear MISO submodels can be assumed at each layer as a function of the possible and actual input signals.

Assume, e.g., a Volterra series model up to memory $m = 4$. Then the model's equation is

$$y(k) = h_0 + h_1(1)u(k-1) + h_1(2)u(k-2) + h_1(3)u(k-3) + h_1(4)u(k-4)$$
$$+ h_2(1,1)u^2(k-1) + h_2(2,2)u^2(k-2) + h_2(3,3)u^2(k-3) + h_2(4,4)u^2(k-4)$$
$$+ h_2(1,2)u(k-1)u(k-2) + h_2(2,3)u(k-2)u(k-3)$$
$$+ h_2(3,4)u(k-3)u(k-4) + h_2(1,3)u(k-1)u(k-3)$$
$$+ h_2(2,4)u(k-2)u(k-4) + h_2(3,4)u(k-3)u(k-4)$$

The number of possible model components is 15 at the first layer. If the submodels are linear MISO systems with 6 inputs (components), then the number of the submodels is

$$\binom{15}{6} = 5005$$

as is seen in Table 5.10.1, row 15, column 6.

There seem to be two ways to overcome the problem:

- All possible model components have to be taken into account at the zero-th layer, but not all possible linear MISO submodels have to be considered;
- The possible components at the zero-th layer should be reduced and nonlinear submodels should be allowed.

It will be shown that not only does the first method imply some heuristics but the second one does also. In the second case let the possible model components be at the zero-th layer only

$$u(k-1), u(k-2), u(k-3), u(k-4)$$

and the structure of the possible submodels in the first layer be

$$y(k) = c_0 + c_1 u(k-i) + c_2 u(k-j) + c_3 u^2(k-i)$$
$$+ c_4 u^2(k-j) + c_5 u(k-i)u(k-j)$$
$$i, j = 1, 2, 3, 4 \qquad\qquad i \ne j$$

In this way the number of submodels and, therefore, the regression procedures needed, reduces to

$$\binom{4}{2} = 6$$

which is shown in Table 5.10.1, row 4, column 2. Through the above procedure all components of the Volterra series model are taken into account, but in fewer combinations than in the previous case when all – also the nonlinear – components were considered at the zero-th layer and, further, only linear submodels were assumed.

For example, the first strategy would consider the submodel

$$y(k) = c_0 + c_1 u(k-1) + c_2 u(k-3) + c_4 u^2(k-2)$$
$$+ c_5 u^2(k-4) + c_6 u(k-1)u(k-4)$$

but this submodel does not appear among the possible submodels in the second strategy. Therefore we can state that the second strategy is a mode of reduction of the number of submodels to be estimated, and it implies many heuristics.

If in the submodels used at the second strategy the cross-product term is omitted, then the Volterra series model restricts itself to the Hammerstein weighting function series model.

It is practical to apply only linear MISO submodels at the second and higher layers, if the model is assumed to be quadratic.

GMDH is then mainly effective when too many parameters would be needed to describe the model by another model structure, e.g., by a model linear-in-parameters. To illustrate this the following examples treat nonparametric modeling.

Example 5.10.2 *Identification of the simple Hammerstein model in nonparametric form (Haber and Perényi, 1987)*

The equation of the static part is

$$v(t) = 2 + u(t) + u^2(t)$$

and the transfer function of the linear dynamic part is

$$G(s) = \frac{1}{1 + 10s}$$

The GMDH method has advantages over other procedures if too many parameters have to be estimated. Therefore not a parametric but a nonparametric Hammerstein model was assumed. To describe the parametric process by a nonparametric process with a few parameters accurately enough, the sampling time was chosen equal to the time constant $\Delta T = T = 10$ [s]. Then the equivalent pulse transfer function of the continuous transfer function is

$$H(q^{-1}) = \frac{0.6321 q^{-1}}{1 - 0.3679 q^{-1}}$$

The weighting function series of $H(q^{-1})$ were truncated after the fourth term

$$H(q^{-1}) = 0.6321 q^{-1} + 0.2325 q^{-2} + 0.08554 q^{-3} + 0.03147 q^{-4}$$

because further terms would have very small values. To eliminate the truncation error the process itself was simulated by the nonparametric model. The equation of the whole model is

$$y(k) = 1.96332 + 0.6321u(k-1) + 0.2325u(k-2)$$

$$+0.08554u(k-3) + 0.03147u(k-4) + 0.31605u^2(k-1)$$

$$+0.11625u^2(k-2) + 0.04277u^2(k-3) + 0.015735u^2(k-4)$$

which is a subset of a Volterra series. The model components ϕ_i and the true parameters θ_i are summarized in Tables 5.10.2 and 5.10.9.

TABLE 5.10.1 The number of submodels at a layer as a function of the possible and actual number of input signals

number of possible inputs \ number of inputs	1	2	3	4	5	6	7	8	9	10	11	12	13	14	15
1	1														
2	2	1													
3	3	3	1												
4	4	6	4	1											
5	5	10	10	5	1										
6	6	15	20	15	6	1									
7	7	21	35	35	21	7	1								
8	8	28	56	70	56	28	8	1							
9	9	36	84	126	126	84	36	9	1						
10	10	45	120	210	252	210	120	45	10	1					
11	11	55	165	330	462	462	330	165	55	11	1				
12	12	66	220	495	792	924	792	495	220	66	12	1			
13	13	78	286	715	1287	1716	1716	1287	715	286	78	13	1		
14	14	91	364	1001	2002	3003	3432	3003	2002	1001	364	91	14	1	
15	15	105	455	1365	3003	5005	6435	6435	5005	3003	1365	455	105	15	1

The test signal was a PRTS with maximum length 26, amplitude 2 and mean value 1. The sampling time was $\Delta T = 2$ [s] and the minimum switching time of the PRTS signal was 5 times more. $N = 26 \cdot 5 = 130$ data pairs were used for the identification.

Before starting the GMDH all possible model components were regressed with the output signal which means that 9 different submodels were estimated, each of them having only one model component. The results are summarized in Table 5.10.3. On the basis of the estimated standard deviations of the residuals σ_{ε_i} the model components ϕ_i could be arranged in series. The column *sequence* shows it starting with the best fitting component. This knowledge about the components was used later to reduce the computations.

Further on, different approximating models are presented.

Common in the procedures (a) to (e) is that all possible components of the models were chosen already at the zero-th layer, and, further on *linear* MISO static models were estimated. This kind of procedure would need too many computations if all possible submodels were to be estimated. Instead of doing that, another strategy was applied: only the neighboring components were used as inputs of the submodels. This *suboptimal* strategy led to acceptable results and required only a few computations.

(a) *model a:*

All possible model components were used. The first layer consists of two submodels, the first one contains the constant term and all linear terms

$$\hat{y}_1^{(1)}(k) = 2.815 + 1.281u(k-1) + 0.465u(k-2) + 0.1711u(k-3) + 0.04088u(k-4)$$

and the second submodel the quadratic terms

$$\hat{y}_2^{(1)}(k) = 0.6751u^2(k-1) + 0.2008u^2(k-2) + 0.07388u^2(k-3) + 0.1597u^2(k-4)$$

The second, final, layer uses the outputs of the first layers

$$\hat{y}(k) \equiv \hat{y}_1^{(2)}(k) = 0.6085\hat{y}_1^{(1)}(k) + 0.4208\hat{y}_2^{(1)}(k)$$

Although only three regressions were used, the approximating model is very good. The good coincidence between the measured (simulated) output signal and the computed output signal – based on the estimated model – is seen in Figure 5.10.2. The whole identification procedure is summarized in Table 5.10.4, where the selected model components of the zero-th layer $\phi_i^{(0)}$, the estimated parameters $\hat{\theta}_i^{(j)}$ at each layer, and the corresponding standard deviation of the residuals $\sigma_{\varepsilon_i}^{(j)}$ are presented. The last column $\hat{\theta}_i$ contains the resulting estimated parameters belonging to the components of the original model of the zero-th layer. They can be calculated by replacing the model inputs of the layers by output signals of the submodels of the lower layers subsequently.

(b) *model b:*
The only difference from the case (a) is that the *worst* model component according to Table 5.10.3 was omitted from the model, i.e., not used as a possible input at the first layer. The estimated equations of the model are as follows:

First layer:

$$\hat{y}_1^{(1)}(k) = 2.832 + 1.275u(k-1) + 0.465u(k-2) + 0.2055u(k-3)$$
$$\hat{y}_2^{(1)}(k) = 0.6751u^2(k-1) + 0.2008u^2(k-2) + 0.07388u^2(k-3) + 0.1597u^2(k-4)$$

Second layer:

$$\hat{y}(k) = \hat{y}_1^{(2)}(k) = 0.6074\hat{y}_1^{(1)}(k) + 0.4220\hat{y}_2^{(1)}(k)$$

The good coincidence between the measured (simulated) output signal and the computed output signal – based on the estimated model – is seen in Figure 5.10.2. The whole identification procedure is summarized in Table 5.10.5.

(c) *model c:*
Similarly to case (b) the first layer has two linear submodels, each of them having four input signals. The difference is, however, that now the model components of the zero-th layer were selected according to the *sequence of importance* given in Table 5.10.3. The *least significant* component $u(k-4)$ was omitted from the model. The model equations are:

First layer:

$$\hat{y}_1^{(1)}(k) = 0.5464u^2(k-1) + 0.2008u^2(k-2) + 0.3576u(k-1) + 0.2278u^2(k-3)$$

$$\hat{y}_2^{(1)}(k) = 1.625u(k-2) + 2.935 + 0.08971u^2(k-4) - 0.07929u(k-3)$$

Second layer:

$$\hat{y}(k) = \hat{y}_1^{(2)}(k) = 0.57654\hat{y}_1^{(1)}(k) + 0.4643\hat{y}_2^{(1)}(k)$$

The measured (simulated) input, output, and computed output signals – based on the estimated model – are seen in Figure 5.10.2. The whole identification procedure is summarized in Table 5.10.6.

(d) *model d:*
The model components of the first layer were selected according to the *sequence of importance* given in Table 5.10.3. In all submodels at the first and subsequent layers only two model components – input signals – were assumed. This fact led to more (4) layers than in cases (a) to (c), where the submodels had up to four model components, and, therefore, two layers gave an accurate approximating model.
    The estimated model equations are as follows:

First layer:

$$\hat{y}_1^{(1)}(k) \equiv u^2(k-1)$$

$$\hat{y}_2^{(1)}(k) = 0.6998u^2(k-2) + 1.047u^2(k-1)$$

$$\hat{y}_3^{(1)}(k) = 0.601u^2(k-3) + 1.202u(k-2)$$

$$\hat{y}_4^{(1)}(k) = 3.144 + 0.4614u^2(k-4)$$

$$\hat{y}_5^{(1)}(k) = 2.177u(k-3) + 1.473u(k-4)$$

Second layer:

$$\hat{y}_1^{(2)}(k) \equiv \hat{y}_1^{(1)}(k)$$

$$\hat{y}_2^{(2)}(k) \equiv \hat{y}_2^{(1)}(k)$$

$$\hat{y}_3^{(2)}(k) \equiv \hat{y}_3^{(1)}(k)$$

$$\hat{y}_4^{(2)}(k) = 0.7959\hat{y}_4^{(1)}(k) + 0.2085\hat{y}_3^{(1)}(k)$$

Third layer:

$$\hat{y}_1^{(3)}(k) = 0.5191\hat{y}_1^{(2)}(k) + 0.5203\hat{y}_2^{(2)}(k)$$

$$\hat{y}_2^{(3)}(k) = 0.5428\hat{y}_3^{(2)}(k) + 0.5238\hat{y}_4^{(2)}(k)$$

Fourth layer:

$$\hat{y}(k) \equiv \hat{y}_1^{(4)}(k) = 0.6837\hat{y}_1^{(3)}(k) + 0.3707\hat{y}_2^{(3)}(k)$$

The resulting model is worse than the previous ones (a) to (c). This fact can be seen in Figure 5.10.2, where the measured (simulated) input, output, and computed output signals – based on the estimated model – are plotted and from the higher standard deviation of the residuals presented in Table 5.10.7.

(e) *model e:*
This model has the same model components in the zero-th layer as model (d). From the four submodel outputs at the first layer only the best three are used for possible inputs to the next (third) layer. This fact led to the reduction of the number of layers and gave a somewhat better standard deviation of the residuals than case (d).
     The estimated equations are as follows.

First layer:

$$\hat{y}_1^{(1)}(k) \equiv u^2(k-1)$$

$$\hat{y}_2^{(1)}(k) = 0.6998u^2(k-2) + 1.047u(k-1)$$

$$\hat{y}_3^{(1)}(k) = 0.601u^2(k-3) + 1.202u(k-2)$$

$$\hat{y}_4^{(1)}(k) = 3.144 + 0.4614u^2(k-4)$$

$$\hat{y}_5^{(1)}(k) = 2.177u(k-3) + 1.473u(k-4)$$

Second layer:

$$\hat{y}_1^{(2)}(k) = 0.5191\hat{y}_1^{(1)}(k) + 0.5203\hat{y}_2^{(1)}(k)$$

$$\hat{y}_2^{(2)}(k) = 0.6007\hat{y}_3^{(1)}(k) + 0.5054\hat{y}_4^{(1)}(k)$$

Third layer:

$$\hat{y}(k) \equiv \hat{y}_1^{(3)}(k) = 0.6399\hat{y}_1^{(2)}(k) + 0.4146\hat{y}_2^{(2)}(k)$$

The measured (simulated) input, output and computed output signals – based on the estimated model – are seen in Figure 5.10.2. The whole identification procedure is summarized in Table 5.10.8.
     The estimated model parameters and standard deviations of the residuals are summarized in Table 5.10.2. Because of the nature of the GMDH the estimated parameters should not be equal to the original parameters since some model components were omitted. The best models are (a) and (b), which have the smallest standard deviation of the residuals. The estimated parameters in the above two cases are not too far from the original ones.
     The efficiency of the GMDH could be increased if *nonlinear* submodels were also allowed. The input signals of the first layers were all combinations of

$$u(k-i), u(k-j) \qquad i = 1, 2, 3, 4 \qquad j = 1, 2, 3, 4 \qquad i \neq j$$

and the submodels of the first layers were static quadratic polynomials with the above input signals. The second, last, layer contained then a linear, static, MISO model. To do

TABLE 5.10.2 True, θ_i, and estimated, $\hat{\theta}_i$, parameters, and estimated standard deviations of the residuals σ_ε at the identification of the nonparametric simple Hammerstein model by quasi-linear submodels at the first layer

No	ϕ_i	θ_i	$\hat{\theta}_i$ (case (a))	$\hat{\theta}_i$ (case (b))	$\hat{\theta}_i$ (case (c))	$\hat{\theta}_i$ (case (d))	$\hat{\theta}_i$ (case (e))
1	1	1.96322	1.713	1.702	1.3627	0.4858	0.6588
2	$u(k-1)$	0.6321	0.7795	0.7744	0.2165	0.3724	0.3486
3	$u(k-2)$	0.2325	0.283	0.2824	0.7545	0.2418	0.2993
4	$u(k-3)$	0.08554	0.1041	0.1248	-0.03681	0.08813	-
5	$u(k-4)$	0.03147	0.02488	-	-	0.05963	-
6	$u^2(k-1)$	0.31605	0.2841	0.2849	0.315	0.3549	0.3322
7	$u^2(k-2)$	0.11625	0.0845	0.08474	0.1158	0.2489	0.2330
8	$u^2(k-3)$	0.04277	0.0311	0.03118	0.1313	0.1209	0.1497
9	$u^2(k-4)$	0.015735	0.0672	0.06739	0.04165	0.0713	0.0967
σ_ε			0.3266	0.3180	0.6612	1.01	0.9175

TABLE 5.10.3 The sequence of importance of the possible model components at the identification of the nonparametric simple Hammerstein model

No	ϕ_i	$\hat{\theta}_i$	σ_{ε_i}	sequence
1	1	4.687 ±0.2783	3.173	6
2	$u(k-1)$	2.5680 ±0.1263	2.770	3
3	$u(k-2)$	2.496 ±0.1422	3.089	5
4	$u(k-3)$	2.294 ±0.1714	3.688	8
5	$u(k-4)$	2.036 ±0.1998	4.258	9
6	$u^2(k-1)$	1.021 ±0.02747	1.660	1
7	$u^2(k-2)$	1.007 ±0.03427	2.047	2
8	$u^2(k-3)$	0.9529 ±0.04736	2.796	4
9	$u^2(k-4)$	0.8804 ±0.05954	3.475	7

real structure identification the cross-product terms $u(k-i), u(k-j), i \neq j$ were first omitted and later they were included in the model. If the model fitting is complete without the cross-product terms then the system is of Hammerstein type.

(f)   *model f:*
The memory of the model was 4. At the first layer 6 quadratic submodels were assumed. The input signals of the submodels were differently delayed input signal values. The cross-product terms were omitted from the quadratic models. The estimated submodels were as follows:

$$\hat{y}_1^{(1)}(k) = 2.050 + 0.6273u(k-1) + 0.3422u(k-2) + 0.3045u^2(k-1) + 0.1632u^2(k-2)$$

$$\hat{y}_2^{(1)}(k) = 1.984 + 0.7473u(k-1) + 0.2316u(k-3) + 0.3714u^2(k-1) + 0.1138u^2(k-3)$$

$$\hat{y}_3^{(1)}(k) = 1.963 + 0.8156u(k-1) + 0.1660u(k-4) + 0.4078u^2(k-1) + 0.0830u^2(k-4)$$

$$\hat{y}_4^{(1)}(k) = 2.383 + 0.8015u(k-2) + 0.05027u(k-3) + 0.3974u^2(k-2)$$
$$+0.02365u^2(k-3)$$

$$\hat{y}_5^{(1)}(k) = 2.413 + 0.8390u(k-2) + 0.001865u(k-4) + 0.4164u^2(k-2)$$
$$-0.0001602u^2(k-4)$$

$$\hat{y}_6^{(1)}(k) = 2.918 + 0.8038u(k-3) - 0.1233u(k-4) + 0.3962u^2(k-3)$$
$$-0.06317u^2(k-4)$$

TABLE 5.10.4 The estimated parameters and standard
deviations of the residuals at the identification of the
nonparametric simple Hammerstein model by structure (a)

No	$\phi_i^{(0)}$	$\hat{\theta}_i^{(1)}$	$\sigma_{\varepsilon_i}^{(1)}$	$\hat{\theta}_i^{(2)}$	$\sigma_{\varepsilon_i}^{(2)}$	$\hat{\theta}_i$
1	1	2.815 ±0.09299				1.713
2	$u(k-1)$	1.281 ±0.0802				0.7795
3	$u(k-2)$	0.465 ±0.1067	0.8537	0.6085 ±0.01486		0.283
4	$u(k-3)$	0.1711 ±0.1067				0.1041
5	$u(k-4)$	0.04088 ±0.08022				0.02488
6	$u^2(k-1)$	0.6751 ±0.04835			0.3266	0.2841
7	$u^2(k-2)$	0.2008 ±0.06685	1.254	0.4208 ±0.01507		0.0845
8	$u^2(k-3)$	0.07388 ±0.06685				0.0311
9	$u^2(k-4)$	0.1597 ±0.04896				0.0672

At the second layer first all output signals of the 6 submodels were considered as possible input signals of a linear static submodel. The parameter estimation resulted in the following equation

$$\hat{y}(k) \equiv \hat{y}_1^{(2)}(k) = -0.6191\hat{y}_1^{(1)}(k) + 1.3890\hat{y}_2^{(1)}(k) + 0.008179\hat{y}_3^{(1)}(k)$$
$$-0.3828\hat{y}_4^{(1)}(k) + 0.9180\hat{y}_5^{(1)}(k) - 0.2690\hat{y}_6^{(1)}(k)$$

The parameters of the estimated model and the standard deviations of the residuals at the regression of the submodels are summarized in Table 5.10.10. The estimated parameters are also shown in Table 5.10.9, where the parameters of the original (simulated) system are also given for comparison. The estimated model is very good; its parameters approximate the true parameters very well, as is seen also on the plot of the computed output signal of the model in Figure 5.10.2.

TABLE 5.10.5 The estimated parameters and standard deviations of
the residuals at the identification of the nonparametric simple
Hammerstein model by structure (b)

No	$\phi_i^{(0)}$	$\hat{\theta}_i^{(1)}$	$\sigma_{\varepsilon_i}^{(1)}$	$\hat{\theta}_i^{(2)}$	$\sigma_{\varepsilon_i}^{(2)}$	$\hat{\theta}_i$
1	1	2.832 ±0.09026				1.7202
2	$u(k-1)$	1.275 ±0.07919				0.7744
3	$u(k-2)$	0.465 ±0.1061	0.8490	0.6074 ±0.01442		0.2824
4	$u(k-3)$	0.2055 ±0.0792				0.1248
5	$u(k-4)$				0.3180	0.0
6	$u^2(k-1)$	0.6751 ±0.04835				0.2849
7	$u^2(k-2)$	0.2008 ±0.06685		0.4220 ±0.01463		0.08474
8	$u^2(k-3)$	0.07388 ±0.06685	1.254			0.03118
9	$u^2(k-4)$	0.1597 ±0.04896				0.06739

TABLE 5.10.6 The estimated parameters and standard deviations of
the residuals at the identification of the nonparametric simple
Hammerstein model by structure (c)

No	$\phi_i^{(0)}$	$\hat{\theta}_i^{(1)}$	$\sigma_{\varepsilon_i}^{(1)}$	$\hat{\theta}_i^{(2)}$	$\sigma_{\varepsilon_i}^{(2)}$	$\hat{\theta}_i$
1	$u^2(k-1)$	0.5464 ±0.06554				0.315
2	$u^2(k-2)$	0.2008 ±0.06742				0.1158
3	$u(k-1)$	0.3576 ±0.1307	1.265	0.5765 ±0.02542		0.2165
4	$u^2(k-3)$	0.2278 ±0.04930				0.1313
5	$u(k-2)$	1.625 ±0.138			0.6612	0.7545
6	1	2.935 ±0.1876				1.3627
7	$u^2(k-4)$	0.08971 ±0.05020	1.483	0.4643 ±0.02573		0.04165
8	$u(k-3)$	-0.07929 ±0.1633				-0.03681
9	$u(k-4)$					0.0

(g) *model g:*
The standard deviations of the residuals at the estimation of the submodels at the first
layer in model (f) were:

$\sigma_{\varepsilon_1}^{(1)} = 0.2377$ $\qquad$ $\sigma_{\varepsilon_2}^{(1)} = 0.3579$ $\qquad$ $\sigma_{\varepsilon_3}^{(1)} = 0.5007$

$\sigma_{\varepsilon_4}^{(1)} = 1.279$ $\qquad$ $\sigma_{\varepsilon_5}^{(1)} = 1.287$ $\qquad$ $\sigma_{\varepsilon_6}^{(1)} = 2.062$

TABLE 5.10.7 The estimated parameters and standard deviations of the residuals at the identification of the nonparametric simple Hammerstein model by structure (d)

No	$\phi_i^{(0)}$	$\hat\theta_i^{(1)}$	$\sigma_{\varepsilon_i}^{(1)}$	$\hat\theta_i^{(2)}$	$\sigma_{\varepsilon_i}^{(2)}$	$\hat\theta_i^{(3)}$	$\sigma_{\varepsilon_i}^{(3)}$	$\hat\theta_i^{(4)}$	$\sigma_{\varepsilon_i}^{(4)}$	$\hat\theta_i$
1	$u^2(k-1)$					0.5191 ±0.06834				0.3549
2	$u^2(k-2)$	0.6998 ±0.04613	1.650			0.5203	1.371	0.6837 ±0.03498		0.2489
3	$u(k-1)$	1.047 ±0.1258				±0.06688				0.3724
4	$u^2(k-3)$	0.601 ±0.06803	2.428			0.5428			1.01	0.1209
5	$u(k-2)$	1.202 ±0.1851				±0.07129				0.2418
6	1	3.144 ±0.3258	2.636	0.7959			2.024	0.3707		0.4858
7	$u^2(k-4)$	0.4614 ±0.06266		±0.06064	2.44	0.5238		±0.03633		0.0713
8	$u(k-3)$	2.177 ±0.3377	3.7	0.2085		±0.07155				0.08813
9	$u(k-4)$	1.473 ±0.3406		±0.04630						0.05963

TABLE 5.10.8 The estimated parameters and standard deviations of the residuals at the identification of the nonparametric simple Hammerstein model by structure (e)

No	$\phi_i^{(0)}$	$\hat\theta_i^{(1)}$	$\sigma_{\varepsilon_i}^{(1)}$	$\hat\theta_i^{(2)}$	$\sigma_{\varepsilon_i}^{(2)}$	$\hat\theta_i^{(3)}$	$\sigma_{\varepsilon_i}^{(3)}$	$\hat\theta_i$
1	$u^2(k-1)$			0.5191 ±0.06834	1.371	0.6399 ±0.03239		0.3322
2	$u^2(k-2)$	0.6998 ±0.04613	1.650	0.5203				0.2330
3	$u(k-1)$	1.047 ±0.1258		±0.06688			0.9175	0.3486
4	$u^2(k-3)$	0.601 ±0.06803	2.428	0.6007				0.1497
5	$u(k-2)$	1.202 ±0.1851		±0.05249	1.846	0.4146		0.2993
6	1	3.144 ±0.3258	2.636	0.5054		±0.03323		0.6588
7	$u^2(k-4)$	0.4614 ±0.06266		±0.05319				0.0967
8	$u(k-3)$	2.177 ±0.3377	3.700					0.0
9	$u(k-4)$	1.473 ±0.3406						0.0

Because the second three standard deviations are much bigger than the first three, the threshold was chosen in such a way that only the first three submodels' outputs were considered as inputs for the linear static submodel of the second layer. The estimated equation of the second layer was then

$$\hat{y}(k) \equiv \hat{y}_1^{(2)}(k) = 0.6219\hat{y}_1^{(1)}(k) + 0.3921\hat{y}_2^{(1)}(k) - 0.01341\hat{y}_3^{(1)}(k)$$

The parameters of the estimated model and the standard deviations of the residuals at the regression of the submodels are summarized in Table 5.10.11. Comparing Tables 5.10.10 and 5.10.11 one can see that the standard deviation of the residuals decreased by omitting the three submodels at the first layer. The excellent coincidence between the simulated and computed output signals – based on model (g) – is seen in Figure 5.10.2. The estimated parameters are not as good as in case (f), as is seen in Table 5.10.9. This is trivial, since some model components were not considered in the model by having omitted the three worst submodels at the first layer.

TABLE 5.10.9 True, θ_i, and estimated, $\hat{\theta}_i$, parameters and estimated standard
deviations of the residuals σ_ε at the identification of the nonparametric simple
Hammerstein model by *nonlinear* submodels at the first layer

No	ϕ_i	θ_i	$\hat{\theta}_i$ (case (f))	$\hat{\theta}_i$ (case (g))	$\hat{\theta}_i$ (case (h))	$\hat{\theta}_i$ (case (i))
1	1	1.96322	1.99386	2.0265	1.9011	2.0251
2	$u(k-1)$	0.6321	0.6563	0.6722	0.6505	0.6716
3	$u(k-2)$	0.2325	0.2515	0.2128	0.2216	0.2138
4	$u(k-3)$	0.08554	0.08626	0.0908	0.04453	0.0909
5	$u(k-4)$	0.03147	0.03624	-0.00223	0.00128	-0.00223
6	$u^2(k-1)$	0.31605	0.3307	0.3295	0.3207	0.3309
7	$u^2(k-2)$	0.11625	0.1292	0.1015	0.1177	0.1029
8	$u^2(k-3)$	0.04277	0.04245	0.0446	0.02194	0.0448
9	$u^2(k-4)$	0.015735	0.01753	-0.00111	0.000516	-0.00112
σ_ε			0.3365	0.2304	0.2093	0.2303

(h) *model h:*
The memory of the model was 4. At the first layer 6 quadratic submodels were assumed.
The input signals of the submodels were differently delayed input signal values. Now
the cross-product terms were not omitted from the quadratic models, which means that it
was not assumed *a priori* that the system was of Hammerstein type. The estimated
submodels were as follows:

$$\hat{y}_1^{(1)}(k) = 2.048 + 0.6266u(k-1) + 0.3436u(k-2) + 0.3066u^2(k-1)$$
$$-0.004478u(k-1)u(k-2) + 0.1653u^2(k-2)$$
$$\hat{y}_2^{(1)}(k) = 1.984 + 0.7473u(k-1) + 0.2321u(k-3) + 0.3718u^2(k-1)$$
$$-0.001164u(k-1)u(k-3) + 0.1143u^2(k-3)$$
$$\hat{y}_3^{(1)}(k) = 1.963 + 0.8156u(k-1) + 0.1660u(k-4) + 0.4078u^2(k-1)$$
$$+0.0000002384u(k-1)u(k-4) + 0.0830u^2(k-4)$$
$$\hat{y}_4^{(1)}(k) = 2.382 + 0.8014u(k-2) + 0.05061u(k-3) + 0.3980u^2(k-2)$$
$$-0.001117u(k-2)u(k-3) + 0.02416u^2(k-3)$$
$$\hat{y}_5^{(1)}(k) = 2.412 + 0.8389u(k-2) + 0.002466u(k-4) + 0.4170u^2(k-2)$$
$$-0.001531u(k-2)u(k-4) - 0.0004616u^2(k-4)$$
$$\hat{y}_6^{(1)}(k) = 2.917 + 0.8034u(k-3) - 0.1224u(k-4) + 0.3975u^2(k-3)$$
$$-0.002828u(k-3)u(k-4) - 0.06187u^2(k-4)$$

At the second layer all output signals of the 6 submodels were first considered as
possible input signals of a linear static submodel. The parameter estimation resulted in
the following equation:

TABLE 5.10.10 The estimated parameters and standard deviations of the residuals at the identification of the nonparametric simple Hammerstein model by structure (f)

No	$y_i^{(0)}$	$\phi_i^{(0)}$	$\hat\theta_i^{(1)}$	$\sigma_{\varepsilon_i}^{(1)}$	$\hat\theta_i^{(2)}$	$\sigma_{\varepsilon_i}^{(2)}$	$\hat\theta_i$	$\hat\theta_i^*$
1		1	2.050±0.03253				-1.2692	1.99386
2		$u(k-1)$	0.6273±0.04264				-0.3884	0.6563
3	$u(k-1)$	$u(k-2)$	0.3422±0.04238	0.2377	-0.6191 ±1.459		-0.2119	0.2515
4	$u(k-2)$	$u^2(k-1)$	0.3045±0.01818				-0.1885	0.3307
5		$u^2(k-2)$	0.1632±0.01818				-0.1010	0.1292
		1	1.984±0.05089				2.7558	-
		$u(k-1)$	0.7473±0.04847				1.0380	-
6	$u(k-1)$	$u(k-3)$	0.2316±0.04786	0.3579	1.389 ±1.399		0.3217	0.08626
	$u(k-3)$	$u^2(k-1)$	0.3714±0.02067				0.5159	-
7		$u^2(k-3)$	0.1138±0.02067				0.1581	0.04245
		1	1.963±0.07458				0.01606	-
		$u(k-1)$	0.8156±0.05972				0.0067	-
8	$u(k-1)$	$u(k-4)$	0.1660±0.05858	0.5007	0.00818 ±0.1921		0.00136	0.03624
	$u(k-4)$	$u^2(k-1)$	0.4078±0.02549				0.00333	-
9		$u^2(k-4)$	0.0830±0.02549			0.3365	0.000679	0.01753
		1	2.838±0.1751				-0.9122	-
		$u(k-2)$	0.8015±0.2294	1.279	-0.3828 ±3.125		-0.3068	-
	$u(k-2)$	$u(k-3)$	0.05027±0.2280				-0.01924	-
	$u(k-3)$	$u^2(k-2)$	0.3974±0.0978				-0.1521	-
		$u^2(k-3)$	0.02365±0.0978				-0.00905	-
		1	2.413±0.1831				2.2151	-
		$u(k-2)$	0.8390±0.1743				0.7702	-
	$u(k-2)$	$u(k-4)$	0.001865±0.1721	1.287	0.9180 ±3.559		0.00171	-
	$u(k-4)$	$u^2(k-2)$	0.4164±0.07436				0.3823	-
		$u^2(k-4)$	-0.00016±0.07436				-0.000147	-
		1	2.918±0.2824				-0.7849	-
		$u(k-3)$	0.8038±0.3698				-0.2162	-
	$u(k-3)$	$u(k-4)$	-0.1233±0.3676	2.062	-0.2690 ±0.2161		0.03317	-
	$u(k-4)$	$u^2(k-3)$	0.3962±0.1577				-0.1066	-
		$u^2(k-4)$	-0.06317±0.1577				0.0170	-

$$\hat y(k) \equiv \hat y_1^{(2)}(k) = -0.5720\hat y_1^{(1)}(k) + 0.4340\hat y_2^{(1)}(k) - 0.03947\hat y_3^{(1)}(k) - 0.1598\hat y_4^{(1)}(k)$$
$$+0.2078\hat y_5^{(1)}(k) - 0.05984\hat y_6^{(1)}(k)$$

The parameters of the estimated model and the standard deviations of the residuals at the regression of the submodels are summarized in Table 5.10.12. The estimated parameters are also shown in Table 5.10.9 in order to compare them with the true and the estimated parameters obtained from the other models (f), (g) and (i). The estimated model is very good, its parameters approximate the true parameters very well. The parameters belonging to the cross-product model components are very small, and the other

TABLE 5.10.11 The estimated parameters and standard deviations of the residuals at the identification of the nonparametric simple Hammerstein model by structure (g)

No	$y_i^{(0)}$	$\phi_i^{(0)}$	$\hat{\theta}_i^{(1)}$	$\sigma_{\varepsilon_i}^{(1)}$	$\hat{\theta}_i^{(2)}$	$\sigma_{\varepsilon_i}^{(2)}$	$\hat{\theta}_i$	$\hat{\theta}_i^*$
1		1	2.050 ±0.03253				1.2749	2.0265
2		$u(k-1)$	0.6273 ±0.04264				0.3901	0.6722
3	$u(k-1)$	$u(k-2)$	0.3422 ±0.04238	0.2377	0.6219 ±0.03976		0.2128	0.2128
4	$u(k-2)$	$u^2(k-1)$	0.3045 ±0.01818				0.1894	0.3295
5		$u^2(k-2)$	0.1632 ±0.01818				0.1015	0.1015
		1	1.984 ±0.05089				0.7779	-
		$u(k-1)$	0.7473 ±0.04847				0.2930	-
6	$u(k-1)$	$u(k-3)$	0.2316 ±0.04786	0.3579	0.3921 ±0.05994	0.2304	0.0908	0.0908
	$u(k-3)$	$u^2(k-1)$	0.3714 ±0.02067				0.1456	-
7		$u^2(k-3)$	0.1138 ±0.02067				0.0446	0.0446
		1	1.963 ±0.07458				-0.0263	-
		$u(k-1)$	0.8156 ±0.05972				-0.0109	-
8	$u(k-1)$	$u(k-4)$	0.1660 ±0.05858	0.5007	-0.01341 ±0.05264		-0.00223	-0.00223
	$u(k-4)$	$u^2(k-1)$	0.4078 ±0.02549				-0.00547	-
9		$u^2(k-4)$	0.08300 ±0.02549				-0.00111	-0.00111
		1	2.838 ±0.1751					-
		$u(k-2)$	0.8015 ±0.2294	1.279				-
	$u(k-2)$	$u(k-3)$	0.05027 ±0.2280					-
	$u(k-3)$	$u^2(k-2)$	0.3974 ±0.0978					-
		$u^2(k-3)$	0.02365 0.0978					-
		1	2.413 ±0.1831					-
		$u(k-2)$	0.8390 ±0.1743	1.287				-
	$u(k-2)$	$u(k-4)$	0.001865 ±0.1721					-
	$u(k-4)$	$u^2(k-2)$	0.4164 ±0.07436					-
		$u^2(k-4)$	-0.000160 ±0.07436					-
		1	2.918 ±0.2824					-
		$u(k-3)$	0.8038 ±0.3698	2.062				-
	$u(k-3)$	$u(k-4)$	-0.1233 ±0.3676					-
	$u(k-4)$	$u^2(k-3)$	0.3962 ±0.1577					-
		$u^2(k-4)$	-0.06317 ±0.1577					-

estimated parameters hardly differ from those of model (f), therefore the system is of Hammerstein type. The excellent coincidence between the simulated and computed output signals – based on model (h) – is seen in Figure 5.10.2.

(i) *model i:*

The standard deviations of the residuals at the estimation of the submodels at the first layer in model (h) were

$$\sigma_{\varepsilon_1}^{(1)} = 0.2386 \qquad \sigma_{\varepsilon_2}^{(1)} = 0.4340 \qquad \sigma_{\varepsilon_3}^{(1)} = 0.5028$$

$$\sigma_{\varepsilon_4}^{(1)} = 1.284 \qquad \sigma_{\varepsilon_5}^{(1)} = 1.292 \qquad \sigma_{\varepsilon_6}^{(1)} = 2.071$$

TABLE 5.10.12 The estimated parameters and standard deviations of the residuals at the identification of the nonparametric simple Hammerstein model by structure (h)

No	$y_i^{(0)}$	$\phi_i^{(0)}$	$\hat{\theta}_i^{(1)}$	$\sigma_{\varepsilon_i}^{(1)}$	$\hat{\theta}_i^{(2)}$	$\sigma_{\varepsilon_i}^{(2)}$	$\hat{\theta}_i$	$\hat{\theta}_i^*$
1		1	2.048 ±0.03298				1.1715	1.9011
2		$u(k-1)$	0.6266 ±0.04284				0.3584	0.6505
3	$u(k-1)$	$u(k-2)$	0.3436 ±0.04272				0.1965	0.2216
4	$u(k-2)$	$u^2(k-1)$	0.3066 ±0.01923	0.2386	0.5720 ±0.03764		0.1754	0.3207
5		$u(k-1)u(k-2)$	-0.004478 ±0.0132				-0.0026	-0.0026
6		$u^2(k-1)$	0.1653 ±0.01923				0.0946	0.1177
		1	1.984 ±0.05205				0.8611	-
		$u(k-1)$	0.7473 ±0.04867				0.3243	-
7	$u(k-1)$	$u(k-3)$	0.2321 ±0.04841				0.1007	0.04453
	$u(k-3)$	$u^2(k-1)$	0.3718 ±0.02162	0.3594	0.4340 ±0.05784		0.1614	-
8		$u(k-1)u(k-3)$	-0.00116 ±0.01488				-0.0005	-0.0005
9		$u^2(k-3)$	0.1143 ±0.02162				0.0496	0.02194
		1	1.963 ±0.07673				-0.07748	-
		$u(k-1)$	0.8156 ±0.06002				-0.0322	-
10	$u(k-1)$	$u(k-4)$	0.1660 ±0.05956				-0.00655	0.00128
	$u(k-4)$	$u^2(k-1)$	0.4078 ±0.02631	0.5028	-0.0394 ±0.04927		-0.0161	-
11		$u(k-1)u(k-4)$	0.0000002 ±0.018				-0.0000	-0.0000
12		$u^2(k-4)$	0.08300 ±0.02631				-0.00328	0.000516
		1	2.382 ±0.1775			0.2093	-0.3806	-
		$u(k-2)$	0.8014 ±0.2306				-0.1281	-
	$u(k-2)$	$u(k-3)$	0.05061 ±0.2299				-0.00809	-
	$u(k-3)$	$u^2(k-2)$	0.3980 ±0.1035	1.284	-0.1598 ±0.04768		-0.0636	-
13		$u(k-2)u(k-3)$	-0.00111 ±0.0711				0.000178	0.000178
		$u^2(k-3)$	0.02416 ±0.1035				-0.00386	-
		1	2.412 ±0.1873				0.5012	-
		$u(k-2)$	0.8389 ±0.175				0.1743	-
	$u(k-2)$	$u(k-4)$	0.002466 ±0.1741				0.00051	-
	$u(k-4)$	$u^2(k-2)$	0.4170 ±0.07777	1.292	0.2078 ±0.05287		0.08665	-
14		$u(k-2)u(k-4)$	-0.00153 ±0.05359				-0.00032	-0.00032
		$u^2(k-4)$	0.000462 ±0.07777				0.000096	-
		1	2.917 ±0.2864				-0.1746	-
		$u(k-3)$	0.8034 ±0.3718				-0.04808	-
	$u(k-3)$	$u(k-4)$	-0.1224 ±0.3708				0.00732	-
	$u(k-4)$	$u^2(k-3)$	0.3975 ±0.1669	2.071	-0.0598±0.01406		-0.0236	-
15		$u(k-3)u(k-4)$	-0.002828 ±0.1147				0.000169	0.000169
		$u^2(k-4)$	-0.06187 ±0.1669				0.0037	-

Because the second three standard deviations are much bigger than the first three, the threshold was chosen in such a way that only the first three submodels' outputs were considered as inputs for the linear static submodel of the second layer. The estimated equation of the second layer was then

$$\hat{y}(k) \equiv \hat{y}_1^{(2)}(k) = 0.6223\hat{y}_1^{(1)}(k) + 0.3917\hat{y}_2^{(1)}(k) - 0.01346\hat{y}_3^{(1)}(k)$$

TABLE 5.10.13 The estimated parameters and standard deviations of the residuals at the identification of the nonparametric simple Hammerstein model by structure (i)

No	$y_i^{(0)}$	$\phi_i^{(0)}$	$\hat\theta_i^{(1)}$	$\sigma_{\varepsilon_i}^{(1)}$	$\hat\theta_i^{(2)}$	$\sigma_{\varepsilon_i}^{(2)}$	$\hat\theta_i$	$\hat\theta_i^*$
1		1	2.048 0.03298				1.2744	2.8251
2		$u(k-1)$	0.6266 0.04284				0.3899	0.6716
3	$u(k-1)$	$u(k-2)$	0.3436 0.04272				0.2138	0.2138
4	$u(k-2)$	$u^2(k-1)$	0.3066 0.01923	0.2386	0.6223 0.03974		0.1908	0.3389
5		$u(k-1)u(k-2)$	-0.004478 0.0132				-0.00279	-0.00279
6		$u^2(k-1)$	0.1653 0.01923				0.1029	0.1029
		1	1.984 0.05205				0.7771	-
		$u(k-1)$	0.7473 0.04867				0.2927	-
7	$u(k-1)$	$u(k-3)$	0.2321 0.04841				0.0909	0.0909
	$u(k-3)$	$u^2(k-1)$	0.3718 0.02162	0.3594	0.3917 0.05989	0.2303	0.1456	-
8		$u(k-1)u(k-3)$	-0.00116 0.01488				-0.00046	-0.00046
9		$u^2(k-3)$	0.1143 0.02162				0.0448	0.0448
		1	1.963 0.07673				-0.02642	-
		$u(k-1)$	0.8156 0.06002				-0.0110	-
10	$u(k-1)$	$u(k-4)$	0.1660 0.05956				-0.00223	-0.00223
	$u(k-4)$	$u^2(k-1)$	0.4078 0.02631	0.5028	-0.01346 0.05261		-0.0055	-
11		$u(k-1)u(k-4)$	0.0000002 0.018				-0.0000	-0.0000
12		$u^2(k-4)$	0.08300 0.02631				-0.00112	-0.00112
		1	2.382 0.1775					-
		$u(k-2)$	0.8014 0.2306					-
	$u(k-2)$	$u(k-3)$	0.05061 0.2299					-
	$u(k-3)$	$u^2(k-2)$	0.3980 0.1035	1.284				-
13		$u(k-2)u(k-3)$	-0.00112 0.0711					-
		$u^2(k-3)$	0.02416 0.1035					-
		1	2.412 0.1873					-
		$u(k-2)$	0.8389 0.175					-
	$u(k-2)$	$u(k-4)$	0.002466 0.1741					-
	$u(k-4)$	$u^2(k-2)$	0.4170 0.07777	1.292				-
14		$u(k-2)u(k-4)$	-0.00153 0.05359					-
		$u^2(k-4)$	0.000462 0.07777					-
		1	2.917 0.2864					-
		$u(k-3)$	0.8034 0.3718					-
	$u(k-3)$	$u(k-4)$	-0.1224 0.3708					-
	$u(k-4)$	$u^2(k-3)$	0.3975 0.1669	2.071				-
15		$u(k-3)u(k-4)$	-0.002828 0.1147					-
		$u^2(k-4)$	-0.06187 0.1669					-

The parameters of the estimated model and the standard deviations of the residuals at the regression of the submodels are summarized in Table 5.10.13. Comparing Tables 5.10.12 and 5.10.13 one can see that the standard deviation of the residuals hardly increased on omitting the three worst submodels at the first layer. The estimated

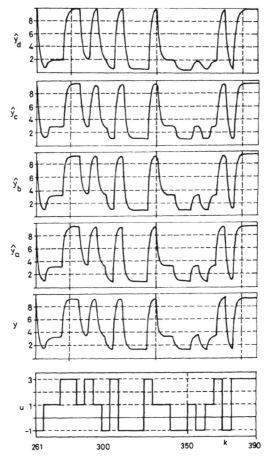

Fig. 5.10.2a Input u and output y signals of the simple Hammerstein model and the computed output $\hat{y}$ signals based on the estimated models

parameters are almost as good as in case (h), as is seen in Table 5.10.9. The parameters belonging to the model components having smaller delayed values of the input signal are accurate enough, but the parameters belonging to the terms $u(k-4)$ and $u^2(k-4)$ are worse. This fact is a consequence of having omitted three submodels at the first layer. The parameters belonging to the cross-product terms are almost zero, which fact newly points to a Hammerstein model. The computed output signal – based on the estimated model – fits to the simulated process output well, as is seen in Figure 5.10.2. ∎

On the basis of the simulation runs (a) to (i) in Example 5.10.2 we can see that the method using quadratic models with few (2) inputs at the first layer is before the method using quasi-linear models with all possible nonlinear (quadratic) terms created from the delayed input signal values already at the first layer.

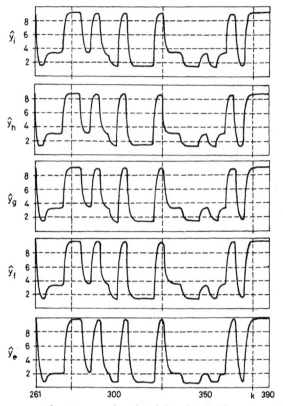

Fig. 5.10.2b Input u and output y signals of the simple Hammerstein model and the computed output $\hat{y}$ signals based on the estimated models

Example 5.10.3 *Identification of the simple Wiener model in nonparametric form (Haber and Perényi, 1987)*

The transfer function of the linear dynamic part is

$$G(s) = \frac{1}{1+10s}$$

and the equation of the static part is

$$y(t) = 2 + v(t) + v^2(t)$$

The GMDH method has advantages over other procedures if too many parameters have to be estimated. Therefore not a parametric but a nonparametric Wiener model was assumed. To describe the parametric process accurately enough by a nonparametric one with few parameters, the sampling time was chosen equal to the time constant $\Delta T = T = 10$ [s]. Then the equivalent pulse transfer function of the continuous transfer function is

$$H\left(q^{-1}\right) = \frac{0.6321q^{-1}}{1 - 0.3679q^{-1}}$$

The weighting function series of $H(q^{-1})$ were truncated after the fourth term

$$H(q^{-1}) = 0.6321q^{-1} + 0.2325q^{-2} + 0.08554q^{-3} + 0.03147q^{-4}$$

because further terms would have very small values. To eliminate the truncation error the process itself was simulated by the nonparametric model. The equation of the whole model is

$$y(k) = 2.0 + 0.6321u(k-1) + 0.2325u(k-2)$$
$$+0.0855u(k-3) + 0.03147u(k-4) + 0.1998u^2(k-1)$$
$$+0.02703u^2(k-2) + 0.00366u^2(k-3) + 0.000495u^2(k-4)$$
$$+0.14696u(k-1)u(k-2) + 0.0199u(k-2)u(k-3)$$
$$+0.00269u(k-3)u(k-4) + 0.05407u(k-1)u(k-3)$$
$$+0.007317u(k-2)u(k-4) + 0.0199u(k-1)u(k-4)$$

which is a subset of a Volterra series. The model components ϕ_i and the true parameters θ_i are listed in Tables 5.10.14 and 5.10.19.

The test signal was a PRTS with maximum length 26, amplitude 2, and mean value 1. The sampling time was $\Delta T = 2$ [s] and the minimum switching time of the PRTS signal was 5 times more. $N = 26 \cdot 5 = 130$ data pairs were used for the identification.

Before starting the GMDH all possible model components were regressed with the output signal which means that 15 different submodels were estimated, each of them having only one model component. The results are summarized in Table 5.10.15. Based on the estimated standard deviations of the residuals σ_{ε_i} the model components ϕ_i can be arranged in series. The column *sequence* shows it starting with the best fitting component. This knowledge about the components was used later to reduce the computations.

Further on different approximating models are presented. Common in the procedures is that all possible components of the models were already chosen at the zero-th layer and, further on quasi-linear MISO static models were estimated. This kind of procedure would need too many computations if all possible submodels were to be estimated. Instead of doing that, another strategy was applied: only the neighboring components were used as inputs of the submodels. This *suboptimal* strategy led to acceptable results and required only a little computation.

(a) *model a:*
All possible model components were used at the zero-th layer. The first layer consists of three submodels, the first one contains the constant term and almost all linear terms

$$\hat{y}_1^{(1)}(k) = 2.681 + 1.211u(k-1) + 0.4907u(k-2) + 0.222u(k-3)$$

and the second submodel consists of the quadratic terms without any cross-product terms

$$\hat{y}_2^{(1)}(k) = 0.6382u^2(k-1) + 0.2095u^2(k-2) + 0.07957u^2(k-3)$$
$$+0.1474u^2(k-4)$$

TABLE 5.10.14 True, θ_i, and estimated, $\hat{\theta}_i$, parameters and estimated standard deviations of the residuals σ_ε at the identification of the nonparametric simple Wiener model by quasi-linear submodels at the first layer

No	ϕ_i	θ_i	$\hat{\theta}_i$ (case (a))	$\hat{\theta}_i$ (case (b))	$\hat{\theta}_i$ (case (c))
1	1	2.0	1.701	1.7377	1.265
2	$u(k-1)$	0.6321	0.7684	0.7561	0.2419
3	$u(k-2)$	0.2325	0.3113	0.43826	0.7057
4	$u(k-3)$	0.0855	0.14086	-	-
5	$u(k-4)$	0.03147	-	-	0.02592
6	$u^2(k-1)$	0.1998	0.252	0.18043	0.217
7	$u^2(k-2)$	0.02703	0.0827	0.06	0.10336
8	$u^2(k-3)$	0.00366	0.03142	0.06213	0.0862
9	$u^2(k-4)$	0.000495	0.0582	-	0.02407
10	$u(k-1)u(k-2)$	0.14696	-	0.1046	0.07947
11	$u(k-2)u(k-3)$	0.0199	-	0.06724	0.01443
12	$u(k-3)u(k-4)$	0.00269	-	-	0.01628
13	$u(k-1)u(k-3)$	0.05407	-	-0.0512	0.01087
14	$u(k-2)u(k-4)$	0.007317	-	-	-
15	$u(k-1)u(k-4)$	0.0199	-	-	0.02358
	σ_ε		0.4451	0.4233	0.6957

and the third one consists of all possible cross-product terms of the differently delayed input values

$$\hat{y}_3^{(1)}(k) = 0.8416u(k-1)u(k-2) - 4.0u(k-2)u(k-3)$$
$$+0.2192u(k-3)u(k-4) + 4.0u(k-1)u(k-3)$$
$$+4.0u(k-2)u(k-4) - 5.0u(k-1)u(k-4)$$

In the first submodel $u(k-4)$ was not considered because it was the *least significant* component according to Table 5.10.15. The standard deviations of the three submodels were

$$\sigma_{\varepsilon_1}^{(1)} = 0.867 \qquad \sigma_{\varepsilon_2}^{(1)} = 1.282 \qquad \sigma_{\varepsilon_3}^{(1)} = 4.851$$

therefore the third submodel's output was not considered any longer as an input signal of the next (second) layer. The equation of the second, final, layer is as follows:

$$\hat{y}(k) \equiv \hat{y}_1^{(2)}(k) = 0.6345\hat{y}_1^{(1)}(k) + 0.3949\hat{y}_2^{(1)}(k)$$

TABLE 5.10.15 The sequence of importance of the possible model components at the identification of the nonparametric simple Wiener model

No	ϕ_i	$\hat{\theta}_i$	σ_{ε_i}	sequence
1	1	0.4538 ±0.2720	3.109	9
2	$u(k-1)$	2.496 ±0.1224	2.685	4
3	$u(k-2)$	2.441 ±0.1359	2.951	8
4	$u(k-3)$	2.247 ±0.1645	3.539	13
5	$u(k-4)$	1.993 ±0.1928	4.109	15
6	$u^2(k-1)$	0.9879 ±0.02773	1.675	1
7	$u^2(k-2)$	0.9779 ±0.03341	1.995	2
8	$u^2(k-3)$	0.9252 ±0.04602	2.718	5
9	$u^2(k-4)$	0.8536 ±0.05799	3.385	10
10	$u(k-1)u(k-2)$	1.013 ±0.03738	2.133	3
11	$u(k-2)u(k-3)$	0.9602 ±0.04956	2.793	6
12	$u(k-3)u(k-4)$	0.8811 ±0.06232	3.466	11
13	$u(k-1)u(k-3)$	0.9976 ±0.05406	2.893	7
14	$u(k-2)u(k-4)$	0.9160 ±0.06695	3.531	12
15	$u(k-1)u(k-4)$	0.9538 ±0.07275	3.616	14

TABLE 5.10.16 The estimated parameters and standard deviations of the residuals at the identification of the nonparametric simple Wiener model by structure (a)

No	$\phi_i^{(0)}$	$\hat{\theta}_i^{(1)}$	$\sigma_{\varepsilon_i}^{(1)}$	$\hat{\theta}_i^{(2)}$	$\sigma_{\varepsilon_i}^{(2)}$	$\hat{\theta}_i$
1	1	2.681 ±0.09217				1.701
2	$u(k-1)$	1.211 ±0.08086	0.867	0.6345 ±0.02101		0.7684
3	$u(k-2)$	0.4907 ±0.1084				0.3113
4	$u(k-3)$	0.2220 ±0.08088				0.14086
5	$u(k-4)$	-			0.4451	
6	$u^2(k-1)$	0.6382 ±0.04941				0.252
7	$u^2(k-2)$	0.2095 ±0.06832	1.282	0.3949 ±0.02134		0.08273
8	$u^2(k-3)$	0.07957 ±0.06832				0.03142
9	$u^2(k-4)$	0.1474 ±0.05004				0.0582
10	$u(k-1)u(k-2)$	0.8416 ±0.2109				0.0
11	$u(k-2)u(k-3)$	$-4.000 ±0.4851 \cdot 10^{11}$				0.0
12	$u(k-3)u(k-4)$	0.2192 ±0.2372	4.851	-		0.0
13	$u(k-1)u(k-3)$	$4.000 ±0.4851 \cdot 10^{11}$				0.0
14	$u(k-2)u(k-4)$	$4.000 ±0.4851 \cdot 10^{11}$				0.0
15	$u(k-1)u(k-4)$	$-5.000 ±0.4851 \cdot 10^{11}$				0.0

Although only three regressions were used, the approximating model is very good. The good coincidence between the measured (simulated) output signal and the computed output signal – based on the estimated model – is seen in Figure 5.10.3. The whole identification procedure is summarized in Table 5.10.16, where the selected model components of the zero-th layer $\phi^{(0)}$, the estimated parameters $\theta^{(j)}$ at each layer, and the corresponding standard deviation of the residuals $\sigma^{(j)}$ are presented. The last column $\hat{\theta}_i$ contains the resulting estimated parameters belonging to the components of the original model of the zero-th layer. They can be calculated by replacing the model inputs of the layers by output signals of the submodels of the lower layers subsequently.

As is seen in Table 5.10.16 the model components belonging to the cross-product terms were omitted from the model during the GMDH. The resulting approximating model is, therefore, a nonparametric Hammerstein model. The computed output signal, based on the estimated model, fits the simulated process output well, as is seen in Figure 5.10.3.

(b) *model b:*
From the possible model components only the best nine – according to Table 5.10.15 – were considered as inputs of the first layer. At the first layer, therefore, three submodels, each of them with three components, were used:

$$\hat{y}_1^{(1)}(k) = 2.728 + 1.187u(k-1) + 0.688u(k-2)$$
$$\hat{y}_2^{(1)}(k) = 0.6302u^2(k-1) + 0.2095u^2(k-2) + 0.217u^2(k-3)$$
$$\hat{y}_3^{(1)}(k) = 0.9095u(k-1)u(k-2) + 0.5847u(k-2)u(k-3) - 0.4452u(k-1)u(k-3)$$

TABLE 5.10.17 The estimated parameters and standard deviations of the residuals at the identification of the nonparametric simple Wiener model by structure (b)

No	$\phi_i^{(0)}$	$\hat{\theta}_i^{(1)}$	$\sigma_{\varepsilon_i}^{(1)}$	$\hat{\theta}_i^{(2)}$	$\sigma_{\varepsilon_i}^{(2)}$	$\hat{\theta}_i$
1	1	2.728 ±0.09248				1.7377
2	$u(k-1)$	1.187 ±0.08234	0.8879	0.6370 ±0.01928		0.7561
3	$u(k-2)$	0.6880 ±0.08236				0.43826
4	$u(k-3)$	-				0.0
5	$u(k-4)$	-				0.0
6	$u^2(k-1)$	0.6302 ±0.05067				0.18043
7	$u^2(k-2)$	0.2095 ±0.07016	1.316	0.2863 ±0.03357	0.4233	0.06
8	$u^2(k-3)$	0.2170 ±0.05127				0.06213
9	$u^2(k-4)$	-				0.0
10	$u(k-1)u(k-2)$	0.9095 ±0.07349				0.1046
11	$u(k-2)u(k-3)$	0.5847 ±0.08271	1.807	0.1150 ±0.02841		0.06724
12	$u(k-3)u(k-4)$	-				0.0
13	$u(k-1)u(k-3)$	-0.4452 ±0.1125				-0.0512
14	$u(k-2)u(k-4)$	-				0.0
15	$u(k-1)u(k-4)$	-				0.0

The second, final, layer also had a three-input static model

$$\hat{y}(k) \equiv \hat{y}_1^{(2)}(k) = 0.6374\hat{y}_1^{(1)}(k) + 0.2863\hat{y}_2^{(1)}(k) + 0.115\hat{y}_3^{(1)}(k)$$

The good coincidence between the measured (simulated) output signal and the computed output signals – based on the estimated model – is seen in Figure 5.10.3. The whole identification procedure is summarized in Table 5.10.17.

The model (b) is better than model (a), which fact can be seen from the smaller standard deviation of the residuals. This fact can be explained by omitting the *least significant* model components at the zero-th layer and that cross-product terms are now also included in the model. The computed output signal – based on the estimated model – fits the simulated process output well, as is seen in Figure 5.10.3.

(c) *model c:*

The model components of the first layer were selected according to the *sequence of importance* given in Table 5.10.15. In all submodels at the first and subsequent layers only two model components – input signals – were assumed. This fact led to more (4) layers than in cases (a) to (c), where the submodels had up to four model components, and, therefore, two layers gave an accurate approximating model.

The estimated model equations are as follows.

First layer:

$$\hat{y}_1^{(1)}(k) \equiv u^2(k-1)$$

$$\hat{y}_2^{(1)}(k) = 0.5809u^2(k-2) + 0.4466u(k-1)u(k-2)$$

$$\hat{y}_3^{(1)}(k) = 1.497u(k-1) + 0.5334u^2(k-3)$$

$$\hat{y}_4^{(1)}(k) = 0.5787u(k-2)u(k-3) + 0.4359u(k-1)u(k-3)$$

$$\hat{y}_5^{(1)}(k) = 1.659u(k-2) + 2.974$$

$$\hat{y}_6^{(1)}(k) = 0.5343u^2(k-4) + 0.3613u(k-3)u(k-4)$$

$$\hat{y}_7^{(1)}(k) = 0.5163u(k-2)u(k-4) + 1.272u(k-3)$$

$$\hat{y}_8^{(1)}(k) = 0.7085u(k-1)u(k-4) + 0.7787u(k-4)$$

Second layer:

$$\hat{y}_1^{(2)}(k) = 0.5674y^{(1)}(k) + 0.4653y^{(1)}(k), \qquad \hat{y}_2^{(2)}(k) = 0.8836y^{(1)}(k) + 0.1363y^{(1)}(k)$$

$$\hat{y}_3^{(2)}(k) = 0.9322y^{(1)}(k) + 0.09872y^{(1)}(k), \qquad \hat{y}_4^{(2)}(k) \equiv \hat{y}_8^{(1)}(k)$$

Third layer:

$$\hat{y}_1^{(3)}(k) = 0.6939\hat{y}_1^{(2)}(k) + 0.3319\hat{y}_2^{(2)}(k)$$

$$\hat{y}_2^{(3)}(k) = 0.9254\hat{y}_3^{(2)}(k) + 0.0675\hat{y}_4^{(2)}(k)$$

Fourth layer:

$$\hat{y}(k) \equiv \hat{y}_1^{(4)}(k) = 0.5511\hat{y}_1^{(3)}(k) + 0.4931\hat{y}_2^{(3)}(k)$$

TABLE 5.10.18 The estimated parameters and standard deviations of the residuals at the identification of the nonparametric simple Wiener model by structure (c)

No	$\phi_i^{(0)}$	$\hat\theta_i^{(1)}$	$\sigma_{\varepsilon_i}^{(1)}$	$\hat\theta_i^{(2)}$	$\sigma_{\varepsilon_i}^{(2)}$	$\hat\theta_i^{(3)}$	$\sigma_{\varepsilon_i}^{(3)}$	$\hat\theta_i^{(4)}$	$\sigma_{\varepsilon_i}^{(4)}$	$\hat\theta_i$
1	$u^2(k-1)$			0.5674 ±0.05987	1.394					0.217
2	$u^2(k-2)$	0.5809 ±0.08340	1.823	0.4653 ±0.06114		0.6939 ±0.06277				0.10336
3	$u(k-1)u(k-2)$	0.4466 ±0.08730					1.273	0.5511 ±0.02830		0.07947
4	$u(k-1)$	1.497 ±0.1145	1.774	0.8836 ±0.06963						0.2419
5	$u^2(k-3)$	0.5334 ±0.04244			1.779	0.3319 ±0.06414				0.0862
6	$u(k-2)u(k-3)$	0.5787 ±0.1229	2.686	0.1363 ±0.07539						0.01443
7	$u(k-1)u(k-3)$	0.4359 ±0.1294								0.01087
8	$u(k-2)$	1.659 ±0.07692	1.443	0.9322 ±0.03926					0.6957	0.7057
9	1	2.974 ±0.1477			1.416	0.9254 ±0.03865				1.265
10	$u^2(k-4)$	0.5343 ±0.1524	3.330	0.09872 ±0.04722						0.02407
11	$u(k-3)u(k-4)$	0.3613 ±0.1599					1.385	0.4931 ±0.02845		0.01628
12	$u(k-2)u(k-4)$	0.5163 ±0.09466	3.195							0.0
13	$u(k-3)$	1.272 ±0.2328								0.0
14	$u(k-1)u(k-4)$	0.7085 ±0.1045	3.524			0.0675 ±0.02819				0.02358
15	$u(k-4)$	0.7787 ±0.2438								0.02592

The resulting model is worse than the previous ones (a) to (c). This fact can be seen in Figure 5.10.3, where the measured (simulated) input, output, and computed output signals – based on the estimated model – are plotted and from the higher standard deviation of the residuals presented in Table 5.10.18. The computed output signal – based on the estimated model – fits the simulated process output well, as is seen in Figure 5.10.3.

The efficiency of the GMDH can be increased if *nonlinear* submodels were also allowed. The input signals of the first layers were all combinations of

$$u(k-i), u(k-j) \qquad i = 1, 2, 3, 4 \qquad j = 1, 2, 3, 4 \qquad i \neq j$$

and the submodels of the first layers were static quadratic polynomials with the above input signals. The second (last) layer then contained a linear (static) MISO model. To do real structure identification first the cross-product terms $u(k-i)u(k-j)$, $i \neq j$, were also considered, and later they were omitted from the model. If the model fitting is complete without the cross-product terms then the system is of Hammerstein type.

(d) *model d:*

The memory of the model was 4. At the first layer 6 quadratic submodels were assumed. The input signals of the submodels were (differently) delayed input signal values. The cross-product terms were also included in the quadratic models. The estimated submodels were as follows:

$$\hat y_1^{(1)}(k) = 2.041 + 0.6326u(k-1) + 0.3510u(k-2) + 0.1916u^2(k-1)$$
$$+ 0.2032u(k-1)u(k-2) + 0.05286u^2(k-2)$$
$$\hat y_2^{(1)}(k) = 2.010 + 0.7493u(k-1) + 0.2345u(k-3) + 0.2845u^2(k-1)$$
$$+ 0.1564u(k-1)u(k-3) + 0.03173u^2(k-3)$$
$$\hat y_3^{(1)}(k) = 2.000 + 0.8156u(k-1) + 0.1660u(k-4) + 0.3416u^2(k-1)$$
$$+ 0.1173u(k-1)u(k-4) + 0.0228u^2(k-4)$$

$$\hat{y}_4^{(1)}(k) = 2.308 + 0.8264u(k-2) + 0.07835u(k-3) + 0.3418u^2(k-2)$$
$$+0.07381u(k-2)u(k-3) - 0.02513u^2(k-3)$$
$$\hat{y}_5^{(1)}(k) = 2.340 + 0.8691u(k-2) + 0.03525u(k-4) + 0.3767u^2(k-2)$$
$$+0.03743u(k-2)u(k-4) - 0.032246u^2(k-4)$$
$$\hat{y}_6^{(1)}(k) = 2.802 + 0.8377u(k-3) - 0.08453u(k-4) + 0.3773u^2(k-3)$$
$$-0.006983u(k-3)u(k-4) - 0.07361u^2(k-4)$$

TABLE 5.10.19 True, θ_i, and estimated, $\hat{\theta}_i$, parameters and estimated standard deviations of the residuals σ_ε at the identification of the nonparametric simple Wiener model by *nonlinear* submodels at the first layer

No	ϕ_i	θ_i	$\hat{\theta}_i$ (case (d))	$\hat{\theta}_i$ (case (e))	$\hat{\theta}_i$ (case (f))	$\hat{\theta}_i$ (case (g))
1	1	2.0	2.0743	2.0343	1.8938	1.9512
2	$u(k-1)$	0.6321	0.6837	0.6815	0.6228	0.6517
3	$u(k-1)$	0.2325	0.2083	0.2091	0.2408	0.2624
4	$u(k-3)$	0.0855	0.10856	0.0973	0.1246	0.1204
5	$u(k-4)$	0.03147	-0.00901	-0.00124	0.06515	-0.0089
6	$u^2(k-1)$	0.1998	0.2291	0.2295	0.2899	0.3076
7	$u^2(k-2)$	0.02703	0.02675	0.0315	0.1210	0.0930
8	$u^2(k-3)$	0.00366	0.00823	0.0132	0.0331	0.0389
9	$u^2(k-4)$	0.000495	0.00495	-0.00017	-0.00251	-0.0024
10	$u(k-1)u(k-2)$	0.14696	0.1245	0.1210	-	-
11	$u(k-2)u(k-3)$	0.0199	0.0120	-	-	-
12	$u(k-3)u(k-4)$	0.00269	0.0000178	-	-	-
13	$u(k-1)u(k-3)$	0.05407	0.0653	0.0649	-	-
14	$u(k-2)u(k-4)$	0.007317	-0.00609	-	-	-
15	$u(k-1)u(k-4)$	0.0199	-0.00246	-0.00088	-	-
	σ_ε		0.1582	0.1588	0.3734	0.3904

At the second layer first all output signals of the 6 submodels were considered as possible input signals of a linear static submodel. The parameter estimation resulted in the following equation:

$$\hat{y}(k) \equiv \hat{y}_1^{(2)}(k) = 0.6129\hat{y}_1^{(1)}(k) + 0.4178\hat{y}_2^{(1)}(k) - 0.0210\hat{y}_3^{(1)}(k)$$
$$+0.1627\hat{y}_4^{(1)}(k) - 0.1626\hat{y}_5^{(1)}(k) - 0.00256\hat{y}_6^{(1)}(k)$$

The parameters of the estimated model and the standard deviations of the residuals at the regression of the submodels are summarized in Table 5.10.20. The estimated parameters are also shown in Table 5.10.19, where the parameters of the original (simulated) system are also given for comparison. The estimated model is very good, its parameters approximate the true parameters well. The excellent coincidence between the simulated and computed output signals – based on model (d) – is seen in Figure 5.10.3.

TABLE 5.10.20 The estimated parameters and standard deviations of the residuals at the identification of the nonparametric simple Wiener model by structure (d)

No	$y_i^{(0)}$	$\phi_i^{(0)}$	$\hat{\theta}_i^{(1)}$	$\sigma_{\varepsilon_i}^{(1)}$	$\hat{\theta}_j^{(2)}$	$\sigma_{\varepsilon_i}^{(2)}$	$\hat{\theta}_i$	$\hat{\theta}_i^*$
1		1	2.041 ±0.04178				1.2509	2.0743
2		$u(k-1)$	0.6326 ±0.05428				0.3877	0.6837
3	$u(k-1)$	$u(k-2)$	0.3510 ±0.05413				0.2151	0.2038
4	$u(k-2)$	$u^2(k-1)$	0.1916 ±0.02436	0.3023	0.6129 ±0.02727		0.1174	0.2291
5		$u(k-1)u(k-2)$	0.2032 ±0.01673				0.1245	0.1245
6		$u^2(k-2)$	0.05286 ±0.02436				0.0324	0.02675
		1	2.010 ±0.05901				0.8398	-
		$u(k-1)$	0.7493 ±0.05518				0.3131	-
7	$u(k-1)$	$u(k-3)$	0.2345 ±0.05489				0.0980	0.10856
	$u(k-3)$	$u^2(k-1)$	0.2845 ±0.02452	0.4074	0.4178 ±0.03612		0.1189	-
8		$u(k-1)u(k-3)$	0.1564 ±0.01687				0.0653	0.0653
9		$u^2(k-3)$	0.03173 ±0.02452				0.0133	0.00823
		1	2.000 ±0.08876				-0.0042	-
		$u(k-1)$	0.8156 ±0.06943				-0.0171	-
10	$u(k-1)$	$u(k-4)$	0.1660 ±0.06890				-0.0035	-0.00901
	$u(k-4)$	$u^2(k-1)$	0.3416 ±0.03044	0.5816	-0.0210 ±0.03107		-0.00717	-
11		$u(k-1)u(k-4)$	0.1173 ±0.02105				-0.00246	-0.00246
12		$u^2(k-4)$	0.02280 ±0.03044				-0.00048	-0.00495
		1	2.308 ±0.1765			0.1582	0.3755	-
		$u(k-2)$	0.8264 ±0.2281				0.1345	-
	$u(k-2)$	$u(k-3)$	0.07835 ±0.2274				0.0127	-
	$u(k-3)$	$u^2(k-2)$	0.3418 ±0.1024	1.270	0.1627 ±0.0952		0.0556	-
13		$u(k-2)u(k-3)$	0.07381 ±0.07033				0.0120	0.0120
		$u^2(k-3)$	-0.02513 ±0.1024				-0.0041	-
		1	2.340 ±0.1859				-0.3805	-
		$u(k-2)$	0.8691 ±0.1738				-0.1413	-
	$u(k-2)$	$u(k-4)$	0.03525 ±0.1729				-0.00573	-
	$u(k-4)$	$u^2(k-2)$	0.3767 ±0.07722	1.283	-0.1626 ±0.09813		-0.06125	-
14		$u(k-2)u(k-4)$	0.03743 ±0.05321				-0.00609	-0.00609
		$u^2(k-4)$	-0.03224 ±0.07722				0.00524	-
		1	2.802 ±0.2819				-0.00717	-
		$u(k-3)$	0.8377 ±0.3661				-0.00214	-
	$u(k-3)$	$u(k-4)$	-0.08453 ±0.3651				0.000216	-
	$u(k-4)$	$u^2(k-3)$	0.3773 ±0.1643	2.039	-0.00256 ±0.0106		-0.000966	-
15		$u(k-3)u(k-4)$	-0.006983 ±0.1129				0.0000178	0.0000178
		$u^2(k-4)$	-0.07361 ±0.1643				0.000188	-

(e) *model e:*

The standard deviations of the residuals at the estimation of the submodels at the first layer in model (d) were

$$\sigma_{\varepsilon_1}^{(1)} = 0.3023 \qquad \sigma_{\varepsilon_2}^{(1)} = 0.4074 \qquad \sigma_{\varepsilon_3}^{(1)} = 0.5816$$

$$\sigma_{\varepsilon_4}^{(1)} = 1.270 \qquad \sigma_{\varepsilon_5}^{(1)} = 1.283 \qquad \sigma_{\varepsilon_6}^{(1)} = 2.039$$

Because the second three standard deviations are much bigger than the first three, the threshold was chosen in such a way that only the first three submodels' outputs were considered as inputs for the linear static submodel of the second layer. The estimated equation of the second layer was then:

$$\hat{y}(k) \equiv \hat{y}_1^{(2)}(k) = 0.5956\hat{y}_1^{(1)}(k) + 0.4148\hat{y}_2^{(1)}(k) - 0.007492\hat{y}_3^{(1)}(k)$$

The parameters of the estimated model and the standard deviations of the residuals at the regression of the submodels are summarized in Table 5.10.21. Comparing Tables 5.10.20 and 5.10.21 one can see that the standard deviation of the residuals are almost the same, even though the three worst submodels were omitted at the first layer. The estimated parameters are not as good as in case (d), as is seen in Table 5.10.19. This is trivial since some model components were not considered in the model by having omitted the three worst submodels at the first layer. The computed output signal – based on the estimated model – fits the simulated process output well, as is seen in Figure 5.10.3.

(f)   *model f:*
The memory of the model was 4. At the first layer 6 quadratic submodels were assumed. The input signals of the submodels were (differently) delayed input signal values. Now the cross-product terms were omitted from the quadratic models which means that the system was assumed of Hammerstein type. The estimated submodels were as follows:

$$\hat{y}_1^{(1)}(k) = 1.970 + 0.6012u(k-1) + 0.4136u(k-2) + 0.2852u^2(k-1)$$
$$+0.1465u^2(k-2)$$
$$\hat{y}_2^{(1)}(k) = 1.906 + 0.7444u(k-1) + 0.2958u(k-3) + 0.3483u^2(k-1)$$
$$+0.09551u^2(k-3)$$
$$\hat{y}_3^{(1)}(k) = 1.892 + 0.8313u(k-1) + 0.2262u(k-4) + 0.3809u^2(k-1)$$
$$+0.06204u^2(k-4)$$
$$\hat{y}_4^{(1)}(k) = 2.282 + 0.8150u(k-2) + 0.1011u(k-3) + 0.3757u^2(k-2)$$
$$+0.008791u^2(k-3)$$
$$\hat{y}_5^{(1)}(k) = 2.315 + 0.8680u(k-2) + 0.04991u(k-4) + 0.3918u^2(k-2)$$
$$-0.01706u^2(k-4)$$
$$\hat{y}_6^{(1)}(k) = 2.805 + 0.8688u(k-3) - 0.08668u(k-4) + 0.3741u^2(k-3)$$
$$-0.0768u^2(k-4)$$

At the second layer first all output signals of the 6 submodels were considered as possible input signals of a linear static submodel. The parameter estimation resulted in the following equation

$$\hat{y}(k) \equiv \hat{y}_1^{(2)}(k) = -0.4166\hat{y}_1^{(1)}(k) + 1.187\hat{y}_2^{(1)}(k) - 0.01241\hat{y}_3^{(1)}(k)$$
$$-0.5698\hat{y}_4^{(1)}(k) + 1.011\hat{y}_5^{(1)}(k) - 0.2013\hat{y}_6^{(1)}(k)$$

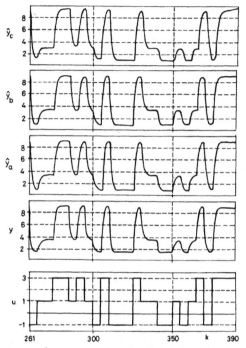

Fig. 5.10.3a Input u and output y signals of the simple Wiener model and the computed output $\hat{y}$ signals based on the estimated models

The parameters of the estimated model and the standard deviations of the residuals at the regression of the submodels are summarized in Table 5.10.22. The estimated parameters are also shown in Table 5.10.19 to compare them with the true and estimated parameters obtained from the other models (d), (e) and (g). The estimated model is worse than models (d) and (e), which fact can be seen from the larger standard deviation of the residuals and from the larger difference between the measured and computed output signals – based on model (f) – in Figure 5.10.3. The reason for the worse model matching is that the cross-product terms fail from the model. As a consequence, the *a priori* assumption that the model was of Hammerstein type must be dropped.

(g) *model g:*
The standard deviations of the residuals at the estimation of the submodels at the first layer in model (f) were

$$\sigma_{\varepsilon_1}^{(1)} = 0.4475 \qquad \sigma_{\varepsilon_2}^{(1)} = 0.5306 \qquad \sigma_{\varepsilon_3}^{(1)} = 0.6499$$

$$\sigma_{\varepsilon_4}^{(1)} = 1.271 \qquad \sigma_{\varepsilon_5}^{(1)} = 1.280 \qquad \sigma_{\varepsilon_6}^{(1)} = 2.030$$

Because the second three standard deviations are much bigger than the first three, the threshold was chosen in such a way that only the first three submodels' outputs were considered as inputs for the linear static submodel of the second layer. The estimated equation of the second layer was then

$$\hat{y}(k) \equiv \hat{y}_1^{(2)}(k) = 0.6345\hat{y}_1^{(1)}(k) + 0.4069\hat{y}_2^{(1)}(k) - 0.03934\hat{y}_3^{(1)}(k)$$

The parameters of the estimated model and the standard deviations of the residuals at the regression of the submodels are summarized in Table 5.10.23. Comparing Tables 5.10.22 and 5.10.23 one can see that the standard deviation of the residuals hardly increased on omitting the three submodels at the first layer. The computed output signal – based on the estimated model – fits well to the simulated process output, as is seen in Figure 5.10.3. ■

On the basis of the simulation runs (a) to (i) in Example 5.10.3 we could see that the method using quadratic models with a few (2) inputs at the first layer is before the method using quasi-linear models already at the first layer with all possible nonlinear (quadratic) terms created from the delayed input signal values.

Several recent papers report on the use of neural nets instead of a fixed static nonlinear structure in GMDH (e.g. Jirina (1994), Ivakhnenko and Müller (1994), Müller (1995), Xue and Watton (1995)).

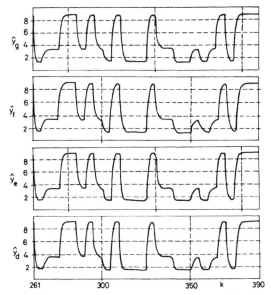

Fig. 5.10.3b Input u and output y signals of the simple Wiener model and the computed output $\hat{y}$ signals based on the estimated models

5.11 REFERENCES

Aguirre, L.A. and S.A. Billings (1995). Improved structure selection for nonlinear models based on term clustering, *Int. Journal of Control*, Vol. 62, 3, pp. 569–587.

Akaike, H. (1970). Statistical predictor identification. *Ann. Inst. Statist. Math.*, Vol. 22, pp. 203–217.

Akaike, H. (1972). Information theory and an extension of the maximum likelihood principle. *Proc. 2nd Int. Symposium on Information Theory, Supp. to Problems of Control and Information Theory*, pp. 267–281.

Aksenova, T.I. (1995). Sufficient condition and convergence rate using different criteria for model selection, *SAMS*, Vol. 20, pp. 69-78.

Bard, Y. and L. Lapidus (1970). Nonlinear system identification. *Ind. Eng. Chem. Fundam.*, Vol. 9, 4, pp. 628–633.

Baumgartner, S.L. and W.S. Rugh (1975). Complete identification of a class of nonlinear systems from steady state frequency response. *IEEE Trans. on Circuits and Systems*, CAS-22, 9, pp. 753–758.

TABLE 5.10.21 The estimated parameters and standard deviations of the residuals at the identification of the nonparametric simple Wiener model by structure (e)

No	$y_i^{(0)}$	$\phi_i^{(0)}$	$\hat{\theta}_i^{(1)}$	$\sigma_{\varepsilon_i}^{(1)}$	$\hat{\theta}_i^{(2)}$	$\sigma_{\varepsilon_i}^{(2)}$	$\hat{\theta}_i$	$\hat{\theta}_i^*$
1		1	2.041 ±0.04178				1.2516	2.0343
2		$u(k-1)$	0.6326 ±0.05428				0.3768	0.6815
3	$u(k-1)$	$u(k-2)$	0.3510 ±0.05413				0.2091	0.2091
4	$u(k-2)$	$u^2(k-1)$	0.1916 ±0.02436	0.3023	0.5956 ±0.02343		0.1141	0.1141
5		$u(k-1)u(k-2)$	0.2032 ±0.01673				0.1210	0.1210
6		$u^2(k-2)$	0.05286 ±0.02436				0.0315	0.8315
		1	2.010 ±0.05901				0.8337	-
		$u(k-1)$	0.7493 ±0.05518				0.3108	-
7	$u(k-1)$	$u(k-3)$	0.2345 ±0.05489				0.0973	0.0973
	$u(k-3)$	$u^2(k-1)$	0.2845 ±0.02452	0.4074	0.4148 ±0.03449	0.1588	0.1180	
8		$u(k-1)u(k-3)$	0.1564 ±0.01687				0.0649	0.0649
9		$u^2(k-3)$	0.03173 ±0.02452				0.0132	0.0132
		1	2.000 ±0.08876				-0.01499	-
		$u(k-1)$	0.8156 ±0.06943				-0.00611	-
10	$u(k-1)$	$u(k-4)$	0.1660 ±0.06890				-0.00124	-0.00124
	$u(k-4)$	$u^2(k-1)$	0.3416 ±0.03044	0.5816	-0.00749 ±0.03038		-0.00256	-
11		$u(k-1)u(k-4)$	0.1173 ±0.02105				-0.00088	-0.00088
12		$u^2(k-4)$	0.02280 ±0.03044				-0.00017	-0.00017
		1	2.308 ±0.1765					-
		$u(k-2)$	0.8264 ±0.2281					-
	$u(k-2)$	$u(k-3)$	0.07835 ±0.2274					-
	$u(k-3)$	$u^2(k-2)$	0.3418 ±0.1024	1.270				-
13		$u(k-2)u(k-3)$	0.07381 ±0.07033					-
		$u^2(k-3)$	-0.02513 ±0.1024					-
		1	2.340 ±0.1859					-
		$u(k-2)$	0.8691 ±0.1738					-
	$u(k-2)$	$u(k-4)$	0.03525 ±0.1729					-
	$u(k-4)$	$u^2(k-2)$	0.3767 ±0.07722	1.283				-
14		$u(k-2)u(k-4)$	0.03743 ±0.05321					-
		$u^2(k-4)$	-0.03224 ±0.07722					-
		1	2.802 ±0.2819					-
		$u(k-3)$	0.8377 ±0.3661					-
	$u(k-3)$	$u(k-4)$	-0.08453 ±0.3651					-
	$u(k-4)$	$u^2(k-3)$	0.3773 ±0.1643	2.039				-
15		$u(k-3)u(k-4)$	-0.006983 ±0.1129					-
		$u^2(k-4)$	-0.07361 ±0.1643					-

Bhansali, R.J. and D.Y. Downham (1977). Some properties of the order of an autoregressive model selected by a generalization of Akaike's FPE criterion. *Biometrica*, Vol. 64, pp. 547–551.

Billings, S.A. and S.Y. Fakhouri (1978a). Identification of a class of nonlinear systems using correlation analysis, *Proc. IEE*, Vol. 125, Pt D, 7, pp. 691–697.

Billings, S.A. and S.Y. Fakhouri (1978b). Theory of separable processes with application to the identification of nonlinear systems, *Proc. IEE*, Vol. 125, Pt D, 9, pp. 1051–1058.

Billings, S.A. and S.Y. Fakhouri (1980). Identification of nonlinear systems using correlation analysis and pseudorandom inputs. *Int. Journal of Systems Science*, Vol. 11, pp.261.

Billings, S.A. and S.Y. Fakhouri (1982). Identification of systems containing linear dynamic and static nonlinear elements, *Automatica*, Vol. 18, 1, pp. 15–26.

TABLE 5.10.22 The estimated parameters and standard deviations of the residuals at the identification of the nonparametric simple Wiener model by structure (f)

No	$y_i^{(0)}$	$\phi_i^{(0)}$	$\hat{\theta}_i^{(1)}$	$\sigma_{\varepsilon_i}^{(1)}$	$\hat{\theta}_i^{(2)}$	$\sigma_{\varepsilon_i}^{(2)}$	$\hat{\theta}_i$	$\hat{\theta}_i^*$
1		1	1.970 ±0.06123				-0.8207	1.8938
2		$u(k-1)$	0.6012 ±0.08026				-0.2505	0.6228
3	$u(k-1)$	$u(k-2)$	0.4136 ±0.07977	0.4475	-0.4166 ±0.5980		-0.1723	0.2408
4	$u(k-2)$	$u^2(k-1)$	0.2852 ±0.03422				-0.1188	0.2899
5		$u^2(k-2)$	0.1465 ±0.03422				-0.0610	0.1210
		1	1.906 ±0.07545				2.2624	-
		$u(k-1)$	0.7444 ±0.07187				0.8836	-
6	$u(k-1)$	$u(k-3)$	0.2958 ±0.07097	0.5306	1.187 ±0.6001		0.3511	0.1246
	$u(k-3)$	$u^2(k-1)$	0.3483 ±0.03065				0.4134	-
7		$u^2(k-3)$	0.09551 ±0.03065				0.1134	0.0331
		1	1.892 ±0.0968				-0.0235	-
		$u(k-1)$	0.8313 ±0.07752				-0.0103	-
8	$u(k-1)$	$u(k-4)$	0.2262 ±0.07604	0.6499	-0.01241 ±0.1379		-0.0028	0.06515
	$u(k-4)$	$u^2(k-1)$	0.3809 ±0.03309				-0.0047	-
9		$u^2(k-4)$	0.06204 ±0.03309			0.3734	-0.00077	-0.00251
		1	2.282 ±0.1739				-1.3003	-
		$u(k-2)$	0.8150 ±0.2279				-0.4644	-
	$u(k-2)$	$u(k-3)$	0.1011 ±0.2265	1.271	-0.5698 ±1.312		-0.0576	-
	$u(k-3)$	$u^2(k-2)$	0.3757 ±0.09718				-0.2141	-
		$u^2(k-3)$	0.008791 ±0.09718				-0.0050	-
		1	2.315 ±0.1822				2.3405	-
		$u(k-2)$	0.8680 ±0.1734				0.8775	-
	$u(k-2)$	$u(k-4)$	0.04991 ±0.1712	1.280	1.011 ±1.476		0.0505	-
	$u(k-4)$	$u^2(k-2)$	0.3918 ±0.07399				0.3961	-
		$u^2(k-4)$	-0.01706 ±0.07399				-0.0172	-
		1	2.805 ±0.2781				-0.5646	-
		$u(k-3)$	0.8388 ±0.3641				-0.1689	-
	$u(k-3)$	$u(k-4)$	-0.08668 ±0.3619	2.030	-0.2013 ±0.1012		0.01745	-
	$u(k-4)$	$u^2(k-3)$	0.3741 ±0.1553				-0.0753	-
		$u^2(k-4)$	-0.0768 ±0.1553				0.01546	-

Billings, S.A. and W.S.F. Voon (1986) A prediction error and stepwise regression algorithm for nonlinear systems, *Int. Journal of Control*, Vol. 44, pp. 803–822.

Cao, S. G. and G. Feng (1995). Modelling of complex control systems. *Proc. 3rd IFAC Symp. on Nonlinear Control Systems Design*, (Tahoe City: CA, USA), Vol. 2, pp. 849–854.

Chen, H.W. (1994). Modeling and identification of parallel and feedback nonlinear systems, *Proc. 33rd IEEE Conf. on Decision and Control*, (Lake Buena Vista: FL, USA), pp. 2267–2272.

Deergha, R.K. and D.C. Reddy (1992). Higher order correlations in the identification of a broad class of nonlinear systems. In: Higher Order Statistics, *Proc. Int. Workshop on Signal Processing*, pp. 313–316.

Desrochers, A.A. (1981). On an improved model reduction technique for nonlinear systems, *Automatica*, Vol. 17, 2, pp. 407–409.

Desrochers, A.A. and S. Mohseni (1984). On determining the structure of a nonlinear system, *Int. Journal of Control*, Vol. 40, 5, pp. 923–938.

Desrochers, A.A. and G.N. Saridis (1978). Identification of nonlinear nondynamic systems with applications to a hot steel rolling mill, *Proc. 7th IFAC World Congress*, (Helsinki: Finland), pp. 191–198.

Desrochers, A.A. and G.N. Saridis (1980). A model reduction technique for nonlinear systems, *Automatica*, Vol. 16, pp. 323–329.

Draper, N.R. and H. Smith (1966). *Applied Regression Analysis*, J. Wiley & Sons, New York.

Duffy, J.J. and M.A. Franklin (1975). A learning identification algorithm and its application to an environmental system, *IEEE Trans. on Systems, Man and Cybernetics*, SMC-5, 2, pp. 226–240.

Dunoyer, A., L. Balmer, K.J. Burnham and D.J.G. James (1996). *Int. Journal of Systems Science*, Vol. 22, 2, pp.43–58.

TABLE 5.10.23 The estimated parameters and standard deviations of the residuals at the identification of the nonparametric simple Wiener model by structure (g)

No	$y_i^{(0)}$	$\phi_i^{(0)}$	$\hat\theta_i^{(1)}$	$\sigma_{\varepsilon_i}^{(1)}$	$\hat\theta_i^{(2)}$	$\sigma_{\varepsilon_i}^{(2)}$	$\hat\theta_i$	$\hat\theta_i^*$
1		1	1.970 ±0.06123				1.2500	1.9512
2		$u(k-1)$	0.6012 ±0.08026				0.3815	0.0517
3	$u(k-1)$	$u(k-2)$	0.4136 ±0.07977	0.4475	0.6345 ±0.06451		0.2624	0.2624
4	$u(k-2)$	$u^2(k-1)$	0.2852 ±0.03422				0.1809	0.3076
5		$u^2(k-2)$	0.1465 ±0.03422				0.0930	0.0930
		1	1.906 ±0.07545				0.7756	-
		$u(k-1)$	0.7444 ±0.07187				0.3029	-
6	$u(k-1)$	$u(k-3)$	0.2958 ±0.07097	0.5306	0.4069 ±0.09722	0.3904	0.1204	0.1204
	$u(k-3)$	$u^2(k-1)$	0.3483 ±0.03065				0.1417	-
7		$u^2(k-3)$	0.09551 ±0.03065				0.0389	0.0389
		1	1.892 ±0.0968				-0.0744	-
		$u(k-1)$	0.8313 ±0.07752				-0.0327	-
8	$u(k-1)$	$u(k-4)$	0.2262 ±0.07604	0.6499	-0.03934 ±0.08558		-0.0089	-0.0089
	$u(k-4)$	$u^2(k-1)$	0.3809 ±0.03309				-0.0150	-
9		$u^2(k-4)$	0.06204 ±0.03309				-0.0024	-0.0024
		1	2.282 ±0.1739					-
		$u(k-2)$	0.8150 ±0.2279					-
	$u(k-2)$	$u(k-3)$	0.1011 ±0.2265	1.271				-
	$u(k-3)$	$u^2(k-2)$	0.3757 ±0.09718					-
		$u^2(k-3)$	0.008791 ±0.09718					
		1	2.315 ±0.1822					-
		$u(k-2)$	0.8680 ±0.1734					-
	$u(k-2)$	$u(k-4)$	0.04991 ±0.1712	1.280				-
	$u(k-4)$	$u^2(k-2)$	0.3918 ±0.07399					-
		$u^2(k-4)$	-0.01706 ±0.07399					-
		1	2.805 ±0.2781					-
		$u(k-3)$	0.8388 ±0.3641					-
	$u(k-3)$	$u(k-4)$	-0.08668 ±0.3619	2.030				-
	$u(k-4)$	$u^2(k-3)$	0.3741 ±0.1553					-
		$u^2(k-4)$	-0.0768 ±0.1553					-

Emara-Shabaik, H.E., K.A.F. Moustafa and J.H.S. Talaq (1995). On identification of parallel block cascade nonlinear models, *Int. Journal of Systems Science*, Vol. 26, 7, 1429–1438.

Farlow, J.S. (1984). (Editor) *Self-organizing methods in modelling*. Marcel Dekker, (New York: USA).

Gardiner, A.B. (1966). Elimination of the effect of nonlinearities on process crosscorrelations, *Electronics Letters*, Vol. 2, pp. 164.

Gardiner, A.B. (1968). Determination of the linear output signal of a process containing single valued nonlinearities, *Electronics Letters*, Vol. 4, pp. 224.

Gardiner, A.B. (1973a). Frequency domain identification of nonlinear systems. *Prepr. 3rd IFAC Symposium on Identification and System Parameter Estimation*, (Hague: The Netherlands), pp. 831–834.

Gardiner, A.B. (1973b). Identification of processes containing single valued nonlinearities, *Int. Journal of Control*, Vol. 5, pp. 1029–1039.

Haber, R. (1985). Structure identification of block oriented models based on the Volterra kernel, *Prepr. 7th IFAC Symposium on Identification and System Parameter Estimation*, (York: UK), (in the Appendix).

Haber, R. (1987). Two step structure identification method: best input–output model fitting from normal operating data and evaluation of its step responses. *Report TUV-IMPA-87/3*, Institute of Machine and Process Automatization, Technical University of Vienna, (Vienna: Austria).

Haber, R. (1989). Structure identification of block oriented models based on the estimated Volterra kernel, *Int. Journal of Systems Science*, Vol. 20, pp. 1355–1380.

Haber, R. and L. Keviczky (1974). Nonlinear structures for identification, *Periodica Polytechnica*, Vol. 18, pp. 393–404.

Haber, R. and L. Keviczky (1976). Identification of nonlinear dynamic systems – Survey paper, *Prepr. 4th IFAC Symposium on Identification and Parameter Estimation*, (Tbilisi: USSR), pp. 62–112.

Haber, R. and L. Keviczky (1985). Identification of linear systems having signal dependent parameters. *Int. Journal of Systems Science*, Vol. 16, pp. 869–884.

Haber, R., L. Keviczky and M. Hilger (1986). Process identification and control in the silicate industry, 4, Discrete time identification of dynamic processes, nonlinear systems (in Hungarian). *Scientific Publications of the Central Research and Design Institute for Silicate Industry*, Vol. 85.

Haber, R. and T. Perényi (1987). Identification of nonlinear systems by GMDH. *Report TUV-IMPA-87/4*, Institute of Machine and Process Automatization, Technical University of Vienna, (Vienna: Austria).

Haber, R. and R. Zierfuss (1987). Structure identification of simple cascade models by correlation method. *Report TUV-IMPA-87/1*, Institute of Machine and Process Automatization, Technical University of Vienna, (Vienna: Austria).

Haber, R. and H. Unbehauen (1990). Structure identification of nonlinear dynamic systems – A Survey on input–output approaches. *Automatica*, Vol. 26, 4, pp.651–677.

Hannan, E.J. and B.G. Quin (1979). The determination of the order of an autoregression. *Journal of the Royal Statistical Society - B*, Vol. 41, 2, pp. 190–195.

Hung, G., L. Stark and P. Eykhoff (1982). On the interpretation of kernels – Computer simulations of responses to impulse pairs, *Prepr. 5th IFAC Symposium on Identification and System Parameter Estimation*, (Washington: USA), pp. 417–420.

Ikeda, S., M. Ochiai and Y. Sawaragi (1976). Sequential GMDH algorithm and its application to river flow prediction, *IEEE Trans. on Systems, Man and Cybernetics*, SMC-6, 7, pp. 473–479.

Ivakhnenko, A.G. (1970). Heuristic self-organization in problems of engineering cybernetics, *Automatica*, Vol. 6, pp. 207–219.

Ivakhnenko, A.G. (1971). Polynomial theory of complex systems, *IEEE Trans. on Systems, Man and Cybernetics*, SMC-1, pp. 364–378.

Ivakhnenko, A.G. and J.A. Müller (1984). *Self-organization of Prediction Models* (in German), Verlag Technik, (Berlin: Germany).

Ivakhnenko, A.G. and J.A. Müller (1994). Present state and new problems of further GMDH development, *SAMS*, Vol. 20.

Janiszowski, K. (1986). Estimation of process models by means of the linear transformation of the measured signals (in German). *Messen, Steuern, Regeln*, Vol. 29, pp. 29–33.

Jirina, M. (1994). The modified GMDH: Sigmoidal and polynomial neural net. *Prepr. 11th IFAC Symp. on Identification and System Parameters Estimation*, (Copenhagen: Denmark), Vol. 2, pp. 309–311.

Kashyap, R.L. (1977). A Bayesian comparison of different classes of dynamic models using empirical data. *IEEE Trans. on Automatic Control*, AC-22, pp. 715–727.

Keulers, M., K. Sepp, K. Breur, A. Reyman and G. Reyman (1993). A simulation study of nonlinear structure identification of a fed-batch baker's yeast process, *Proc. American Control Conference*, (San Francisco: CA, USA), Vol. 3, pp. 2256-2260.

Korenberg, M.J. (1985). Identifying noisy cascades of linear and static nonlinear systems, *Proc. 7th IFAC Symposium on Identification and System Parameter Estimation*, (York: UK), pp. 421–426.

Korenberg, M.J., S.A. Billings, Y.P. Liu and P.J. McIlroy (1988). Orthogonal parameter estimation algorithm for nonlinear stochastic systems. *Int. Journal of Control*, Vol. 48, pp. 193–210.

Kortmann, M. and H. Unbehauen (1987). A new algorithm for automatic selection of the optimal structure at the identification of nonlinear systems (in German). *Automatisierungstechnik*, Vol. 12, pp. 491–498.

Kortmann, M. and H. Unbehauen (1988a). Structure detection in the identification of nonlinear systems. *Automatique Productique Informatique Industrielle*, Vol. 22, pp. 5–25.

Kortmann, M. and H. Unbehauen (1988b). Two algorithms for model structure determination of nonlinear dynamic systems with applications to industrial processes. *Prepr. 8th IFAC Symposium on Identification and System Parameter Estimation*, (Beijing: China), pp. 939–946.

Leontaritis, I.J. and S.A. Billings (1985a). Input–output parametric models for nonlinear systems. Part I.: Deterministic nonlinear systems. *Int. Journal of Control*, Vol. 41, pp. 303–328.

Leontaritis, I.J. and S.A. Billings (1985b). Input–output parametric models for nonlinear systems. Part II.: Stochastic nonlinear systems. *Int. Journal of Control*, Vol. 41, pp. 329–341.

Leontaritis, I.J. and S.A. Billings (1987). Model selection and validation methods for nonlinear systems. *Int. Journal of Control*, Vol. 45, pp. 311–341.

Li, C.J. and Jeon, Y.C. (1993). Genetic algorithm in identifying nonlinear auto-regressive with exogenous input models for nonlinear systems, *Proc. American Control Conference*, (San Francisco: CA, USA), Vol. 3, pp. 2305-2309.

Marmarelis, P.Z. and V.Z. Marmarelis (1978). *Analysis of Physiological Systems – The White Noise Approach*, Plenum Press, (New York: USA).

Marmarelis, P.Z. and K.I. Naka (1974). Identification of multi-input biological systems, *IEEE Trans. on Biomedical Eng.*, BME-21, pp. 88–101.

Mehra, R.K. (1977). Group method of data handling (GMDH): Review and Experience. (manuscript).

Mital, A. (1984). Prediction of human static and dynamic strengths by modified basic GMDH algorithm, *IEEE Trans. on Systems, Man and Cybernetics*, SMC-14, 5, pp. 773–776.

Müller, J.A. (1995). Self-organizing modelling. Present state and new problems, *SAMS*, Vol. 18–19, pp. 87–92.

Ravindra, H.V., M. Raghunandan, Y.G. Srinivasa and R. Krishnamurthy (1994). Tool wear estimation by group method of data handling in turning, *Int. Journal of Production Research*, Vol. 32, pp. 1295–1312.

Schetzen, M. (1965a). Measurements of the kernels of nonlinear systems of finite degree. *Int. Journal of Control*, Vol. 1, pp. 251–263.

Schetzen, M. (1965b). Synthesis of a class of nonlinear systems. *Int. Journal of Control*, Vol. 1, pp. 251–263.

Shimizu, S. and T. Nishikawa (1984). An analysis of mechanisms to occur the threshold in the GMDH, *Prepr. 9th IFAC World Congress*, (Budapest: Hungary).

Simeu, E. (1995). Application of NARMAX modelling to eddy current brake process. *Proc. 4th IEEE Conf. on Control Applications*, pp. 444–449.

Singh, Y.P. and S. Subramanian (1980). Frequency response identification of structure of nonlinear systems, *IEE Proc.*, Vol. 127, Pt. D, 3, pp. 77–82.

Sriniwas, R.G. and Y. Arkun, I.L. Chien and B.A. Ogunnaike (1995). Nonlinear identification and control of a high-purity distillation column: a case study. *Journal of Process Control*, Vol. 5, 3, pp. 149–162.

Thouvarez, F. and L. Jezequel (1996). Identification of NARMAX models on a modal base, *Journal of Sound and Vibration*, Vol. 189, 2, pp. 193–213.

Xue, Y. and J. Watton (1995). A self-organizing neural network approach to data-based modelling of fluid power systems dynamics using the GMDH algorithm, *Proc. Institute Mechanical Engineers*, Vol. 209, pp. 229–240.

Yao, L. (1996). Genetic algorithm based identification of nonlinear systems by sparse Volterra filters, *Proc. IEEE Conference on Emerging Technologies and Factory Automation*, Vol. 1, pp. 327–333.

Yoshimura, T., D. Deepack and H. Takagi (1985). Track/vehicle system identification by a revised group method of data handling (GMDH), *Int. Journal of Systems Science*, Vol. 16, 1, pp. 131–144.

Yoshimura, T., R. Kiyozumi, K. Nishino and T. Soeda (1982a). Prediction of air pollutant concentrations by revised GMDH algorithms in Tokushima Prefecture, *IEEE Trans. on Systems, Man and Cybernetics*, SMC-12, 1, pp. 50–55.

Yoshimura, T., U.S. Pandey, T. Takagi and T. Soeda (1982b). Prediction of the peek food using revised GMDH algorithm, *Int. Journal of Systems Science*, Vol. 13, 5, pp. 547–557.

Vandersteen, G. (1996). Nonparametric identification of the linear dynamic parts of nonlinear systems containing one static nonlinearity, *Proc. 35th IEEE Conf. on Decision and Control*, (Kobe: Japan), Vol. 1, pp. 1099–1102.

Zierfuss, R. and R. Haber (1987). Structure identification of linear-in-parameters nonlinear dynamic systems with stepwise regression and with forward regression using orthogonal model components (in German). *Report*, Institute of Machine and Process Automation, Technical University of Vienna, (Vienna: Austria).

6. Model Validity Tests

6.1 INTRODUCTION

It is of great importance to decide whether the model obtained by structure identification and parameter estimation is an adequate one or not. Since the true parameters of the process are unknown, we cannot compare the estimated parameters with them. However the residuals and the model's output signal can be calculated by means of the estimated parameters from the measured input and output signals.

A trivial and very illustrative test is to plot the measured and computed model output signals in the same coordinate system.

The estimation methods assume that the source noise, and therefore the computed residuals,
- are white noise,
- have zero mean value,
- are independent of the input signal and of the model components, and
- usually have a normal distribution.

The following methods are recommended for checking the residuals (Haber, 1993):
- checking the mean value for being zero;
- chi-square test;
- time sequence plot;
- normality test;
- run test of randomness;
- Durbin–Watson test for possible fitting of a first-order linear autoregressive model to the residuals;
- checking the auto-correlation function whether it has a Dirac function shape;
- plot of the residuals against the previous input, output, and residual values and against the actual input and computed noise-free model output signals;
- plot of the conditional mean value of the residuals according to the previous input, output, and residual values and against the actual input and computed noise-free model output signals;
- checking the cross-dispersional function of the residual and the previous input, output and residual values and the actual input and the computed noise-free model output signals for being zero;
- checking the cross-correlation function of the residual and the previous input, output and residual values and the actual input and computed noise-free model output signals for being zero.

The model validity test should be extended by checking the estimated model parameters for
- significance, and
- physical interpretation.

6.2 TIME SEQUENCE PLOT OF THE COMPUTED MODEL OUTPUT SIGNAL

Usually the aim of the identification is to deliver a good process model. A first glance at the plots of the measured and computed output signals based on the estimated parameters shows how good the model's fit is. The model output can be computed
- by neglecting the effect of the residuals,

- by taking into account the effect of the residuals.

Although the latter case helds only when the noise model is also considered correct, it is practical to simulate the noise-free output signal, as well. If the fitting is correct then the computed noise-free output signal fits the measured output signal but does not follow the high frequency fluctuations caused by the noises.

A qualitative measure of the goodness of fit is the mean square error (MSE) of the residuals

$$\text{MSE} = \frac{1}{N} \sum_{k=1}^{N} \left[y(k) - \hat{y}(k) \right]^2$$

which has to be normalized by the variance of the output signal. The relative mean square error (RMSE) of the residuals

$$\text{RMSE} = \frac{\dfrac{1}{N} \sum_{k=1}^{N} \left[y(k) - \hat{y}(k) \right]^2}{\dfrac{1}{N} \sum_{k=1}^{N} \left[y(k) - \bar{y}(k) \right]^2}$$

or its square root, is then a measure of the fitting. The smaller the RMSE the better the fitting.

Example 6.2.1   *Approximation of the noise-free simple Wiener model by simple linear and nonlinear structures (Haber and Zierfuss, 1987)*

The transfer function of the linear part is

$$\frac{V(s)}{U(s)} = \frac{1}{1 + 10s}$$

and the equation of the nonlinear static term is

$$Y = 2 + V + 0.5V^2$$

The test signal was a PRTS with maximum length 26, amplitude ±2, and mean value 0. The sampling time was 2 [s] and the minimum switching time of the PRTS signal was 5 times more. $N = 130$ data pairs were used for the identification. The following models were fitted by LS parameter estimation to the input–output data:

- *second-order linear model (L2):*

$$y(k) = (1.398 \pm 0.06697)y(k-1) - (0.4735 \pm 0.05563)y(k-2)$$
$$+ (0.1607 \pm 0.04092) + (0.3312 \pm 0.01561)u(k-1)$$
$$- (0.1794 \pm 0.02882)u(k-2)$$

- *second-order quadratic generalized Hammerstein model (H2):*

$$y(k) = (1.354 \pm 0.06704)y(k-1) - (0.4367 \pm 0.5541)y(k-2)$$
$$+(0.1195 \pm 0.03747) + (0.2133 \pm 0.02702)u(k-1)$$
$$-(0.1068 \pm 0.03118)u(k-2) + (0.05870 \pm 0.1153)u^2(k-1)$$
$$-(0.02832 \pm 0.01224)u^2(k-2)$$

- *second-order quadratic parametric Volterra model* (V2):

$$y(k) = (1.458 \pm 0.02812)y(k-1) - (0.5242 \pm 0.02325)y(k-2)$$
$$+(0.1281 \pm 0.01553) + (0.2228 \pm 0.01121)u(k-1)$$
$$-(0.1556 \pm 0.01308)u(k-2) + (0.211 \pm 0.005024)u^2(k-1)$$
$$-(0.07344 \pm 0.005405)u^2(k-2) + (0.08429 \pm 0.003487)u(k-1)u(k-2)$$

- *second-order simple bilinear model* (B2):

$$y(k) = (1.284 \pm 0.4964)y(k-1) - (0.4027 \pm 0.03649)y(k-2)$$
$$+(0.2644 \pm 0.3114) + (0.03576 \pm 0.008591)u(k-1)$$
$$+(0.04137 \pm 0.01028)u(k-2) + (0.08037 \pm 0.002031)u(k-1)y(k-1)$$
$$-(0.05727 \pm 0.003662)u(k-2)y(k-2)$$

- *second-order quadratic general polynomial model* (P2):

$$y(k) = (1.136 \pm 0.07565)y(k-1) - (0.2412 \pm 0.07194)y(k-2)$$
$$+(0.2781 \pm 0.02992) + (0.01655 \pm 0.009245)u(k-1)$$
$$+(0.01056 \pm 0.02235)u(k-2) + (0.07804 \pm 0.002057)u(k-1)y(k-1)$$
$$-(0.01449 \pm 0.01628)u(k-2)y(k-2) + (0.01382 \pm 0.003437)u^2(k-1)$$
$$-(0.001989 \pm 0.004902)u^2(k-2) - (0.024 \pm 0.0107)y^2(k-1)$$
$$+(0.009712 \pm 0.00536)y^2(k-2)$$

The input and output signals of the Wiener model and the computed output signals based on the identified models are drawn in Figure 6.2.1. As is seen, the outputs of following models fit well to the process output:
- second-order quadratic parametric Volterra model;
- second-order simple bilinear model;
- second-order quadratic polynomial model.   ∎

Example 6.2.2 *Approximation of the noisy simple Wiener model by simple linear and nonlinear structures (Haber and Zierfuss, 1987)*
The noise-free process and the test signal are the same as in Example 6.2.1. White noise with mean value of zero and standard deviation of 0.2 was superposed on the output signal. $N = 130$ data pairs were used for the identification. The extended LS parameter estimation method was applied with 10 iteration steps. The following models resulted:

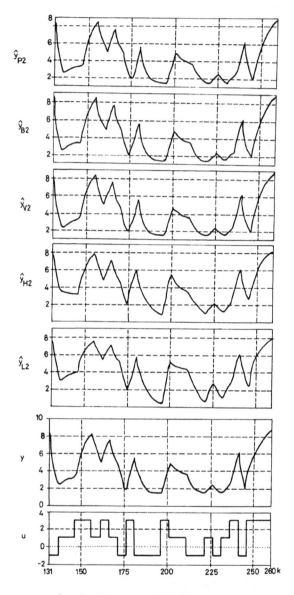

Fig. 6.2.1 The measured
input u and output y signals
of the noise-free simple
Wiener model, and the
computed output signals of
the estimated linear $(\hat{y}_{L2})$,
Hammerstein $(\hat{y}_{H2})$,
parametric Volterra $(\hat{y}_{V2})$,
bilinear $(\hat{y}_{B2})$ and general
polynomial $(\hat{y}_{P2})$ models

- *second-order linear model* (L2):

$$y(k) = -(0.02587 \pm 0.5536)y(k-1) + (0.6875 \pm 0.4526)y(k-2)$$
$$+(0.6415 \pm 0.1979) + (0.3825 \pm 0.02393)u(k-1)$$
$$+(0.3277 \pm 0.2151)u(k-2) + (0.8412 \pm 0.5583)\varepsilon(k-1)$$
$$+(0.05077 \pm 0.074)\varepsilon(k-2)$$

- *second-order quadratic generalized Hammerstein model* (H2):

$$y(k) = (0.3880 \pm 0.6571)y(k-1) + (0.3441 \pm 0.5352)y(k-2)$$
$$+ (0.4059 \pm 0.1926) + (0.3077 \pm 0.04774)u(k-1)$$
$$+ (0.09791 \pm 0.2016)u(k-2) + (0.03602 \pm 0.01965)u^2(k-1)$$
$$+ (0.03399 \pm 0.03457)u^2(k-2) + (0.3656 \pm 0.6615)\varepsilon(k-1)$$
$$+ (0.02393 \pm 0.08352)\varepsilon(k-2)$$

- *second-order quadratic parametric Volterra model* (V2):

$$y(k) = (1.478 \pm 0.1331)y(k-1) - (0.5425 \pm 0.1093)y(k-2)$$
$$+ (0.1254 \pm 0.0592) + (0.2936 \pm 0.03921)u(k-1)$$
$$- (0.2277 \pm 0.05441)u(k-2) + (0.003069 \pm 0.01687)u^2(k-1)$$
$$- (0.04835 \pm 0.01856)u^2(k-2) + (0.07449 \pm 0.01122)u(k-1)u(k-2)$$
$$+ (0.8674 \pm 0.1603)\varepsilon(k-1) + (0.03165 \pm 0.07699)\varepsilon(k-2)$$

- *second-order simple bilinear model* (B2):

$$y(k) = (0.9334 \pm 0.3417)y(k-1) - (0.1527 \pm 0.2489)y(k-2)$$
$$+ (0.4989 \pm 0.2137) + (0.07716 \pm 0.03391)u(k-1)$$
$$+ (0.03118 \pm 0.05708)u(k-2) + (0.06961 \pm 0.006469)u(k-1)y(k-1)$$
$$- (0.02088 \pm 0.02184)u(k-2)y(k-2) - (0.7180 \pm 0.3476)\varepsilon(k-1)$$
$$+ (0.08784 \pm 0.1875)\varepsilon(k-2)$$

- *second-order quadratic general polynomial model* (P2):

$$y(k) = (0.5189 \pm 0.2517)y(k-1) + (0.3552 \pm 0.2174)y(k-2)$$
$$+ (0.4007 \pm 0.1173) + (0.08995 \pm 0.03745)u(k-1)$$
$$- (0.04997 \pm 0.05417)u(k-2) + (0.06277 \pm 0.005915)u(k-1)y(k-1)$$
$$+ (0.06035 \pm 0.01942)u(k-2)y(k-2) + (0.00866 \pm 0.01295)u^2(k-1)$$
$$+ (0.01042 \pm 0.01338)u^2(k-2) - (0.2197 \pm 0.2517)y^2(k-1)$$
$$+ (0.0004812 \pm 0.008761)y^2(k-2) - (0.2197 \pm 0.2517)\varepsilon(k-1)$$
$$- (0.4370 \pm 0.1910)\varepsilon(k-2)$$

The measured input and output signals of the Wiener model and the computed noise-free output signals – based on the identified models – are drawn in Figure 6.2.2. As is seen, the outputs of following models fit well to the process output:
- second-order quadratic parametric Volterra model;
- second-order simple bilinear model;
- second-order quadratic polynomial model.   ■

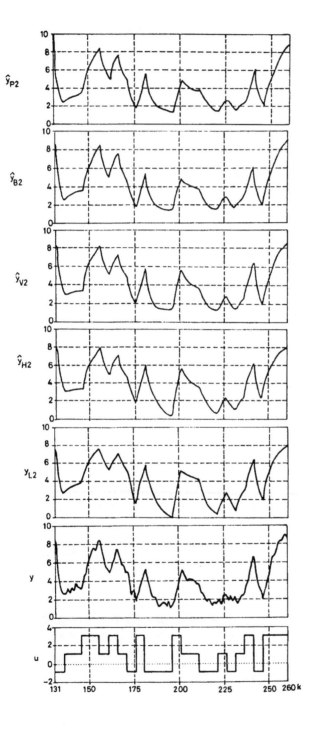

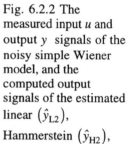

Fig. 6.2.2 The measured input u and output y signals of the noisy simple Wiener model, and the computed output signals of the estimated linear $(\hat{y}_{L2})$, Hammerstein $(\hat{y}_{H2})$, parametric Volterra $(\hat{y}_{V2})$, bilinear $(\hat{y}_{B2})$ and general polynomial $(\hat{y}_{P2})$ models

6.3 CHECKING THE MEAN VALUE OF THE RESIDUAL FOR ZERO VALUE

The expected value of the residuals

$$E\{\varepsilon(k)\} = \frac{1}{N}\sum_{k=1}^{N}\varepsilon(k)$$

and its standard deviation

$$\sigma_\varepsilon = E\left\{\left[\varepsilon(k) - E\{\varepsilon(k)\}\right]^2\right\} = \frac{1}{N-1}\sum_{k=1}^{N}\left[\varepsilon(k) - E\{\varepsilon(k)\}\right]^2$$

can be calculated from the residual sequence $\{\varepsilon(k)\}$. In the sequel we assume that the sequence of the residuals is stationary and ergodic. Then the expected and mean values coincide:

$$\bar{\varepsilon} = \overline{\varepsilon(k)} = E\{\varepsilon(k)\}$$

The mean value of the residuals has to be zero, otherwise either the structure or the parameter estimations are wrong. Any model structure that contains also a constant term leads to zero mean value if the parameter estimation is correct. There is no need for a constant term if the output signal is zero for zero input signal in the steady-state. To decide whether the calculated mean value of the residuals can be considered zero, the standard deviation also has to be calculated.

The standard deviation of the mean value of the residual is:

$$\sigma_{\bar{\varepsilon}} = \frac{\sigma_\varepsilon}{\sqrt{N}}$$

because

$$\sigma_{\bar{\varepsilon}}^2 = E\left\{\left[\frac{1}{N}\sum_{k=1}^{N}\varepsilon(k) - \bar{\varepsilon}\right]^2\right\} = \frac{1}{N^2}E\left\{\left[\sum_{k=1}^{N}\varepsilon(k) - N\bar{\varepsilon}\right]^2\right\}$$

$$= \frac{1}{N^2}E\left\{\left[\sum_{k=1}^{N}[\varepsilon(k) - \bar{\varepsilon}]\right]^2\right\} = \frac{1}{N^2}E\left\{\sum_{k=1}^{N}[\varepsilon(k) - \bar{\varepsilon}]^2\right\} = \frac{\sigma_\varepsilon^2}{N}$$

(We have used the assumption that the residuals reduced by their mean value are uncorrelated to each other.) The mean value of the residuals can be considered zero with 95 percent confidence if the following inequality is fulfilled:

$$|\bar{\varepsilon}| < 1.96\frac{\sigma_\varepsilon}{\sqrt{N}} \approx 2\frac{\sigma_\varepsilon}{\sqrt{N}}$$

The above inequality is valid if the distribution of the residuals is Gaussian. Then the mean value normalized by its estimated standard deviation has a Student (t) distribution.

Generally the standard deviation of the residuals does not have to be computed additionally from the residuals. It can be obtained from the normalized loss function of the parameter estimation, which quantity is usually computed with the estimated parameters:

$$\sigma_\varepsilon = E\left\{\left[\varepsilon(k) - \bar{\varepsilon}\right]^2\right\} \approx E\left\{\left[\varepsilon(k)\right]^2\right\} = \sqrt{\frac{1}{N-1}\sum_{k=1}^{N}\varepsilon^2(k)} \approx \sqrt{\frac{2}{N-1}J(\hat{\theta})}$$

Here $J(\hat{\theta})$ is the not normalized loss function

$$J(\hat{\theta}) = \frac{1}{2}\sum_{k=1}^{N}\varepsilon^2(k)$$

6.4 CHI-SQUARE TEST OF THE RESIDUALS

The residuals generally have a Gaussian distribution, because owing to the *Central Limit Theorem* the sum of a large amount of random data has a normal distribution independently of the distribution of the individual noise components. Since the sum of independent normalized Gaussian signals has a chi-square distribution we can test the residuals for whether they have the above feature.

Define the normalized residual values by

$$\varepsilon'(k) = \frac{\varepsilon(k) - E\{\varepsilon(k)\}}{\sigma_\varepsilon} \qquad k = 1, \ldots, N$$

The N-times sum of squares of the normalized residuals has to be less than the tabulated chi-square value at degree of freedom N with $100(1 - \alpha)$ percent probability:

$$\sum_{k=1}^{N}\varepsilon'^2(k) = \varepsilon'^T\varepsilon' \leq \Xi^2(N, \alpha)$$

6.5 TIME SEQUENCE PLOT OF THE RESIDUALS

As stated in Section 6.3 the values of the residuals should be within the domain

$$-1.95\sigma_\varepsilon \leq \varepsilon(k) \leq 1.95\sigma_\varepsilon$$

with probability 95 percent if the residuals have a normal distribution, which is usually the case.

Many consequences can be drawn from the time sequence plot of the residuals. If the model fit is correct then the followings can be observed (Draper and Smith, 1966):
- the residuals oscillate round the zero value;
- the residuals have no drift;
- almost all residuals are within the $\pm 2\sigma_\varepsilon$ range;
- the band of the residuals, i.e., a horizontal band in which most of residuals are, is symmetric about the zero line and has a constant width;
- the outliers, which are the values outside the $\pm 3\sigma_\varepsilon$, occur very seldom and are equally distributed, and are not correlated to any change in the input signal or in the model components;

- there seems to be no relation (correlation) between the residuals and the input signal;
- there seems to be no relation (correlation) between the residuals and the previous output signal values;
- there seems to be no relation (correlation) between the residuals and the computed noise-free model output.

The observation of the plot of the residuals simultaneously with the measured input and output signals is a very efficient tool of the model validity test. Box and Jenkins (1976) wrote in their famous classical book *Time Series Analysis: Forecasting and Control* the following: "It cannot be too strongly emphasized that visual inspection of a plot of the residuals themselves is an indispensable first step in the checking process".

Example 6.5.1 *Approximation of the noise-free simple Wiener model by simple linear and nonlinear structures (Haber and Zierfuss, 1987) (Continuation of Example 6.2.1)*

The plot of the residuals is seen in Figure 6.5.1. The computed standard deviation of the residuals for the different approximations was as follows:
- *second-order linear model* (L2): $\sigma_\varepsilon = 0.1673$
- *second-order quadratic generalized Hammerstein model* (H2): $\sigma_\varepsilon = 0.1502$
- *second-order quadratic parametric Volterra model* (V2): $\sigma_\varepsilon = 0.06224$
- *second-order simple bilinear model* (B2): $\sigma_\varepsilon = 0.045$
- *second-order quadratic general polynomial model* (P2): $\sigma_\varepsilon = 0.04175$

As is seen, the goodness of the approximation improves in the order of listing of the models. The linear and Hammerstein models are much worse than the others. There is very little difference between the bilinear and the general polynomial models.

If the model fit is good then the residuals should be within the band $|\varepsilon(k)| \leq 1.95\sigma_\varepsilon$.

This is also marked on the plots. The order of goodness of the models is supported also by this test, mainly if not only the number of occurrences when the residual leaves the limits is counted but also the sizes of the peaks.

Example 6.5.2 *Approximation of the noisy simple Wiener model by simple linear and nonlinear structures (Haber and Zierfuss, 1987) (Continuation of Example 6.2.2)*

The plot of the residuals is seen in Figures 6.5.2a and 6.5.2b. The computed standard deviation of the residuals for the different approximations was as follows:
- *second-order linear model* (L2): $\sigma_\varepsilon = 0.3295$
- *second-order quadratic generalized Hammerstein model* (H2): $\sigma_\varepsilon = 0.3198$
- *second-order quadratic parametric Volterra model* (V2): $\sigma_\varepsilon = 0.2640$
- *second-order simple bilinear model* (B2): $\sigma_\varepsilon = 0.2310$
- *second-order quadratic general polynomial model* (P2): $\sigma_\varepsilon = 0.2044$

As is seen, the goodness of the approximation improves in the order of listing of the models. The linear and Hammerstein models are much worse than the others. There is not too much difference between all other models. With the best, polynomial model the standard deviation of the residuals is almost equal to that of the source noise.

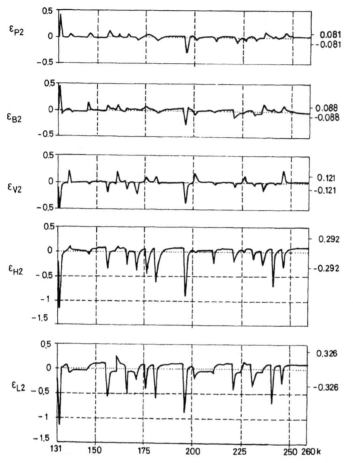

Fig. 6.5.1 The residuals of the estimated linear (ε_{L2}), Hammerstein (ε_{H2}), parametric Volterra (ε_{V2}), bilinear (ε_{B2}) and general polynomial (ε_{P2}) models at the identification of the noise-free simple Wiener model

The band $\pm 1.95\sigma_\varepsilon$ is also marked on the plots. Most of the residuals fall within the bounds with every model fitted, no relevant difference can be seen from the plots in this respect.   ∎

6.6 TIME SEQUENCE PLOT OF CERTAIN MEAN VALUES OF THE DATA TO THE ACTUAL TIME POINT

Cleveland and Kleiner (1975) suggested plotting certain mean values of the residuals at the actual time point. The following running mean values are to be plotted:

- mid-mean: the average value of all data up to the actual time point;
- lower semi-mid-mean: the average value of all data less than the mid-mean value up to the actual time point;
- upper semi-mid-mean: the average value of all data greater than the mid-mean value up to the actual point.

The time sequence plots of the mean values tend to a constant value if the residuals are independent of the input signal and have no drift. Of course, the mid-mean itself should tend to zero.

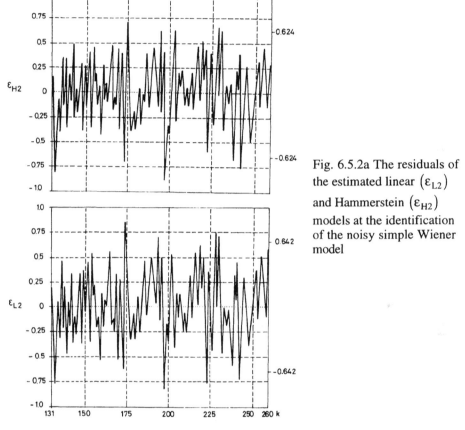

Fig. 6.5.2a The residuals of the estimated linear $\left(\varepsilon_{L2}\right)$ and Hammerstein $\left(\varepsilon_{H2}\right)$ models at the identification of the noisy simple Wiener model

6.7 PLOTTING OF THE HISTOGRAM OF THE RESIDUALS

A plot of the amplitude histogram of the residuals shows how random the residuals are. If the shape of the empirical density histogram distribution differs from the known bell curve of the Gaussian distribution we can guess that the estimated model is not correct. The reason for the high probability that the residuals commonly have a normal distribution is that the sum of large number of random noises has a Gaussian distribution independently of their distribution.

It can happen of course that the source noise has a different distribution from Gaussian, and then the residuals will have the same non-Gaussian distribution, although the identification is correct.

In any case the median of the distribution should fall into the zero abscissa because this fact shows that the residuals have a zero mean value.

650   Chapter 6

6.8 NORMALITY TEST OF THE RESIDUALS

As explained in the previous section, the residuals usually have a Gaussian distribution if the estimated model is correct.

There are different normality tests that can show whether the residuals have a normal (Gaussian) distribution or not:

- plot of the empirical probability density or distribution function;
- plot of the inverse Gaussian cumulative distribution function of the empirical cumulative distribution function;
- chi-square test;
- calculation of the skewness;
- calculation of the kurtosis.

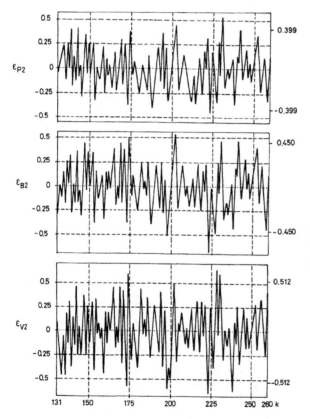

Fig. 6.5.2b The residuals of the estimated parametric Volterra $\left(\varepsilon_{V2}\right)$, bilinear $\left(\varepsilon_{B2}\right)$ and general polynomial $\left(\varepsilon_{P2}\right)$ models at the identification of the noisy simple Wiener model

It is practical to perform the test on the normalized residuals

$$\varepsilon'(k) = \frac{\varepsilon(k) - \mathrm{E}\{\varepsilon(k)\}}{\sigma_\varepsilon} \approx \frac{\varepsilon(k)}{\sigma_\varepsilon}$$

The normality test methods have already been described in Section 4.5 in detail.

6.9 FREQUENCY DOMAIN ANALYSIS OF THE RESIDUALS

If the residuals form a white noise sequence then its frequency spectrum is uniform over the frequency range $-\infty < \omega < \infty$. The plot of the residual frequency spectrum should also show approximately equal amplitude values at all frequencies. The occasional rises show modeling errors at the frequencies where they occur. The errors can be divided into three main types:
- low frequency error;
- middle frequency error;
- high frequency error.

The low frequency error is a consequence of the modeling error in the steady state relation. The middle frequency error occurs in the middle of the frequency bandwidth of the process and shows modeling errors in the dominant time constants of the linearized model. The high frequency errors show modeling errors in the small time constants (relative to the dominant). The high frequency errors are less critical because the very small time constants that would cause too large memory vector of the parametric model (e.g., of order larger than three) are usually neglected at the identification.

6.10 RUN TEST OF THE RANDOMNESS OF THE RESIDUALS

The run test presents a measure of the randomness of the residuals. To perform the test check every residual value whether it is greater or less than the mean value, i.e., whether it is positive or negative. Define the following values:
- N^+:   the number of positive values;
- N^-:   the number of negative values;
- run: a group where the sign of the consecutive residual values is the same;
- N_r: the number of runs.

Obviously $N = N^+ + N^-$.

The randomness of the residuals is minimum if only two runs exist, i.e., the first group contains only positive (or negative) residuals and all further data (which build the second group) have the opposite sign. One would think that the maximal randomness of the residuals happens if all neighboring values have different signs, i.e.,

$$N^+ = N^- = \frac{N_r}{2} = \frac{N}{2} \rightarrow N_r = N$$

if the number of data is even. This is obviously not correct, because in this case the residuals have a dominant high frequency component with half period time equal to the sampling time. The maximal randomness is characterized by a number between 2 and N and lying far away from both limits:

$$2 << N_{\text{ropt}} << N$$

As an example of what has been stated above consider a PRBS test signal that was planned for maximal randomness, since its auto-correlation function is similar to that of white noise.

Example 6.10.1 *PRBS test signal*

Take a sequence of maximal length with period 15 and lower and upper values -2 and 2 (Figure 2.4.1c):

$$2, -2, -2, -2, 2, -2, -2, 2, 2, -2, 2, -2, 2, 2, 2$$

Replace the values by their signs and separate the runs by brackets

$$+), (-, -, -), (+), (-, -), (+, +), (-), (+), (-), (+, +, +$$

The number of runs is $N_r = 8$, a number which is far from either 1 or 15. Observe that $N^+ = 8$ and $N^- = 7$ and $8 \approx 15/2$. ∎

From the above example and from physical considerations it follows that the optimal number of the runs is half the number of the data. However, this is correct exactly only if the number of positive and negative signs is equal. In practice the number of positive and negative signs may differ from each other. Even if they were to be equal, and therefore the maximal randomness of the residuals would be characterized by $N_r = N/2$, we need a measure of whether the observed N_r is close enough to the optimal value or not.

The question is whether the residuals can be considered random enough or not in the knowledge of N^+, N^- and N_r. The mean value of the number of runs $\bar{N}_r$ and its standard deviation σ_{N_r} is (Draper and Smith, 1966; Bendat and Piersol, 1971):

$$\bar{N}_r = \frac{2N^+N^-}{N} + 1 \tag{6.10.1}$$

$$\sigma_{N_r} = \sqrt{\frac{2N^+N^-\left(2N^+N^- - N\right)}{N^2(N-1)}} \tag{6.10.2}$$

The above formulas are valid if $N^+ > 10$ and $N^- > 10$. The residuals can be considered random enough if

$$\bar{N}_r - 2\sigma_{N_r} < N_r < \bar{N}_r + 2\sigma_{N_r}$$

In the special case where the number of positive and negative signs is equal, i.e.,

$$N^+ = N^- = \frac{N}{2}$$

then (6.10.1) and (6.10.2) reduce to

$$\bar{N}_r = \frac{N}{2} + 1$$

$$\sigma_{N_r} = \sqrt{\frac{N}{4} \frac{N-2}{N-1}}$$

For a number of data that is large they tend to

$$\left. \begin{array}{l} \overline{N}_r \approx \dfrac{N}{2} + 1 \\[4mm] \sigma_{N_r} \approx \dfrac{\sqrt{N}}{2} \end{array} \right\} \quad \text{if } N \to \infty$$

We see that our intuition that the maximal randomness is characterized by $N_r = N/2$ if $N^+ = N^-$ was correct.

It should be noted that from the residuals having zero mean one does not conclude that the number of the positive and the negative terms is equal.

6.11 DURBIN–WATSON TEST OF THE RESIDUALS

The Durbin–Watson test (Durbin and Watson, 1951; Draper and Smith, 1966) can be used for checking whether a first-order autoregressive model can be fitted to the residuals, i.e.,

$$\varepsilon(k) + a_1 \varepsilon(k-1) = \varepsilon_0(k)$$

where $\varepsilon_0(k)$ is Gaussian white noise signal. The following statistic measure has to be calculated

$$\rho = \frac{E\left\{[\varepsilon(k) - \varepsilon(k-1)]^2\right\}}{E\left\{[\varepsilon(k)]^2\right\}}$$

Durbin and Watson (1951) presented the critical values $d_U = d_U(N, n_\theta, \alpha)$ and $d_L = d_L(N, n_\theta, \alpha)$ for deciding whether the residuals are uncorrelated or not. (The significance level $100(1-\alpha) = 95$ percent is recommended.) Based on the calculated test values one can decide as follows (Draper and Smith, 1966):
- if $\rho > d_U$ the residuals are uncorrelated;
- if $\rho < d_L$ the residuals are correlated

The following trends can be observed (Draper and Smith, 1966):
- d_L increases with the number of data;
- d_U decreases with the number of data;
- d_L decreases with the accuracy;
- d_U decreases with the accuracy;
- d_L decreases with the number of parameters;
- d_U increases with the number of parameters.

Some values are given for information (Draper and Smith, 1966):
- $d_L(N = 100, n_\theta = 1, \alpha = 0.05) = 1.65$
- $d_L(N = 100, n_\theta = 5, \alpha = 0.05) = 1.57$
- $d_U(N = 100, n_\theta = 1, \alpha = 0.05) = 1.69$

• $d_U(N = 100, n_\theta = 5, \alpha = 0.05) = 1.78$.

The auto-correlation function analysis, presented in the next section, checks whether any linear autoregressive model can be fitted to the residuals. That test makes the Durbin–Watson test superfluous. In the nonlinear case, neither the general nor the special test based on the auto-correlation function can show whether the model fit is perfect. This problem will be treated in Sections 6.12 and 6.16.

6.12 ANALYSIS OF THE AUTO-CORRELATION FUNCTION OF THE RESIDUALS

If the model fit is good then the residuals are uncorrelated and their auto-correlation function is zero for any shifting times except zero

$$r_{\varepsilon\varepsilon}(\kappa) = \begin{cases} \sigma_\varepsilon^2 & \text{if} & \kappa = 0 \\ 0 & \text{if} & \kappa \neq 0 \end{cases}$$

It is practical to calculate the auto-correlation function of the normalized residuals $\varepsilon'(k)$. Their auto-correlation function should be

$$r_{\varepsilon'\varepsilon'}(\kappa) = \begin{cases} 1 & \text{if} & \kappa = 0 \\ 0 & \text{if} & \kappa \neq 0 \end{cases}$$

if the model is good. The standard deviation of the auto-correlation function is (e.g., Box and Jenkins, 1976)

$$\hat{\sigma}_{r_{\varepsilon'\varepsilon'}} = \sqrt{\hat{var}\{r_{\varepsilon'\varepsilon'}(\kappa)\}} = \frac{1}{\sqrt{N}}$$

The residuals are uncorrelated if the auto-correlation function for any non-zero shifting time is

$$-\frac{2}{\sqrt{N}} \approx -\frac{1.96}{\sqrt{N}} < r_{\varepsilon'\varepsilon'}(\kappa) < \frac{1.96}{\sqrt{N}} \approx \frac{2}{\sqrt{N}} \qquad \forall \kappa \neq 0$$

The plot of the auto-correlation function is a useful tool for model validity test, because:
• if the auto-correlation function has a periodic trend then the residuals have the periodic trend with the same frequency;
• the deviation of the auto-correlation function from zero for large shifting times means steady-state process modeling error;
• the deviation of the auto-correlation function from zero in the small shifting range area (except zero) means a process modeling error in the high frequency domain.

Box and Pierce (1970) have shown that for linear systems the sum of squares of the normalized auto-correlation functions from shifting time 1 to κ_{max}

$$\sum_{\kappa=1}^{\kappa_{max}} r_{\varepsilon'\varepsilon'}^2(\kappa) \tag{6.12.1}$$

is approximately chi-square distributed with $\left(\kappa_{max} - n_\theta \approx \kappa_{max}\right)$ degree of freedom, where n_θ is the number of parameters. In the case of a satisfactory fitting the sum (6.12.1) should be less than the chi-square value $\Xi^2(\kappa_{max}, \alpha)$ with the above degree of freedom and $100(1-\alpha) = 95$ percent probability. The above test is based on two facts. First, that the sum of squares of independent Gaussian random variables with zero mean and unit standard deviation is chi-squared distributed with degree of freedom equal to the number of squares. Second, that the auto-correlation function values at any shifting time differing from zero can be considered as owing to the *Central Limit Theorem* random variables with Gaussian distribution .

The auto-correlation function test checks only whether a linear relation between the actual and previous values of the residuals exists or not. It can happen, however, that there exists a nonlinear relation between two residual values and, nevertheless, the auto-correlation function has a Dirac function form. The above problem will be dealt with in Section 6.16 in details.

Example 6.12.1 *Approximation of the noise-free simple Wiener model by simple linear and nonlinear structures (Haber and Zierfuss, 1987) (Continuation of Examples 6.2.1 and 6.5.1)*

The plots of the normalized auto-correlation functions of the residuals are seen in Figures 6.12.1a and 6.12.1b. The band $\pm 2/\sqrt{N} = \pm 2/\sqrt{130} = \pm 0.175$ is also marked. The number of places where the auto-correlation function leaves the bound is given below:

- *second-order linear model* (L2): 4;
- *second-order quadratic generalized Hammerstein model* (H2): 6;
- *second-order quadratic parametric Volterra model* (V2): 5;
- *second-order simple bilinear model* (B2): 3;
- *second-order quadratic general polynomial model* (P2):4.

The difference is not significant. However, the magnitude of the outliers is much better with the linear, Hammerstein, and parametric Volterra models than with the bilinear and general polynomial models.   ■

Example 6.12.2 *Approximation of the noisy simple Wiener model by simple linear and nonlinear structures (Haber and Zierfuss, 1987) (Continuation of Examples 6.2.2 and 6.5.2)*

The plots of the normalized auto-correlation functions of the residuals are seen in Figures 6.12.2a and 6.12.2b. The band $\pm 2/\sqrt{N} = \pm 2/\sqrt{130} = \pm 0.175$ is also marked. The number of places where the auto-correlation function leaves the bound is given below:

- *second-order linear model* (L2): 4;
- *second-order quadratic generalized Hammerstein model* (H2): 5;
- *second-order quadratic parametric Volterra model* (V2): 3;
- *second-order simple bilinear model* (B2): 2;
- *second-order quadratic general polynomial model* (P2):1.

The difference seems to be significant. However, the magnitudes of the outliers hardly exceeds the limits. If another limit were taken into account the distribution of the number of outliers would be different. Therefore, a model validity test based only on the normalized auto-correlation function of the residuals is not a reliable method.   ∎

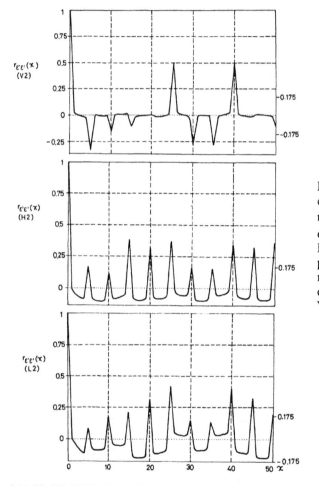

Fig. 6.12.1a The auto-correlation functions of the residuals $r_{\varepsilon'\varepsilon'}(\kappa)$ of the estimated linear (L2), Hammerstein (H2) and parametric Volterra (V2) models at the identification of the noise-free simple Wiener model

6.13 PLOT OF THE RESIDUALS AGAINST THE INPUT, OUTPUT AND PREVIOUS RESIDUAL VALUES

If the structure and parameter estimation were performed correctly then the residuals build a white noise sequence and they do not depend on the:
- actual and previous input signal values;
- the previous measured output signals;
- the actual computed noise-free output signal;
- the previous residual values.

Consequently a scatter plot of the residuals against all the above signal values should show that there is no relation between them and the residual.

The input signal and the source noise are independent from each other and the output

signal is a function of the two signals. Therefore it is enough to plot the residuals either against
- the actual and previous input signal values, and
- the previous residual values,

or against
- the actual and previous input signal values, and
- the previous measured output signals.

It is difficult to plot multi-dimensional scatter plots, therefore the residuals are usually plotted only against one signal value (component).

If it turns out that the residuals obviously have a functional relationship to a signal component investigated, then the above component should be tried to be used as a model component in addition to the ones previously used.

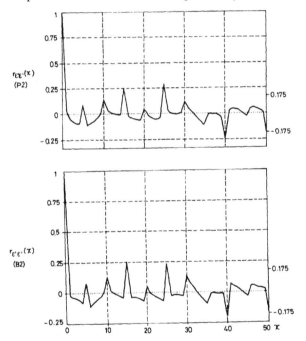

Fig. 6.12.1b The auto-correlation functions of the residuals $r_{\varepsilon'\varepsilon'}(\kappa)$ of the estimated bilinear (B2) and general polynomial (P2) models at the identification of the noise-free simple Wiener model

6.14 COMPUTATION AND PLOT OF THE CONDITIONAL MEAN VALUE OF THE RESIDUALS AGAINST THE INPUT, OUTPUT, AND THE PREVIOUS RESIDUAL VALUES

When evaluating the scatter plots of the residuals against the actual and previous input, output, and residual values defined in Section 6.13, respectively, we, indeed, examine the conditional mean value of the residuals according to the above variables. The conditional mean value is, however, a more exact, quantifiable function than the averaged value obtained from the scatter plot by eye.

The multi-dimensional conditional mean value of the residuals against the signals $x_i(k)$, $i = 1, ..., M$ is denoted by

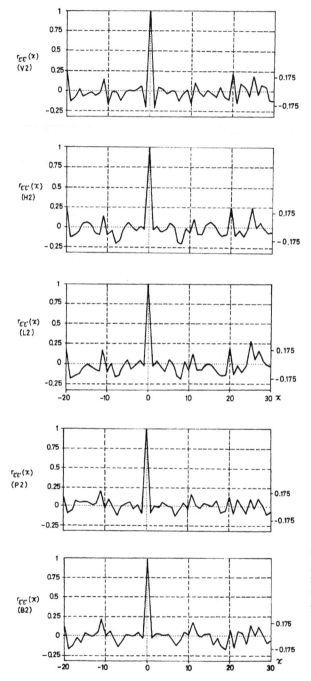

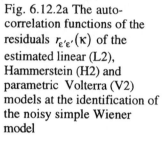

Fig. 6.12.2a The auto-correlation functions of the residuals $r_{\varepsilon'\varepsilon'}(\kappa)$ of the estimated linear (L2), Hammerstein (H2) and parametric Volterra (V2) models at the identification of the noisy simple Wiener model

Fig. 6.12.2b The auto-correlation functions of the residuals $r_{\varepsilon'\varepsilon'}(\kappa)$ of the estimated bilinear (B2) and general polynomial (P2) models at the identification of the noisy simple Wiener model

$$E\{\varepsilon(k)|\mathbf{x}(k)\} = E\{\varepsilon(k)|x_1(k), \ldots, x_M(k)\}$$

where the signals $x_i(k)$, $i = 1, \ldots, M$

$$\mathbf{x}(k) = [x_1(k), \ldots, x_M(k)]^T$$

can be the same signal values as defined in Section 6.13 for independent variables of the scatter plots:

- the actual and previous input signal values: $u(k - \kappa)$, $\kappa = 0, 1, \ldots$;
- the measured previous output signals: $y(k - \kappa)$, $\kappa = 1, 2, \ldots$;
- the actual computed noise-free output signal: $\hat{w}(k)$
- the previous residual values: $\varepsilon(k - \kappa)$, $\kappa = 1, 2, \ldots$.

According to the remark in Section 6.13 it is enough to calculate the conditional mean values at different values of either

- the actual and previous input signal values, and
- the previous residual values,

or

- the actual and previous input signal values, and
- the previous measured output signals.

To compute the conditional mean value of the residuals the amplitude range of each component $x_i(k)$, $i = 1, \ldots, M$ has to be divided into N_q intervals. The lower and upper bounds of the jth domain of signal $x_i(k)$ are the elements of the vector

$$\mathbf{x}_{i\min} = \left[x_{i1\min}, \ldots, x_{iN_q\min}\right]^T$$

$$\mathbf{x}_{i\max} = \left[x_{i1\max}, \ldots, x_{iN_q\max}\right]^T$$

respectively.

The one-dimensional conditional mean value of the residual is defined as the mean value of those residual values which correspond to the discrete time points k where the following conditions are fulfilled

$$x_{ij_i\min} \le x_i(k) < x_{ij_i\max} \quad i = 1, \ldots, M \tag{6.14.1}$$

Rajbman and Chadeev (1980) recommended the total number of intervals for one signal as

$$N_q \approx 1 + 3.32\lg(N)$$

This value is less than the total number of intervals recommended for the chi-square test of normality (Table 4.5.1).

If the dependence of the residuals on M signals is searched then each signal $x_i(k)$, $i = 1, \ldots, M$, has to be divided into N_q intervals. The total number of domains is then

$$N_{qM} = N_q M$$

Now an M-dimensional conditional mean value has to be computed. Collect the lower and upper limits of the domains into the vectors

$$x_{min} = \left[x_{min}(1), ..., x_{min}\left(N_{qM}\right) \right]^T$$

$$x_{max} = \left[x_{max}(1), ..., x_{max}\left(N_{qM}\right) \right]^T$$

respectively. The model fit is perfect if all conditional mean values are zero. This fact can be checked if the following quantities are computed:
- the conditional mean value in the jth interval

$$\hat{\bar{\varepsilon}}_j = \frac{1}{N} \sum_{k=1}^{N} \varepsilon(k) \left| x_{min}(j) \le x(k) < x_{max}(j) \right.$$

- the standard deviation of the conditional mean in the jth interval

$$\hat{\sigma}_{\varepsilon_j} = \sqrt{\frac{1}{N_j - 1} \sum_{k=1}^{N} \left[\varepsilon(k) - \hat{\bar{\varepsilon}}_j \right]^2 \left| x_{min}(j) \le x(k) < x_{max}(j) \right.}$$

If the residuals have a Gaussian distribution then the random variable

$$\frac{\varepsilon(k) - \hat{\bar{\varepsilon}}_j}{\hat{\sigma}_{\varepsilon_j}} \sqrt{N_j}$$

has a Student (t) distribution with $\left(N_j - 1\right)$ degree of freedom. The computed conditional mean value can be considered zero if it falls into the confidence interval

$$-\frac{\hat{\sigma}_{\varepsilon_j}}{\sqrt{N_j}} t\left(N_j - 1, \alpha\right) \le \hat{\bar{\varepsilon}}_i < \frac{\hat{\sigma}_{\varepsilon_j}}{\sqrt{N_j}} t\left(N_j - 1, \alpha\right)$$

where $t\left(N_j - 1, \alpha\right)$ is the Student (t) value at $100(1-\alpha)$ percent probability level. If the number of samples that fall in the jth interval N_j is larger than 10, then the model fitting is perfect with 95 percent probability if

$$-2\frac{\hat{\sigma}_{\varepsilon_j}}{\sqrt{N_j}} \le \hat{\bar{\varepsilon}}_i < 2\frac{\hat{\sigma}_{\varepsilon_j}}{\sqrt{N_j}}$$

is satisfied.
    A weaker test can be defined when not each domain will be investigated separately, but a cumulative quantity will be computed and checked.
    For convenience often only the one-dimensional conditional mean values of the

residuals are computed. This is the case if the conditional mean values are computed only according to one signal component, e.g., according to $x_i(k)$

$$E\{\epsilon(k)|x_i(k)\} = E\{\epsilon(k)|x_{ij,\min} \le x_i(k) < x_{ij,\max}\}$$

The one-dimensional conditional mean value is the mean value of all multi-dimensional conditional mean values corresponding to the same interval in the variable investigated. Therefore the same test can be applied as in the multi-dimensional case. However, it can happen that the multi-dimensional conditional mean differs significantly from zero in some domains, but the one-dimensional conditional mean values can be considered zero because of the compensation effect of other domains belonging to the same interval of the investigated signal component. Consequently the acceptance of the one-dimensional conditional mean value test is only a necessary, but not sufficient, condition for the correct modeling. In spite of all these factors it is a useful test.

6.15 ANALYSIS OF THE CROSS-DISPERSIONAL FUNCTIONS OF THE RESIDUALS AND THE INPUT, OUTPUT, AND PREVIOUS RESIDUAL VALUES

In addition to the conditional mean value $E\{\epsilon(k)|x(k)\}$ being zero, its standard deviation around zero has to be small enough to ensure that the conditional mean value is zero with a significance enough level. The cross-dispersion coefficient of the normalized variables of $\epsilon'(k)$ and $x'(k)$

$$D_{x'\epsilon'} = \text{var}\{E_{x'}\{E_{\epsilon'}\{\epsilon'(k)|x'(k)\}\}\}$$

$$= E_{x'}\{[E_{\epsilon'}\{\epsilon'(k)|x'(k)\} - E_{\epsilon'}\{\epsilon'(k)\}]^2\}$$

is a measure for the variance of the conditional mean values about the mean value of the normalized residuals. (The latter is zero if there is no modeling error.)

If both the process and the noise model are correct then all conditional mean values are zero and the cross-dispersional coefficient is zero.

Similarly to the analysis of the conditional mean values the cross-dispersional coefficients have to be computed and checked between the above signals
- the actual and previous normalized input signal values: $u'(k - \kappa)$, $\kappa = 0, 1, \ldots$;
- the measured normalized previous output signals: $y'(k - \kappa)$, $\kappa = 1, 2, \ldots$;
- the actual computed noise-free normalized output signal: $\hat{w}'(k)$;
- the previous normalized residual values: $\epsilon'(k - \kappa)$, $\kappa = 1, 2, \ldots$.

and between the normalized residuals $\epsilon'(k)$.

According to the remark in Section 6.13 it is enough to calculate the cross-dispersional coefficients of either
- the actual and previous normalized input signal values, and
- the previous normalized residual values,

or

- the actual and previous normalized input signal values, and
- the previous measured normalized output signals,

and of the normalized residuals.

According to its definition the multi-dimensional cross-dispersional function contains cross-dispersional coefficients belonging to series of shifted values of the independent variable

$$D_{x'\varepsilon'}(\kappa_1, ..., \kappa_M) = E_{x'}\left\{\left[E_{\varepsilon'}\{\varepsilon'(k)|x'_1(k - \kappa_1), ..., x'_M(k - \kappa_M)\} - E_{\varepsilon'}\{\varepsilon'(k)\}\right]^2\right\}$$

If the signals $x'_1(k - \kappa_1), ..., x'_M(k - \kappa_M)$ and $\varepsilon'(k)$ are independent then the corresponding cross-dispersional function is zero,

$$D_{x'\varepsilon'}(\kappa_1, ..., \kappa_M) = 0, \quad \kappa_i \geq 1, \quad \forall i$$

If $x_i(k)$ is the input signal or the computed noise-free model output signal then zero shifting time $\kappa_i = 0$ is also allowed.

In addition to the normalization it is practical to whiten both the normalized signal components $x'_i(k)$ and the normalized residuals $\varepsilon'(k)$. The whitened normalized signals are denoted by $x_i'^w(k)$ and $\varepsilon'^w(k)$. The reason for the whitening and its technique was explained in Section 4.9.

The normalized cross-dispersional function is the variance of the conditional mean values at a given shifting time. As stated in Section 6.14, the conditional mean values have a Student distribution that can be well approximated for a large number of data by a Gaussian distribution. The sum of squares of independent Gaussian random variables with zero mean and unit standard deviation have a chi-square distribution. Similar tests to the model validity tests based on correlation functions can be performed also in this case.

The computation of a one-dimensional cross-dispersional function is much simpler and takes much less time than the computation of a multi-dimensional cross-dispersional function.

Similarly to the case of the residual test based on the one- and multi-dimensional conditional mean values, the one-dimensional cross-dispersional test is only a necessary but not a sufficient test for the bias-free identification.

6.16 ANALYSIS OF DIFFERENT CORRELATION FUNCTIONS OF THE RESIDUALS

Two model validity tests based on correlation functions are known for model validity test with linear systems (e.g., Box and Jenkins, 1976):
   • auto-correlation function based test of the residuals, and
   • cross-correlation function based test of the input signal and the residuals

If the linear model fitting is perfect then both correlation functions are zero, or at least a hypothesis that they are zero can be accepted with an acceptable probability. With nonlinear systems the above tests fail, as shown by the following example.

Example 6.16.1 *The residual is the square of the input signal that is a Gaussian white noise*
Assume that the input signal is a white noise Gaussian signal with zero mean and unit

standard deviation and the residual is the square of the input signal,

$$\varepsilon(k) = u^2(k) = u'^2(k)$$

The auto-correlation function of the input signal is

$$r_{u'u'}(\kappa) = \begin{cases} 1 & \kappa = 0 \\ 0 & \kappa \neq 0 \end{cases}$$

Then the normalized cross-correlation function is zero since any even order auto-correlation function of the input signal is zero

$$r_{u'\varepsilon}(\kappa) = E\left\{u'(k-\kappa)\frac{\varepsilon(k) - E\{\varepsilon\}}{\sigma_\varepsilon}\right\} = \frac{1}{\sigma_\varepsilon} E\{u'(k-\kappa)\varepsilon(k)\}$$

$$= \frac{1}{\sigma_\varepsilon} E\{u'(k-\kappa)u'^2(k)\} = 0 \quad \forall \kappa$$

Therefore, a test based on the cross-correlation function cannot detect that the model fitting is not perfect. Calculate, however, the normalized (nonlinear) cross-correlation function of the normalized square of the input signal

$$x(k) = \frac{u'^2(k) - E\{u'^2(k)\}}{\mathrm{var}\{u'^2(k)\}} = \frac{u'^2(k) - \sigma_u^2}{\sigma_u^2}$$

and of the residual. This correlation function differs from zero

$$r_{x\varepsilon}(\kappa) = E\left\{\frac{u'^2(k-\kappa) - \sigma_u^2}{\sigma_u^2} \frac{\varepsilon(\kappa) - \bar{\varepsilon}}{\sigma_\varepsilon}\right\}$$

$$= \frac{1}{\sigma_u^2}\frac{1}{\sigma_\varepsilon} E\{u'^2(k-\kappa)u'^2(k)\} = \frac{r_{u'u'}^2(\kappa) + 2\sigma_u^4}{\sigma_u^2\sigma_\varepsilon} \neq 0 \quad \forall \kappa$$

The special (second-order) cross-correlation function can detect the modeling error. ∎

From the above example we can see that the linear cross-correlation function cannot detect any systematic error, but the cross-correlation function of the normalized square of the normalized input signal and of the residual can detect a systematic error.

There are two questions that arise:
1. What kind of correlation functions can detect all modeling errors?
2. What kind of statistical test shows that the correlation function under investigation can be considered zero?

The recommended special correlation functions are cross-correlation functions of a function of the residual and of a function of one of the following signals :
- input signal;
- measured output signal values except the actual one;
- residual values except the actual one;
- computed (noise-free) output signal.

Based on references, theoretical considerations and simulations the following normalized correlation functions seem to be practical for the analysis of the residuals with nonlinear systems

- correlation functions used also for residual analysis of linear systems (Box and Jenkins, 1976):
    - auto-correlation function of the residuals $r_{\varepsilon'\varepsilon'}(\kappa)$, $\kappa \neq 0$;
    - cross-correlation function of the input signal and the residuals $r_{u'\varepsilon'}(\kappa)$;

- correlation functions used especially for residual analysis of nonlinear systems:
    - second-order auto-correlation function of the residuals $r_{\left(\varepsilon'^2\right)'\varepsilon'}(\kappa)$

      recommended for the nonlinearity test of a process in the form of $r_{y'y'^2}(\kappa)$ by Billings and Voon (1983);
    - cross-correlation function of the square of the normalized input signal and the residual $r_{\left(u'^2\right)'\varepsilon'}(\kappa)$ (Billings and Voon, 1986);
    - cross-correlation function of the square of the normalized input signal and the square of the residuals $r_{\left(u'^2\right)'\left(\varepsilon'^2\right)'}(\kappa)$ (Billings and Voon, 1986);
    - cross-correlation function of the product of the normalized residual and the normalized input signal and of the residual $r_{(u'\varepsilon')\varepsilon'}(\kappa)$ (based on Billings and Voon, 1983).

The two components of the correlation function are normalized to obtain the correlation functions between 0 and 1 at least at the zero shifting time.

We shall investigate whether the mentioned correlation function can detect a modeling error or not.

Assumption 6.16.1   The following assumptions are made:
- the input signal $u(k)$ and the source noise $e(k)$ are independent;
- the source noise $e(k)$ has zero mean value;
- the residual $\varepsilon(k)$ has also zero mean value (otherwise the model fitting would be wrong).   ∎

In the examples worked out we will assume that both the input signal and the source noise have Gaussian distributions if the type of the distribution is needed.

If the process and noise model fittings are correct then the computed residuals are equal to the source noise

$$\varepsilon(k) = e(k)$$

In any other cases the residuals are functions of the input signal and/or the source noise. The functional dependence can be described by a two-variable Volterra operator

$$\begin{aligned}
\varepsilon(k) &= V\big[u'(k), e(k)\big] = x(k) + e(k) \\
&= V^u\big[u'(k)\big] + V^{ue}\big[u'(k), e(k)\big] + V^e\big[e(k)\big] + e(k)
\end{aligned} \tag{6.16.1}$$

where $V^u[\ldots]$ and $V^e[\ldots]$ contain only functions of the input signal and the residuals, respectively, and the $V^{ue}[\ldots]$ operator contains product terms between functions of the above signals. The mean value of $x(k)$ is equal to zero, unless the residual does not have a zero expected value, and the estimated model can not be perfect. The component $x(k)$ is called the systematic error. The input signal was considered by its normalized value $u'(k)$. We investigate some correlation functions in turn, whether they would detect the different residual types or not.

The systematic error is a function only of the input signal
First we treat the case when the systematic errors are functions only of the input signal

$$\varepsilon(k) = V^u[u'(k)] + e(k) = x(k) + e(k) \tag{6.16.2}$$

with

$$E\{x(k)\} = 0$$

Lemma 6.16.1
(Haber, 1993) $r_{\varepsilon'\varepsilon'}(\kappa)$ differs from zero for at least one shifting time except zero
  • if the systematic errors are even functions of the source noise only, or
  • if the systematic errors are odd functions of the source noise only and the input signal is not a white noise.

Proof.   The auto-correlation function of the normalized residuals is proportional to $r_{xx}(\kappa)$ and $r_{\varepsilon\varepsilon}(\kappa)$:

$$r_{\varepsilon'\varepsilon'}(\kappa) \sim r_{xx}(\kappa) + r_{xe}(\kappa) + r_{ex}(\kappa) + r_{ee}(\kappa) = r_{xx}(\kappa) + r_{ee}(\kappa) = r_{xx}(\kappa) \quad \text{for } \kappa \neq 0$$

1. *Case when the systematic errors are even functions of the input signal*
   Then $r_{xx}(\kappa)$ differs from zero for $\kappa \neq 0$. As an example take, e.g.,
   $$x(k) = u'^2(k) - 1$$

The above signal has a zero mean value because

$$E\{u'^2(k)\} = E\left\{\left(\frac{u'(k) - E\{u'(k)\}}{\sigma_u}\right)^2\right\} = \frac{1}{\sigma_u^2} E\left\{[u(k) - E\{u(k)\}]^2\right\} = 1$$

Let $u(k)$ a Gaussian white noise. Then

$$\begin{aligned}
r_{x'x'}(\kappa) &= E\left\{[u'^2(k-\kappa) - 1][u'^2(k) - 1]\right\} \\
&= E\{u'^2(k-\kappa)u'^2(k)\} - 2E\{u'^2(k)\} + 1 \tag{6.16.3} \\
&= [r_{u'u'}^2(\kappa) + 2] - 2 + 1 = r_{u'u'}^2(\kappa) + 1
\end{aligned}$$

which differs from zero for all shifting times.

2. *Case when the systematic errors are odd functions of the input signal*
Then

$$r_{\varepsilon'\varepsilon'}(\kappa) \sim r_{xx}(\kappa)r_{ee}(\kappa)$$

and $r_{xx}(\kappa)$ differs from zero only when $u(k)$ is a color noise. Let first the systematic errors a linear function of the input signal

$$x(k) = u'(k)$$

and thus

$$r_{xx}(\kappa) = r_{u'u'}(\kappa)$$

which differs from a Dirac function if $u(k)$ is a color noise. Now let the systematic errors a cubic function of the input signal

$$x(k) = u'^3(k)$$

then

$$r_{xx}(\kappa) = r_{(u'^3)(u'^3)}(\kappa) = E\{u'^3(k-\kappa)u'^3(k)\} = 9\sigma_u^4 r_{u'u'}(\kappa) + 6r_{u'u'}^3(\kappa)$$

which differs from a Dirac function if the input signal is not a white noise.   ■

Lemma 6.16.2
(Haber, 1993) $r_{(\varepsilon'^2)\,\varepsilon'}(\kappa)$ differs from zero for at least one shifting time

• if the systematic errors are even functions of the input signal only, or
• if the systematic errors are odd functions of the input signal only and the odd order moments of either the normalized input signal and or the systematic errors are not zero.

Proof.   The second-order auto-correlation function is

$$r_{(\varepsilon'^2)\,\varepsilon'}(\kappa) \sim E\{\varepsilon'^2(k-\kappa)\varepsilon'(k)\} = E\{[x(k-\kappa)+e(k-\kappa)]^2[x(k)+e(k)]\}$$

$$= E\{[x^2(k-\kappa)+2x(k-\kappa)e(k-\kappa)+e^2(k-\kappa)][x(k)+e(k)]\}$$

$$= E\{x^2(k-\kappa)x(k)\} + 2E\{x(k-\kappa)x(k)e(k-\kappa)\} \qquad (6.16.4)$$

$$+E\{e^2(k-\kappa)x(k)\} + E\{x^2(k-\kappa)e(k)\} + 2E\{x(k-\kappa)e(k-\kappa)e(k)\}$$

$$+E\{e^2(k-\kappa)e(k)\}$$

Terms 2 and 4 are zero because they are linear functions of the source noise. Terms 3 and 5 vanish because $x(k)$ and $e(k)$ are independent signals and $x(k)$ has a zero mean value. The last term is zero if the odd order moments of the source noise are zero. If $x(k)$ is an odd function of u' then the first term is also zero. Otherwise it does not vanish. As an example for an even function of the systematic error take

$$x(k) = u'^2(k) - 1$$

Then

$$E\{x^2(k-\kappa)x(k)\}\big|_{\kappa=0} = E\left\{\left[u'^2(k) - 1\right]^3\right\} = E\{u'^6\} - 3E\{u'^4\} + 3E\{u'^2\} + 1 = 10 \neq 0 \blacksquare$$

Lemma 6.16.3
(Haber, 1993) $r_{u'\varepsilon'}(\kappa)$ differs from zero for at least one shifting time
- if the systematic errors are odd functions of the input signal only, or
- if the systematic errors are even functions of the input signal only and the odd order moments of the normalized input signal are not zero.

Proof.   The linear cross-correlation function is

$$r_{u'\varepsilon'}(\kappa) \sim E\{u'(k-\kappa)\varepsilon'(k)\} = E\{u'(k-\kappa)[x(k) + e(k)]\}$$
$$= E\{u'(k-\kappa)x(k)\}$$

(6.16.5)

If $x(k)$ is an odd function of $u'(k)$ then the cross-correlation function is not zero for all shifting times. If $x(k)$ is an even function of $u'(k)$ then $r_{u'\varepsilon'}(\kappa)$ is zero for all shifting times if the odd order moments of $u'(k)$ are zero. As an illustration for an even degree function consider

$$x'(k) = u'^2(k) - 1$$

Then

$$r_{u'\varepsilon'}(\kappa) \sim E\{u'(k-\kappa)[u'^2(k) - 1]\} = E\{u'(k-\kappa)u'^2(k)\}$$

is zero if the odd order moments of $u'(k)$ are zero. $\blacksquare$

Lemma 6.16.4
(Haber, 1993) $r_{(u'^2)'\varepsilon'}(\kappa)$ differs from zero for at least one shifting time
- if the systematic errors are even functions of the input signal only, or
- if the systematic errors are odd functions of the input signal only and the odd order moments of the normalized input signal are not zero

Proof.   The nonlinear cross-correlation function under investigation is

$$r_{\left(u'^2\right)'\varepsilon'}(\kappa) \sim E\left\{\left[u'^2(k)-1\right]\left[x(k)+e(k)\right]\right\}$$

$$= E\left\{u'^2(k)x(k)\right\} - E\left\{x(k)\right\} + E\left\{u'^2(k)e(k)\right\} - E\left\{e(k)\right\} \qquad (6.16.6)$$

$$= E\left\{u'^2(k)x(k)\right\}$$

A comparison of (6.16.6) with (6.16.5) shows that $r_{\left(u'^2\right)'\varepsilon'}(\kappa)$ behaves for odd (even) functions of $u'(k)$ like $r_{u'\varepsilon'}(\kappa)$ for even (odd) functions of $u'(k)$. ∎

Lemma 6.16.5
(Haber, 1993) $r_{u'\left(\varepsilon'^2\right)}(\kappa)$ differs from zero for at least one shifting time if the systematic errors are functions only of the input signal and the odd order moments of the normalized input signal are not zero.

Proof. The nonlinear cross-correlation function under investigation

$$r_{u'\left(\varepsilon'^2\right)}(\kappa) \sim E\left\{u'(k-\kappa)\left[\left[x(k)+e(k)\right]^2 - E\left\{\left[x(k)+e(k)\right]^2\right\}\right]\right\}$$

$$= E\left\{u'(k-\kappa)x^2(k)\right\} + 2E\left\{u'(k-\kappa)x(k)\right\}E\left\{e(k)\right\} + E\left\{u'(k)\right\}E\left\{e^2(k)\right\}$$

$$= E\left\{u'(k-\kappa)x^2(k)\right\}$$

$$(6.16.7)$$

which expression has features similar to (6.16.6) if in (6.16.6) $x(k)$ is an odd function of $u'(k)$. Consequently $r_{u'\left(\varepsilon'^2\right)}(\kappa)$ is zero if the odd order moments of $u'(k)$ are zero. ∎

Lemma 6.16.6
(Haber, 1993) $r_{(u'\varepsilon')e'}(\kappa)$ differs from zero for at least one shifting time except zero if the systematic errors are functions only of the input signal and the odd order moments of the normalized input signal are not zero.

Proof. The nonlinear cross-correlation function under investigation is

$$r_{(u'\varepsilon')e'}(\kappa) \sim E\left\{u'(k-\kappa)\left[x(k-\kappa)+e(k-\kappa)\right]\left[x(k)+e(k)\right]\right\}$$

$$= E\left\{u'(k-\kappa)x(k-\kappa)x(k)\right\} + E\left\{u'(k-\kappa)x(k-\kappa)\right\}E\left\{e(k)\right\}$$

$$+ E\left\{u'(k-\kappa)x(k)\right\}E\left\{e(k)\right\} + E\left\{u'(k)\right\}E\left\{e(k-\kappa)e(k)\right\} \qquad (6.16.8)$$

$$= E\left\{u'(k-\kappa)x(k-\kappa)x(k)\right\}$$

(6.16.8) has the same structure as (6.16.7) therefore the conditions when $r_{(u'\varepsilon')e'}(\kappa)$ and $r_{u'\left(\varepsilon'^2\right)}(\kappa)$ are zero are the same. ∎

Lemma   6.16.7

(Haber, 1993) $r_{\left(u'^2\right)'\left(\varepsilon'^2\right)'}(\kappa)$ differs from zero for at least one shifting time if the systematic errors are functions only of the input signal.

Proof.   The nonlinear cross-correlation function under investigation is

$$r_{\left(u'^2\right)'\left(\varepsilon'^2\right)'}(\kappa) \sim \mathrm{E}\left\{\left[u'^2(k-\kappa)-1\right]\left[\left[x(k)+e(k)\right]^2 - \mathrm{E}\left\{\left[x(k)+e(k)\right]^2\right\}\right]\right\}$$

$$= \mathrm{E}\left\{\left[u'^2(k-\kappa)-1\right]\left[x^2(k)+2x(k)e(k)+e^2(k)\right]\right\}$$

$$= \mathrm{E}\left\{u'^2(k-\kappa)x^2(k)\right\} - \mathrm{E}\left\{x^2(k)\right\} + 2\mathrm{E}\left\{\left[u'^2(k-\kappa)-1\right]x(k)e(k)\right\}$$

$$+\mathrm{E}\left\{\left[u'^2(k-\kappa)-1\right]e^2(k)\right\}$$

$$(6.16.9)$$

In (6.16.9) term 4 is zero because the multiplier of $e^2(k)$ is independent of $e(k)$ and has a zero mean value. Term 3 is also zero because $e(k)$ is independent of the other functions. Thus (6.16.9) reduces to

$$r_{\left(u'^2\right)'\left(\varepsilon'^2\right)'}(\kappa) \sim \mathrm{E}\left\{u'^2(k-\kappa)x^2(k)\right\} - \mathrm{E}\left\{x^2(k)\right\}$$

The remaining terms are never zero for all shifting times. As an example, for an odd (linear) function consider

$$x(k) = u'(k)$$

thus (6.16.9) becomes

$$r_{\left(u'^2\right)'\left(\varepsilon'^2\right)'}(\kappa)\Big|_{\kappa=0} \sim \mathrm{E}\left\{u'^4(k)\right\} - \mathrm{E}\left\{u'^2(k)\right\} = 3-1 = 2 \neq 0$$

In an other example let $x(k)$ an even (quadratic) function of $u'(k)$

$$x(k) = u'^2(k)-1$$

thus (6.16.9) becomes

$$r_{\left(u'^2\right)'\left(\varepsilon'^2\right)'}(\kappa)\Big|_{\kappa=0} \sim \mathrm{E}\left\{u'^2(k)\left[u'^2(k)-1\right]^2\right\} - \mathrm{E}\left\{\left[u'^2(k)-1\right]^2\right\}$$

$$= \mathrm{E}\left\{u'^6(k)\right\} - 2\mathrm{E}\left\{u'^4(k)\right\} + \mathrm{E}\left\{u'^2(k)\right\} - \mathrm{E}\left\{u'^4(k)\right\} + 2\mathrm{E}\left\{u'^2(k)\right\} - 1$$

$$= 15-6+1-3+2-1 = 8 \neq 0$$

∎

TABLE 6.16.1 Features of different correlation functions when the
systematic errors in the residuals are functions of the input signal only
(o.o.m: odd order moments)

Correlation function	κ	$\varepsilon(k) = V^u[u'(k)] + e(k)$	
		$V^u[u'(k)]$ odd function	$V^u[u'(k)]$ even function
$r_{\varepsilon'\varepsilon'}(\kappa)$	≠0	=0 if u' is a white noise ≠0 otherwise	≠0
$r_{(\varepsilon'^2)'\varepsilon'}(\kappa)$	∀	=0 if $\begin{cases} \text{o.o.m. of } e \text{ are zero and} \\ \text{o.o.m. of } u' \text{ are zero} \end{cases}$ ≠0 otherwise	≠0
$r_{u'\varepsilon'}(\kappa)$	∀	≠0	=0 if o.o.m. of u' are zero ≠0 otherwise
$r_{(u'^2)'\varepsilon'}(\kappa)$	∀	=0 if o.o.m. of u' are zero ≠0 otherwise	≠0
$r_{u'(\varepsilon'^2)'}(\kappa)$	∀	=0 if o.o.m. of u' are zero ≠0 otherwise	=0 if o.o.m. of u' are zero ≠0 otherwise
$r_{(u'\varepsilon')\varepsilon'}(\kappa)$	∀	=0 if o.o.m. of u' are zero ≠0 otherwise	=0 if o.o.m. of u' are zero ≠0 otherwise
$r_{(u'^2)'(\varepsilon'^2)'}(\kappa)$	∀	≠0	≠0

Theorem 6.16.1   (Haber, 1993) Table 6.16.1 summarizes under what circumstances the normalized correlation functions detect the systematic errors if they are a function of only the input signal as shown in (6.16.2), e.g.,

- auto-correlation function of the residuals $r_{\varepsilon'\varepsilon'}(\kappa)$, $\kappa \neq 0$;
- second-order auto-correlation function of the residuals $r_{(\varepsilon'^2)'\varepsilon'}(\kappa)$;
- cross-correlation function of the input signal and the residuals $r_{u'\varepsilon'}(\kappa)$;
- cross-correlation function of the square of the input signal and the residual $r_{(u'^2)'\varepsilon'}(\kappa)$;
- cross-correlation function of the input signal and the square of the residual $r_{u'(\varepsilon'^2)'}(\kappa)$;
- cross-correlation function of the product of the residual and the input signal and of the residual $r_{(u'\varepsilon')\varepsilon'}(\kappa)$;
- cross-correlation function of the square of the input signal and the square of the residuals $r_{(u'^2)'(\varepsilon'^2)'}(\kappa)$.

Proof.   The proof is a summary of the results of the Lemmas 6.16.1 to 6.16.7.   ∎

The systematic error is a function only of the source noise
Second, we treat the case when the systematic errors are functions only of the source noise,

$$\varepsilon(k) = V^e[e(k)] + e(k) = x(k) + e(k) \qquad (6.16.10)$$

with

$$E\{x(k)\} = 0$$

Lemma 6.16.8

(Haber, 1993) $r_{\varepsilon'\varepsilon'}(\kappa)$ differs from zero for at least one shifting time except zero
- if the systematic errors are only odd function of the source noise, or
- if the systematic errors are only even functions of the source noise and either the above function is dynamic (not static) or the odd order moments of the systematic errors are not zero.

Proof.   The auto-correlation function of the residuals is

$$r_{\varepsilon'\varepsilon'}(\kappa) \sim E\{[x(k-\kappa)+e(k-\kappa)][x(k)+e(k)]\}$$
$$= E\{x(k-\kappa)x(k)\} + E\{x(k-\kappa)e(k)\} + E\{e(k-\kappa)x(k)\} + r_{ee}(\kappa) \qquad (6.16.11)$$

If $x(k)$ is an odd function of $e(k)$ then all terms exist beside $r_{ee}(\kappa)$ and they contribute to the resulting $r_{\varepsilon'\varepsilon'}(\kappa)$. Let, e.g.,

$$x(k) = e(k - \kappa_1)$$

then

$$r_{\varepsilon'\varepsilon'}(\kappa) \sim r_{ee}(\kappa) + r_{ee}(\kappa + \kappa_1) + r_{ee}(\kappa - \kappa_1) + r_{ee}(\kappa)$$

and $r_{\varepsilon'\varepsilon'}(\kappa)$ differs from zero for κ equal to 0 and $\pm\kappa_1$. (The case $\kappa_1 = 0$ would not be detected by the test.) As a second example let

$$x(k) = e^3(k - \kappa_1)$$

then

$$r_{\varepsilon'\varepsilon'}(\kappa) \sim E\{e^3(k-\kappa_1-\kappa)e^3(k-\kappa_1)\} + E\{e^3(k-\kappa_1-\kappa)e(k)\}$$
$$+ E\{e(k-\kappa)e^3(k-\kappa_1)\} + r_{ee}(\kappa)$$
$$= 9\sigma_e^4 r_{ee}(\kappa) + 6r_{ee}^3(\kappa) + 3\sigma_e^2 r_{ee}(\kappa+\kappa_1) + 3\sigma_e^2 r_{ee}(\kappa-\kappa_1) + r_{ee}(\kappa)$$

which differs from zero for $\kappa = \kappa_1$. However, if $\kappa_1 = 0$ then

$$r_{\varepsilon'\varepsilon'}(\kappa) \sim [1 + 6\sigma_e^2 + 9\sigma_e^4 + 6r_{ee}^2(\kappa)]r_{ee}(\kappa)$$

and $r_{\varepsilon'\varepsilon'}(\kappa)$ has a Dirac function shape because there is no systematic error. Consequently if the systematic errors are odd static functions of the source noise then we cannot see from $r_{\varepsilon'\varepsilon'}(\kappa)$ whether the static function is linear or nonlinear. ∎

Lemma 6.16.9

(Haber, 1993) $r_{(\varepsilon'^2)\varepsilon'}(\kappa)$ differs from zero for at least one shifting time

- if the systematic errors are even functions of the source noise only, or
- if the systematic errors are odd functions of the source noise only and the odd order moments of the source noise are not zero.

Proof.   The second-order auto-correlation function is:

$$r_{(\varepsilon'^2)'\varepsilon'}(\kappa) \sim E\left\{\left[\varepsilon'^2(k-\kappa) - E\{\varepsilon'^2(k)\}\right]\varepsilon'(k)\right\}$$

$$= E\left\{\left[\varepsilon'^2(k-\kappa)\varepsilon'(k)\right]\right\} = E\left\{\left[x(k-\kappa) + e(k-\kappa)\right]^2\left[x(k) + e(k)\right]\right\}$$

$$= E\left\{\left[x^2(k-\kappa) + 2x(k-\kappa)e(k-\kappa) + e^2(k-\kappa)\right]\left[x(k) + e(k)\right]\right\} \qquad (6.16.12)$$

$$= E\left\{x^2(k-\kappa)x(k)\right\} + 2E\left\{x(k-\kappa)x(k)e(k-\kappa)\right\} + E\left\{e^2(k-\kappa)x(k)\right\}$$

$$+ E\left\{x^2(k-\kappa)e(k)\right\} + 2E\left\{x(k-\kappa)e(k-\kappa)e(k)\right\} + E\left\{e^2(k-\kappa)e(k)\right\}$$

If $x(k)$ is an odd function of $e(k)$ then all terms in (6.16.12) are even order auto-correlation functions of the residuals. They are zero only if the odd order moments of the source noise are zero. As an example, for an even function of the systematic error of the source noise take the quadratic function

$$x(k) = e^2(k) - \sigma_e^2$$

In (6.16.12) the terms that are odd functions of $e(k)$ or its shifted values vanish if the odd order moments of $e(k)$ are zero. Thus (6.16.12) reduces to

$$r_{(\varepsilon'^2)'\varepsilon'}(\kappa) \sim E\left\{x^2(k-\kappa)x(k)\right\} + E\left\{e^2(k-\kappa)x(k)\right\} + 2E\left\{x(k-\kappa)e(k-\kappa)e(k)\right\}$$

which at $\kappa = 0$ is

$$r_{(\varepsilon'^2)'\varepsilon'}(\kappa)\Big|_{\kappa=0} \sim E\left\{x^3(k)\right\} + 3E\left\{x(k)e^2(k)\right\}$$

$$= E\left\{\left[e^2(k) - \sigma_e^2\right]^3\right\} + 3E\left\{\left[e^2(k) - \sigma_e^2\right]e^2(k)\right\}$$

$$= E\left\{e^6(k)\right\} - 3\sigma_e^2 E\left\{e^4(k)\right\} + 3\sigma_e^4 E\left\{e^2(k)\right\}$$

$$- \sigma_e^6 + 3E\left\{e^4(k)\right\} - 3\sigma_e^2 E\left\{e^2(k)\right\}$$

$$= 15\sigma_e^6 - 3\sigma_e^2 3\sigma_e^4 + 3\sigma_e^4\sigma_e^2 - \sigma_e^6 + 3\cdot 3\sigma_e^4 - 3\sigma_e^2\sigma_e^2 = 8\sigma_e^6 + 6\sigma_e^4 \neq 0$$

This example shows that if $x(k)$ is an even function of $e(k)$ then the test detects it. ∎

Lemma 6.16.10
(Haber, 1993) $r_{u'\varepsilon'}(\kappa)$ is zero for all shifting times and consequently is unable to detect any systematic error in the residuals if the systematic error is a function of the source noise only.

Proof.   The linear cross-correlation function $r_{u'\varepsilon'}(\kappa)$ is zero for all shifting times because the two components of the correlation function are independent of each other and the first multiplier $u'(k)$ has a zero mean value.   ∎

Lemma 6.16.11

(Haber, 1993) $r_{(u'^2)\varepsilon'}(\kappa)$ is zero for all shifting times and consequently is unable to detect any systematic error in the residuals if the systematic error is function of the source noise only.

Proof.   The nonlinear cross-correlation function $r_{(u'^2)\varepsilon'}(\kappa)$ is zero for all shifting times because the two components of the correlation function are independent of each other and the first multiplier $u'^2(k)$ has a zero mean value.   ∎

Lemma 6.16.12

(Haber, 1993) $r_{u'(\varepsilon'^2)}(\kappa)$ is zero for all shifting times and consequently is unable to detect any systematic error in the residuals if the systematic error is function of the source noise only.

Proof.   The nonlinear cross-correlation function $r_{u'(\varepsilon'^2)}(\kappa)$ is zero for all shifting times because the two components of the correlation function are independent of each other and the first multiplier $u'(k)$ has a zero mean value.   ∎

Lemma 6.16.13

(Haber, 1993) $r_{(u'\varepsilon')\varepsilon'}(\kappa)$ is zero for all shifting times and consequently is unable to detect any systematic error in the residuals if the systematic error is function of the source noise only.

Proof.   The nonlinear cross-correlation function has the form

$$r_{(u'\varepsilon')\varepsilon'}(\kappa) \sim \mathrm{E}\left\{\left[u'(k-\kappa)\varepsilon'(k-\kappa)-\mathrm{E}\{u'(k-\kappa)\varepsilon'(k)\}\right]\varepsilon'(k)\right\}$$

$$= \mathrm{E}\{u'(k-\kappa)\varepsilon'(k-\kappa)\varepsilon'(k)\} - \mathrm{E}\{u'(k-\kappa)\varepsilon'^2(k)\} \qquad (6.16.13)$$

$$= \mathrm{E}\left\{u'(k-\kappa)\left[\varepsilon'(k-\kappa)\varepsilon'(k)-\varepsilon'^2(k)\right]\right\}$$

which is zero for all shifting times because the two components of the correlation function are independent of each other and the first multiplier $u'(k)$ has a zero mean value.   ∎

Lemma 6.16.14

(Haber, 1993) $r_{(u'^2)(\varepsilon'^2)}(\kappa)$ is zero for all shifting times and consequently is unable to detect any systematic error in the residuals if the systematic error is function of the source noise only.

Proof.   The nonlinear cross-correlation function $r_{\left(u'^2\right)'\left(\varepsilon'^2\right)'}(\kappa)$ is zero for all shifting times because the two components of the correlation function are independent of each other and the first multiplier $\left(u'^2(k)\right)'$ has a zero mean value.   ∎

TABLE 6.16.2 Features of different correlation functions when the systematic
errors in the residuals are functions of the source error only
(o.o.m: odd order moments)

Correlation function	κ	$\varepsilon(k) = V^e\left[e(k)\right] + e(k)$	
		$V^e\left[e(k)\right]$ *odd function*	$V^e\left[e(k)\right]$ *even function*
$r_{\varepsilon'\varepsilon'}(\kappa)$	$\neq 0$	$\neq 0$	$=0$ if $\begin{cases} u' \text{ is a white noise} \\ \text{and } V^e[e(k)] \text{ is static function} \end{cases}$ $\neq 0$ otherwise
$r_{\left(\varepsilon'^2\right)'\varepsilon'}(\kappa)$	$\forall$	$=0$ if o.o.m. of e are zero $\neq 0$ otherwise	$\neq 0$
$r_{u'\varepsilon'}(\kappa)$	$\forall$	$=0$	$=0$
$r_{\left(u'^2\right)'\varepsilon'}(\kappa)$	$\forall$	$=0$	$=0$
$r_{u'\left(\varepsilon'^2\right)'}(\kappa)$	$\forall$	$=0$	$=0$
$r_{(u'\varepsilon')\varepsilon'}(\kappa)$	$\forall$	$=0$	$=0$
$r_{\left(u'^2\right)'\left(\varepsilon'^2\right)'}(\kappa)$	$\forall$	$=0$	$=0$

Theorem 6.16.2   (Haber, 1993) Table 6.16.2 summarizes under what circumstances the normalized correlation functions:
- auto-correlation function of the residuals $r_{\varepsilon'\varepsilon'}(\kappa)$, $\kappa \neq 0$,
- second-order auto-correlation function of the residuals $r_{\left(\varepsilon'^2\right)'\varepsilon'}(\kappa)$,
- cross-correlation function of the input signal and the residuals $r_{u'\varepsilon'}(\kappa)$,
- cross-correlation function of the square of the input signal and the residuals $r_{\left(u'^2\right)'\varepsilon'}(\kappa)$,
- cross-correlation function of the input signal and the square of the residuals $r_{u'\left(\varepsilon'^2\right)'}(\kappa)$,
- cross-correlation function of the product of the residual and the input signal and of the residuals $r_{(u'\varepsilon')\varepsilon'}(\kappa)$,
- cross-correlation function of the square of the input signal and the square of the residuals $r_{\left(u'^2\right)'\left(\varepsilon'^2\right)'}(\kappa)$,

detect the systematic errors if they are a function of only the source noise, as shown in (6.16.10).

Proof.   The proof is a summary of the results of the Lemmas 6.16.8 to 6.16.14.   ■

The systematic error is a function of both the input signal and source noise
Lastly we treat the case when the systematic error is a function of both the input signal and the source noise

$$\varepsilon(k) = V^{ue}\big[u'(k), e(k)\big] + e(k) = x(k) + e(k) \tag{6.16.14}$$

with

$$E\{x(k)\} = 0$$

We distinguish four cases depending on what kind of function the systematic error of the normalized input signal and the source noise is:
1. an odd function of both the normalized input signal and the source noise;
2. an odd function of the normalized input signal and even function of the source noise;
3. an even function of the normalized input signal and odd function of the source noise;
4. an even function of both the normalized input signal and the source noise.

In the sequel product terms of the normalized input signal and the source noise or their powers in the form $u^i(k - \kappa_1)e^j(k - \kappa_2)$, $(i, j, \kappa_1, \kappa_2$ integers) will be treated. Any analytical nonlinear function can be described by its Taylor series where the product terms occur. The signal $x(k)$ now has the form

$$x(k) = f(u'(k))g(e(k)) - E\{f(u'(k))\}E\{g(e(k))\} \tag{6.16.15}$$

where both functions $f(...)$ and $g(...)$ may be odd or even functions and may include time shifts (delays) of the arguments. Introduce the following short notations

$$E\{f(u(k))\} = \bar{f}(u'(k)) = \bar{f}(u') = \bar{f} \tag{6.16.16}$$

$$E\{g(u(k))\} = \bar{g}(e'(k)) = \bar{g}(e') = \bar{g} \tag{6.16.17}$$

$$r_{ff}(\kappa) = E\{f(u(k - \kappa))f(u(k))\}$$

$$r_{gg}(\kappa) = E\{g(u(k - \kappa))g(u(k))\}$$

Lemma 6.16.15
(Haber, 1993) $r_{\varepsilon'\varepsilon'}(\kappa)$ differs from zero for at least one shifting time except zero
• if the systematic errors are odd functions of both the input signal and the source noise and neither the odd order moments of the normalized input signal nor the source noise are zero, and either both functions $f(...)$ and $g(...)$ consist of the sum of at least two differently shifted terms, and the delays for $f(...)$ and $g(...)$ are the same or the input signal is a color noise and $g(...)$ consists of the sum

of at least two differently shifted terms;

- if the systematic errors are odd functions of the input signal and even functions of the source noise and the odd order moments of the normalized input signal are not zero and either the input signal is a colored signal or $f(u'(k))$ is a function of at least two differently shifted input terms;

- if the systematic errors are even functions of the input signal and odd functions of the source noise and the odd order moments of the source noise are not zero and the source noise and $g(e(k))$ is not a static function of $e(k)$;

- if the systematic errors are even functions of both the input signal and the source noise and the odd order moments of the source noise are not zero and both functions $f(...)$ and $g(...)$ do not consist of the product of two differently shifted terms

Proof.   The auto-correlation function of the systematic errors is

$$r_{\varepsilon'\varepsilon'}(\kappa) \sim E\{[f(u'(k-\kappa))g(e(k-\kappa)) - \bar{f}(u')\bar{g}(e) + e(k-\kappa)]$$

$$\times [f(u'(k))g(e(k)) - \bar{f}(u')\bar{g}(e) + e(k)]\}$$

$$= E\{f(u'(k-\kappa))f(u'(k))\}E\{g(e(k-\kappa))g(e(k))\}$$

$$-\bar{f}^2(u')\bar{g}^2(e) + \bar{f}(u')E\{g(e(k-\kappa))g(e(k))\} - \bar{f}^2(u')\bar{g}^2(e) + \bar{f}^2(u')\bar{g}^2(e) -$$

$$-\bar{f}(u')\bar{g}(e)E\{e(k)\} + \bar{f}(u')E\{g(e(k))e(k-\kappa)\} - \bar{f}(u')\bar{g}(e)E\{e(k)\} + r_{ee}(\kappa)$$

$$= E\{f(u'(k-\kappa))f(u'(k))\}E\{g(e(k-\kappa))g(e(k))\} - \bar{f}^2(u')\bar{g}^2(e)$$

$$+\bar{f}(u')[E\{g(e(k-\kappa))e(k)\} + E\{g(e(k))e(k-\kappa)\}] + r_{ee}(\kappa)$$

$$(6.16.18)$$

Different cases will be investigated in the sequel individually.

1. *Case when both $f(u'(k))$ and $g(e(k))$ are odd functions:*
Further on we assume that all odd order moments of $u'(k)$ and $e(k)$ are zero. Because $f(...)$ is an odd function (6.16.18) reduces to

$$r_{\varepsilon'\varepsilon'}(\kappa) \sim E\{f(u'(k-\kappa))f(u'(k))\}E\{g(e(k-\kappa))g(e(k))\} + r_{\varepsilon'\varepsilon'}(\kappa) \qquad (6.16.19)$$

If the source noise is a white noise then the systematic error is not always detectable as shown in detail:

(a) *Assume that $g(...)$ is a linear function of $e(k-\kappa_1)$ and $e(k-\kappa_2)$:*

$$g(e(k)) = c_1 e(k-\kappa_1) + c_2 e(k-\kappa_2)$$

The auto-correlation function of $g(e(k))$ is

$$r_{gg}(\kappa) = E\{g(e(k-\kappa))g(e(k))\}$$
$$= c_1^2 r_{ee}(\kappa) + c_1 c_2 r_{ee}(\kappa + \kappa_1 - \kappa_2) + c_1 c_2 r_{ee}(\kappa + \kappa_2 - \kappa_1) + c_2^2 r_{ee}(\kappa)$$

which differs from zero for $\kappa = 0$, $\kappa = \kappa_2 - \kappa_1$ and $\kappa = \kappa_1 - \kappa_2$. If $\kappa_1 = \kappa_2$ then $r_{gg}(\kappa)$ differs from zero only at $\kappa = 0$.

(b) *Assume that* $g(...)$ *is a one-term cubic function:*

$$g(e(k)) = e(k - \kappa_1)e^2(k - \kappa_2)$$

The auto-correlation function $r_{gg}(\kappa)$ becomes

$$
\begin{aligned}
r_{gg}(\kappa) &= E\{e(k - \kappa_1 - \kappa)e^2(k - \kappa_2 - \kappa)e(k - \kappa_1)e^2(k - \kappa_2)\} \\
&= 2\sigma_e^2 r_{ee}(\kappa_1 - \kappa_2)r_{ee}(\kappa - \kappa_1 + \kappa_2) + 4r_{ee}^2(\kappa_1 - \kappa_2)r_{ee}(\kappa) \\
&\quad + \sigma_e^4 r_{ee}(\kappa) + 2r_{ee}^3(\kappa) + 2\sigma_e^2 r_{ee}(\kappa_1 - \kappa_2 + \kappa)r_{ee}(\kappa_1 - \kappa_2) \\
&\quad + 4r_{ee}(\kappa + \kappa_1 - \kappa_2)r_{ee}(\kappa - \kappa_1 + \kappa_2)r_{ee}(\kappa)
\end{aligned}
$$

(6.16.20)

If $\kappa_1 = \kappa_2$ then (6.16.20) reduces to

$$r_{gg}(\kappa) = 9\sigma_e^4 r_{ee}(\kappa) + 6r_{ee}^3(\kappa)$$

which is zero for $\kappa \neq 0$. If $\kappa_1 \neq \kappa_2$ then (6.16.20) reduces to

$$r_{gg}(\kappa) = r_{ee}(\kappa)\left[\sigma_e^4 + 2r_{ee}^2(\kappa) + 4r_{ee}(\kappa + \kappa_1 - \kappa_2)r_{ee}(\kappa - \kappa_1 + \kappa_2)\right]$$

which is zero for $\kappa \neq 0$. If $\kappa_2 = 0$ and $\kappa_1 \neq 0$ then (6.16.20) reduces to

$$
\begin{aligned}
r_{gg}(\kappa) &= 2\sigma_e^2 r_{ee}(\kappa_1)r_{ee}(\kappa - \kappa_1) + 4r_{ee}^2(\kappa_1)r_{ee}(\kappa) + \sigma_e^4 r_{ee}(\kappa) \\
&\quad + 2r_{ee}^3(\kappa) + 2\sigma_e^2 r_{ee}(\kappa_1 + \kappa)r_{ee}(\kappa_1) + 4r_{ee}(\kappa_1 + \kappa)r_{ee}(\kappa - \kappa_1)r_{ee}(\kappa) \\
&= \sigma_e^4 r_{ee}(\kappa) + 2r_{ee}^3(\kappa) + 4r_{ee}(\kappa_1 + \kappa)r_{ee}(\kappa - \kappa_1)r_{ee}(\kappa)
\end{aligned}
$$

(6.16.21)

because with a white noise $r_{ee}(\kappa_1) = 0$ if $\kappa_1 \neq 0$. The expression (6.16.21) is zero for all shifting times except zero.

(c) *Assume that* $g(...)$ *is a two-term cubic function:*

For example, let

$$g(e(k)) = c_1 e^3(k - \kappa_1) + c_2 e^3(k - \kappa_2)$$

The auto-correlation function $g(e'(k))$

$$r_{gg}(\kappa) = c_1^2 E\{e^3(k - \kappa_1 - \kappa)e^3(k - \kappa_1)\} + c_1 c_2 E\{e^3(k - \kappa_1 - \kappa)e^3(k - \kappa_2)\}$$

$$+ c_1 c_2 E\{e^3(k - \kappa_2 - \kappa)e^3(k - \kappa_1)\} + c_2^2 E\{e^3(k - \kappa_2 - \kappa)e^3(k - \kappa_2)\}$$

$$= c_1^2[9\sigma_e^4 r_{ee}(\kappa) + 6r_{ee}^3(\kappa)] + c_1 c_2[9\sigma_e^4 r_{ee}(\kappa + \kappa_1 - \kappa_2) + 6r_{ee}^3(\kappa + \kappa_1 - \kappa_2)]$$

$$+ c_1 c_2[9\sigma_e^4 r_{ee}(\kappa + \kappa_2 - \kappa_1) + 6r_{ee}^3(\kappa + \kappa_2 - \kappa_1)] + c_2^2[9\sigma_e^4 r_{ee}(\kappa) + 6r_{ee}^3(\kappa)]$$

which differs from zero only at $\kappa = 0$, $\kappa = \kappa_2 - \kappa_1$ and $\kappa = \kappa_1 - \kappa_2$. If $\kappa_1 = \kappa_2$ then $r_{gg}(\kappa)$ differs from zero only at $\kappa = 0$.

From the three examples we can draw the conclusion that $r_{gg}(\kappa)$ differs from zero for at least one shifting time $\kappa \neq 0$ if $g(...)$ is the sum of at least two odd functions shifted differently.

What was mentioned above about $r_{gg}(\kappa)$ is valid for $r_{ff}(\kappa)$, which means that $r_{ff}(\kappa)$ differs from zero for at least one shifting time $\kappa \neq 0$ if $f(...)$ is the sum of at least two odd functions shifted differently. The expression (6.16.19) differs from zero if both multiplicative terms $r_{ff}(\kappa)$ and $r_{gg}(\kappa)$ differ from zero at the same shifting time(s). That is fulfilled if both functions $f(...)$ and $g(...)$ consist of the same shifted terms, e.g.,

$$\varepsilon'(k) = [c_1 u'(k - 2) + c_2 u'(k - 3)][c_3 e^3(k - 2) + c_4 e^3(k - 3)] + e(k)$$

In contrast to the source noise, the input signal can be a color signal. Assume that the auto-correlation function $r_{uu}(\kappa)$ differs from zero for any shifting time. Investigate the case when $f(...)$ is a linear function

$$f(u'(k)) = c_1 u'(k - \kappa_1) + c_2 u'(k - \kappa_2)$$

The auto-correlation function of $f(u'(k))$ is

$$r_{ff}(\kappa) = E\{f(u'(k - \kappa))f(u'(k))\}$$

$$= c_1^2 r_{u'u'}(\kappa) + c_1 c_2 r_{u'u'}(\kappa + \kappa_1 - \kappa_2) + c_1 c_2 r_{u'u'}(\kappa + \kappa_2 - \kappa_1) + c_2^2 r_{u'u'}(\kappa)$$

which differs from zero for $\kappa \neq 0$ independently from the relation between the shifting times κ_1 and κ_2.

The final conclusion is that the error can be detected if both functions $f(...)$ and $g(...)$ are odd functions and:

- either both functions $f(...)$ and $g(...)$ consist of the sum of at least two differently shifted terms, and the delays for $f(...)$ and $g(...)$ are the same, e.g.,

$$f(u'(k)) = c_1 u'(k - \kappa_1) + c_2 u'(k - \kappa_2)$$
$$g(u'(k)) = c_3 u'(k - \kappa_1) + c_4 u'(k - \kappa_2)$$

or

- the input signal is a color noise with $r_{uu}(\kappa) \neq 0$, $\forall \kappa$, and $g(\ldots)$ consist of the sum of at least two differently shifted terms.

2. *Case when $f(u'(k))$ is an odd function and $g(e(k))$ is an even function:*
Further on assume that all odd order moments of $u'(k)$ are zero. Because $f(\ldots)$ is an odd function (6.16.18) reduces to (6.16.19). Assume now that $g(\ldots)$ is a quadratic function, e.g.,

$$g(e(k)) = e^2(k - \kappa_1)$$

The auto-correlation function of $g(e(k))$ is

$$r_{gg}(\kappa) = E\{e^2(k - \kappa_1 - \kappa)e^2(k - \kappa_1)\} = 2r_{ee}^2(\kappa) + \sigma_e^4$$

which differs from zero for all shifting times.

The final conclusion is that the error can be detected if $f(\ldots)$ is an odd and $g(\ldots)$ is an even function and

- either $f(\ldots)$ consists of at least two differently shifted terms, or
- the input signal is a color noise that $r_{uu}(\kappa) \neq 0$, $\forall \kappa$.

3. *Case when $f(u'(k))$ is an even and $g(e(k))$ is an odd function:*
Assume all odd order moments of the source noise are zero. Then the second term in (6.16.18) becomes zero and (6.16.18) reduces to

$$r_{\varepsilon'\varepsilon'}(\kappa) \sim E\{f(u'(k-\kappa))f(u'(k))\}E\{g(e(k-\kappa))g(e(k))\}$$
$$+ \bar{f}(u')\left[E\{g(e(k-\kappa))e(k)\} + E\{g(e(k))e(k-\kappa)\}\right] + r_{e'e'}(\kappa) \qquad (6.16.22)$$
$$= r_{ff}(\kappa)r_{gg}(\kappa) + \bar{f}(u')\left[r_{eg}(-\kappa) + r_{eg}(\kappa)\right] + r_{ee}(\kappa)$$

In (6.16.22) the notation

$$r_{eg}(\kappa) = E\{e(k - \kappa)g(e(k))\}$$

was introduced.

Assume, for example,

$$g(e(k)) = e(k - \kappa_3)$$

The second term of (6.16.22) is

$$\bar{f}(u')\big[r_{eg}(-\kappa)+r_{eg}(\kappa)\big]=\bar{f}(u')\big[r_{ee}(\kappa+\kappa_3)+r_{ee}(\kappa-\kappa_3)\big]$$

which differs from zero for $\kappa=\kappa_3\neq0$ or $\kappa=-\kappa_3\neq0$ independently of the type of input signal. If $\kappa_3=0$ then the second term of (6.16.22) vanishes for $\kappa\neq0$ and the first term $r_{ff}(\kappa)r_{gg}(\kappa)$ has to be investigated whether it detects the systematic error or not. The auto-correlation functions $r_{ff}(\kappa)$ and $r_{gg}(\kappa)$ have been already investigated under Case 1 and $g(\ldots)$ should be a properly delayed function of the source noise to detect the error.

The conclusion is that the error can be detected if $f(\ldots)$ is an even and $g(\ldots)$ an odd function and $g(\ldots)$ is a delayed function of the source noise.

Examples for detectable residuals are

$$\varepsilon(k)=cu'(k-\kappa_1)e(k-\kappa_3)+e(k),\qquad \kappa_3\geq1$$
$$\varepsilon(k)=\big[c_1u'^2(k-\kappa_1)+c_2u'^2(k-\kappa_2)\big]e^3(k-\kappa_3)+e(k),\qquad \kappa_3\geq1$$

The residuals

$$\varepsilon(k)=c_1u'^2(k-\kappa_1)e(k)+e(k)$$
$$\varepsilon(k)=\big[c_1u'^2(k-\kappa_1)+c_2u'^2(k-\kappa_2)\big]e^3(k)$$

are not detectable by the auto-correlation function test.

4. *Case when both $f(u'(k))$ and $g(e'(k))$ are even functions:*
The third term in (6.16.18) becomes zero and (6.16.18) reduces to

$$r_{\varepsilon'\varepsilon'}(\kappa)\sim \mathrm{E}\big\{f(u'(k-\kappa))f(u'(k))\big\}\mathrm{E}\big\{g(e(k-\kappa))g(e(k))\big\}-\bar{f}^2(u')\bar{g}^2(e')+r_{ee}(\kappa)$$
$$=r_{ff}(\kappa)r_{gg}(\kappa)-\bar{f}^2(u')\bar{g}^2(e')+r_{ee}(\kappa)$$

$$(6.16.23)$$

(a) *Assume that both $f(\ldots)$ and $g(\ldots)$ are functions of one delayed square term*

$$f(u'(k))=u'^2(k-\kappa_1)$$
$$g(e(k))=e^2(k-\kappa_2)$$

Assuming a Gaussian distribution the auto-correlation functions are

$$r_{ff}(\kappa)=2r_{u'u'}^2(\kappa)+1$$
$$r_{gg}(\kappa)=2r_{ee}^2(\kappa)+\sigma_e^4$$

The auto-correlation function of the residuals is

$$r_{\varepsilon'\varepsilon'}(\kappa) \sim \left[2r_{u'u'}^2(\kappa)+1\right]\left[2r_{ee}^2(\kappa)+\sigma_e^4\right]-\sigma_e^4+r_{ee}(\kappa) \tag{6.16.24}$$

If both the input signal and the source noise are white noises then (6.16.24) becomes zero for $\kappa \neq 0$ and the systematic error cannot be detected. If, however, the input signal is a color signal then for $\kappa \neq 0$ (6.16.24) becomes

$$r_{\varepsilon'\varepsilon'}(\kappa) \sim 2\sigma_e^4 r_{u'u'}^2(\kappa)$$

which would detect the modeling error.

(b) *Assume that both functions $f(\ldots)$ and $g(\ldots)$ are the sum of two differently delayed square terms*

$$f(u'(k)) = c_1 u'^2(k-\kappa_1) + c_2 u'^2(k-\kappa_2)$$
$$g(e'(k)) = c_3 e'^2(k-\kappa_3) + c_4 e'^2(k-\kappa_4)$$

Assuming a Gaussian distribution the auto-correlation functions $r_{ff}(\kappa)$ and $r_{gg}(\kappa)$ become

$$r_{ff}(\kappa) = E\{f(u'(k-\kappa))f(u'(k))\} = c_1^2\left[2r_{u'u'}^2(\kappa)+1\right]$$
$$+c_1 c_2\left[2r_{u'u'}^2(\kappa+\kappa_1-\kappa_2)+1\right]+c_1 c_2\left[2r_{u'u'}^2(\kappa+\kappa_2-\kappa_1)+1\right] \tag{6.16.25}$$
$$+c_2^2\left[2r_{u'u'}^2(\kappa)+1\right]$$

$$r_{gg}(\kappa) = E\{f(e(k-\kappa))f(e(k))\} = c_3^2\left[2r_{ee}^2(\kappa)+\sigma_e^4\right]$$
$$+c_3 c_4\left[2r_{ee}^2(\kappa+\kappa_3-\kappa_4)+\sigma_e^4\right]+c_3 c_4\left[2r_{ee}^2(\kappa+\kappa_4-\kappa_3)+\sigma_e^4\right] \tag{6.16.26}$$
$$+c_4^2\left[2r_{ee}^2(\kappa)+\sigma_e^4\right]$$

The expression (6.16.23) becomes with (6.16.25) and (6.16.26)

$$r_{\varepsilon'\varepsilon'}(\kappa) \sim \left[c_1^2\left[2r_{u'u'}^2(\kappa)+1\right]+c_1 c_2\left[2r_{u'u'}^2(\kappa+\kappa_1-\kappa_2)+1\right]\right.$$
$$+c_1 c_2\left[2r_{u'u'}^2(\kappa+\kappa_2-\kappa_1)+1\right]+c_2^2\left[2r_{u'u'}^2(\kappa)+1\right]\right]$$
$$\times\left[c_3^2\left[2r_{ee}^2(\kappa)+\sigma_e^4\right]+c_3 c_4\left[2r_{ee}^2(\kappa+\kappa_3-\kappa_4)+\sigma_e^4\right]\right. \tag{6.16.27}$$
$$+c_3 c_4\left[2r_{ee}^2(\kappa+\kappa_4-\kappa_3)+\sigma_e^4\right]+c_4^2\left[2r_{ee}^2(\kappa)+\sigma_e^4\right]\right]$$
$$-(c_1+c_2)^2\sigma_e^4(c_3+c_4)^2\sigma_e^4+r_{ee}(\kappa)$$

If both the input signal and the source noise are white noises then (6.16.27) has to be investigated at $\kappa = \kappa_2 - \kappa_1$ or $\kappa = \kappa_1 - \kappa_2$ and $\kappa = \kappa_4 - \kappa_3$ or $\kappa = \kappa_3 - \kappa_4$. If the shifting time is none of the listed ones and also differs from zero then

(6.16.27) becomes

$$r_{\varepsilon'\varepsilon'}(\kappa) \sim \left[c_1^2 + 2c_1c_2 + c_2^2\right]\left[c_3^2\sigma_e^4 + 2c_3c_4\sigma_e^4 + c_4^2\sigma_e^4\right] - (c_1+c_2)^2(c_3+c_4)^2\sigma_e^4$$

$$= [c_1+c_2]^2\left[(c_3+c_4)\sigma_e^2\right]^2 - (c_1+c_2)^2(c_3+c_4)^2\sigma_e^4 = 0$$

If the shifting time is one of the combinations
- $\kappa = \kappa_1 - \kappa_2 = \kappa_3 - \kappa_4 \neq 0$ or
- $\kappa = \kappa_1 - \kappa_2 = \kappa_4 - \kappa_3 \neq 0$ or
- $\kappa = \kappa_2 - \kappa_1 = \kappa_3 - \kappa_4 \neq 0$ or
- $\kappa = \kappa_2 - \kappa_1 = \kappa_4 - \kappa_3 \neq 0$

then

$$r_{\varepsilon'\varepsilon'}(\kappa) \sim \left[(c_1+c_2)^2 + 2c_1c_2\right]\left[(c_3+c_4)^2 + 2c_3c_4\right]\sigma_e^4 - (c_1+c_2)^2(c_3+c_4)^2\sigma_e^4$$

$$= \sigma_e^4\left[2c_1c_2(c_3+c_4)^2 + 2c_3c_4(c_1+c_2)^2 + 4c_1c_2c_3c_4\right]$$

the value of which depends on the coefficients c_1, c_2, c_3 and c_4.
If the shifting time is $\kappa = \kappa_1 - \kappa_2 \neq 0$ and $\kappa \neq \kappa_3 - \kappa_4$ then

$$r_{\varepsilon'\varepsilon'}(\kappa) \sim \left[(c_1+c_2)^2 + 2c_1c_2\right](c_3+c_4)^2\sigma_e^4 - (c_1+c_2)^2(c_3+c_4)^2\sigma_e^4$$

$$= 2c_1c_2(c_3+c_4)^2\sigma_e^4$$

(6.16.28)

differs from zero for non-zero coefficients c_1, c_2, c_3 and c_4.

(c) *Assume quadratic functions in the form of the product terms*

$$f(u'(k)) = c_1 u'(k-\kappa_1)u'(k-\kappa_2)$$
$$g(e(k)) = c_3 e(k-\kappa_3)e(k-\kappa_4)$$

The auto-correlation functions $r_{ff}(\kappa)$ and $r_{gg}(\kappa)$ become

$$r_{ff}(\kappa) = c_1^2 E\{u'(k-\kappa_1-\kappa)u'(k-\kappa_2-\kappa)u'(k-\kappa_1)u'(k-\kappa_2)\}$$

$$= c_1^2\left[r_{u'u'}^2(\kappa_1-\kappa_2) + r_{u'u'}^2(\kappa) + r_{u'u'}(\kappa+\kappa_1-\kappa_2)r_{u'u'}(\kappa+\kappa_2-\kappa_1)\right]$$

$$r_{gg}(\kappa) = c_3^2 E\{e(k-\kappa_3-\kappa)e(k-\kappa_4-\kappa)e(k-\kappa_3)e(k-\kappa_4)\}$$

$$= c_3^2\left[r_{ee}^2(\kappa_1-\kappa_2) + r_{ee}^2(\kappa) + r_{ee}(\kappa+\kappa_3-\kappa_4)r_{ee}(\kappa+\kappa_4-\kappa_3)\right]$$

The expression (6.16.23) becomes

$$r_{\varepsilon'\varepsilon'}(\kappa) \sim c_1^2\left[r_{u'u'}^2(\kappa_1-\kappa_2) + r_{u'u'}^2(\kappa) + r_{u'u'}(\kappa+\kappa_1-\kappa_2)r_{u'u'}(\kappa+\kappa_2-\kappa_1)\right]$$

$$\times c_3^2 \left[r_{ee}^2(\kappa_1 - \kappa_2) + r_{ee}^2(\kappa) + r_{ee}(\kappa + \kappa_3 - \kappa_4) r_{ee}(\kappa + \kappa_4 - \kappa_3) \right]$$

$$-c_1^2 r_{u'u'}^2(\kappa_1 - \kappa_2) c_3^2 r_{ee}^2(\kappa_1 - \kappa_2) + r_{ee}(\kappa)$$

$$= c_1^2 \left[r_{u'u'}^2(\kappa_1 - \kappa_2) + r_{u'u'}^2(\kappa) + r_{u'u'}(\kappa + \kappa_1 - \kappa_2) r_{u'u'}(\kappa + \kappa_2 - \kappa_1) \right]$$

$$\times c_3^2 \left[r_{e'e'}^2(\kappa) + r_{e'e'}(\kappa + \kappa_3 - \kappa_4) r_{ee}(\kappa + \kappa_4 - \kappa_3) \right] + r_{ee}(\kappa)$$

(6.16.29)

because the source noise is a white noise. If both the input signal and the source noise are white noises then (6.16.29) has to be investigated at
- $\kappa = \kappa_2 - \kappa_1$ or $\kappa = \kappa_1 - \kappa_2$ and
- $\kappa = \kappa_4 - \kappa_3$ or $\kappa = \kappa_3 - \kappa_4$

If the shifting time is none of those listed and also differs from zero then (6.16.29) becomes zero. If the shifting time is $\kappa = \kappa_2 - \kappa_1 \neq 0$, and $\kappa \neq \kappa_4 - \kappa_3$, then

$$r_{\varepsilon'\varepsilon'}(\kappa) \sim c_1^2 \sigma_u^4 \cdot 0 = 0$$

If the shifting time is $\kappa = \kappa_2 - \kappa_1 = \kappa_4 - \kappa_3 \neq 0$ then

$$r_{\varepsilon'\varepsilon'}(\kappa) \sim c_1^2 \sigma_u^4 c_3^2 \sigma_e^4 \neq 0$$

(6.16.30)

The conclusion is that the systematic errors can be detected if both functions $f(...)$ and $g(...)$ do not consist of the product of two differently shifted terms. ∎

Lemma 6.16.16
(Haber, 1993) $r_{(\varepsilon'^2)'\varepsilon'}(\kappa)$ differs from zero for at least one shifting time:
- if the systematic errors are odd functions of both the input signal and the source noise and the odd order moments of the source noise are not zero; or
- if the systematic errors are odd functions of the input signal and even functions of the source noise and the odd order moments of the normalized input signal or the source noise are not zero, or
- if the systematic errors are even functions of the input signal and odd function of the source noise and the odd order moments of the source noise are not zero;
- if the systematic errors are even functions of both the input signal and the source noise.

Proof.   The second-order auto-correlation function is

$$r_{(\varepsilon'^2)'\varepsilon'}(\kappa) \sim E\left\{ \left[\varepsilon'^2(k-\kappa) - E\{\varepsilon'^2(k)\} \right] \varepsilon'(k) \right\} = E\left\{ \varepsilon'^2(k-\kappa)\varepsilon'(k) \right\}$$

$$= E\left\{ \left[f(u'(k-\kappa))g(e(k-\kappa)) - \bar{f}(u')\bar{g}(e) + e(k-\kappa) \right]^2 \right.$$

$$\times \left[f(u'(k))g(e(k)) - \bar{f}(u')\bar{g}(e) + e(k) \right] \right\}$$

$$= E\left\{ \left[f^2(u'(k-\kappa))g^2(e(k-\kappa)) + \bar{f}^2(u')\bar{g}^2(e) + e^2(k-\kappa) \right. \right.$$

$$-2\bar{f}(u')\bar{g}(e)f(u'(k-\kappa))g(e(k-\kappa))-2\bar{f}(u')\bar{g}(e)e(k-\kappa)$$

$$+2f(u'(k-\kappa))g(e(k-\kappa))e(k-\kappa)\big][f(u'(k))g(e(k))-\bar{f}(u')\bar{g}(e)+e(k)]\big\}$$

$$=\mathrm{E}\big\{f^2(u'(k-\kappa))f(u'(k))\big\}\mathrm{E}\big\{g^2(e(k-\kappa))g(e(k))\big\}$$

$$-f(u')g(e)\mathrm{E}\big\{f^2(u'(k))\big\}\mathrm{E}\big\{g^2(e(k))\big\}$$

$$+\mathrm{E}\big\{f^2(u'(k-\kappa))\big\}\mathrm{E}\big\{g^2(e(k-\kappa))e(k)\big\}+\bar{f}^3(u')\bar{g}^3(e)-\bar{f}^3(u')\bar{g}^3(e)$$

$$+\bar{f}^2(u')\bar{g}^2(e)\mathrm{E}\big\{e(k)\big\}+\bar{f}(u')\mathrm{E}\big\{e^2(k-\kappa)g(e(k))\big\}$$

$$-\bar{f}(u')\bar{g}(e)\mathrm{E}\big\{e^2(k)\big\}+\mathrm{E}\big\{e^2(k-\kappa)e(k)\big\}$$

$$-2\bar{f}(u')\bar{g}(e)\mathrm{E}\big\{f(u'(k-\kappa))f(u'(k))\big\}\mathrm{E}\big\{g(e(k-\kappa))g(e(k))\big\}$$

$$+2\bar{f}^3(u')\bar{g}^3(e)-2\bar{f}^2(u')\bar{g}(e)\mathrm{E}\big\{g(e(k-\kappa))e(k)\big\}$$

$$-2\bar{f}^2(u')\bar{g}(e)\mathrm{E}\big\{e(k-\kappa)g(e(k))\big\}$$

$$+2\bar{f}^2(u')\bar{g}^2(e)\mathrm{E}\big\{e(k)\big\}-2\bar{f}^2(u')\bar{g}(e)\mathrm{E}\big\{e(k-\kappa)e(k)\big\}$$

$$+2\mathrm{E}\big\{f(u'(k-\kappa))f(u'(k))\big\}\mathrm{E}\big\{g(e(k-\kappa))e(k-\kappa)g(e(k))\big\}$$

$$-2\bar{f}^2(u')\bar{g}(e)\mathrm{E}\big\{g(e(k-\kappa))e(k-\kappa)\big\}+2\bar{f}(u')\mathrm{E}\big\{g(e(k-\kappa))e(k-\kappa)e(k)\big\}$$

$$(6.16.31)$$

1. Case when g(...) is an odd function:
Assume all odd order moments of the source noise are zero. Then all terms are zero and the systematic error is undetectable.

2. Case when f(...) is odd and g(...) is an even function:
Assume all odd order moments of the source noise are zero. Then the following terms become zero in (6.16.31): 3, 6, 9, 12, 13, 14, 16, 17. Assume, furthermore, that all odd order moments of the normalized input signal are zero. Then the following terms in (6.16.31) become also zero: 1, 2, 4, 5, 6, 7, 8, 10, 11, 12, 13, 14, 15, 17, 18. If both assumptions are valid then all terms in (6.16.31) vanish and the systematic error is undetectable.

3. Case when f(...) and g(...) are even functions:
Assume, furthermore, that all odd order moments of the normalized input signal and of the source noise are zero. Because of g(...) is an even function the terms 3, 6, 9, 12, 13, 14, 16, 17 become zero in (6.16.31). The terms 4 and 5 vanish because they are identical to each other with different signs. No further term is missing in (6.16.31) because of f(...) is an even function. Then (6.16.31) becomes

$$r_{(\varepsilon'^2)'\varepsilon'}(\kappa)\sim\mathrm{E}\big\{f^2(u'(k-\kappa))f(u'(k))\big\}\mathrm{E}\big\{g^2(e(k-\kappa))g(e(k))\big\}$$

$$-\bar{f}_0(u')\bar{g}_0(e)\mathrm{E}\big\{f^2(u'(k))\big\}\mathrm{E}\big\{g^2(e(k))\big\}$$

$$+\bar{f}(u')\mathrm{E}\big\{e^2(k-\kappa)g(e(k))\big\}-\bar{f}(u')\bar{g}(e)\mathrm{E}\big\{e^2(k)\big\}$$

$$-2\bar{f}(u')\bar{g}(e)\mathrm{E}\{f(u'(k-\kappa))f(u'(k))\}\mathrm{E}\{g(e(k-\kappa))g(e(k))\}$$
$$+2\bar{f}^3(u')\bar{g}^3(e)-2\bar{f}(u')\bar{g}(e)\mathrm{E}\{e(k-\kappa)e(k)\}$$
$$+2\bar{f}(u')\mathrm{E}\{g(e(k-\kappa))e(k-\kappa)e(k)\} \qquad\qquad (6.16.32)$$

(6.16.32) differs from zero for any modeling error. The following example illustrates it:

$$\varepsilon(k)=u'^2(k-\kappa_1)e^2(k-\kappa_2)+e(k)$$

which means that

$$f(u'(k))=u'^2(k-\kappa_1)$$
$$g(e(k))=e^2(k-\kappa_2)$$

The components of (6.16.32) are as follows:

$$\bar{f}(u')=\mathrm{E}\{u'^2(k)\}=1$$
$$\bar{g}(e)=\mathrm{E}\{e^2(k)\}=\sigma_e^2$$
$$\bar{f}^2(u')=\mathrm{E}\{u'^4(k)\}=3$$
$$\bar{g}^2(e)=\mathrm{E}\{e^4(k)\}=3\sigma_e^4$$
$$\mathrm{E}\{e^2(k)\}=\sigma_e^2$$
$$\mathrm{E}\{e^2(k-\kappa)g(e(k))\}=\mathrm{E}\{e^2(k-\kappa)e^2(k-\kappa_2)\}=2r_{ee}^2(k-\kappa_2)+\sigma_e^4$$
$$\mathrm{E}\{f(u'(k-\kappa))f(u'(k))\}=\mathrm{E}\{u'^2(k-\kappa_1-\kappa)u'^2(k-\kappa_1)\}=2r_{u'u'}^2(\kappa)+1$$
$$\mathrm{E}\{g(e(k-\kappa))g(e(k))\}=\mathrm{E}\{e^2(k-\kappa_2-\kappa)e^2(k-\kappa_2)\}=2r_{ee}^2(\kappa)+\sigma_e^4$$
$$\mathrm{E}\{f^2(u'(k-\kappa))f(u'(k))\}=\mathrm{E}\{u'^4(k-\kappa_1-\kappa)u'^2(k-\kappa_1)\}=12r_{u'u'}^2(\kappa)+3$$
$$\mathrm{E}\{g^2(e(k-\kappa))g(e(k))\}=\mathrm{E}\{e^4(k-\kappa_2-\kappa)e^2(k-\kappa_2)\}=12\sigma_e^2r_{ee}^2(\kappa)+3\sigma_e^6$$
$$\mathrm{E}\{e(k)\}=0$$
$$\mathrm{E}\{e(k-\kappa)e(k)\}=r_{ee}(\kappa)$$
$$\mathrm{E}\{g(e(k-\kappa))e(k-\kappa)e(k)\}=\mathrm{E}\{e^2(k-\kappa)e(k-\kappa)e(k)\}=3\sigma_e^2r_{ee}^2(\kappa)$$

Replacing of all these components into (6.16.31) leads to

$$r_{(\varepsilon'^2)'\varepsilon'}(\kappa)\sim\left[12r_{u'u'}^2(\kappa)+3\right]\left[12r_{ee}^2(\kappa)+3\sigma_e^2\right]-\sigma_e^2 3\cdot 3\sigma_e^4+\left[2r_{ee}^2(\kappa-\kappa_2)+\sigma_e^4\right]$$

$$-\sigma_e^2\sigma_e^2+2\sigma_e^2\left[2r_{u'u'}^2(\kappa)+1\right]\left[2r_{ee}^2(\kappa)+\sigma_e^4\right]+2\sigma_e^6-2\sigma_e^2 r_{ee}(\kappa)+2\cdot 3\sigma_e^2 r_{ee}^2(\kappa)$$

$$= 144\sigma_e^2 r_{u'u'}^2(\kappa) r_{ee}^2(\kappa) + 36\sigma_e^6 r_{u'u'}^2(\kappa) + 36\sigma_e^2 r_{ee}^2(\kappa) + 9\sigma_e^6 - 9\sigma_e^6$$
$$+2r_{ee}^2(\kappa - \kappa_2) + \sigma_e^4 - \sigma_e^4 - 8\sigma_e^2 r_{u'u'}^2(\kappa) r_{ee}^2(\kappa) - 4\sigma_e^6 r_{u'u'}^2(\kappa)$$
$$-4\sigma_e^2 r_{ee}^2(\kappa) - 2\sigma_e^6 + 2\sigma_e^6 - 2\sigma_e^2 r_{ee}(\kappa) + 6\sigma_e^2 r_{ee}^2(\kappa) = 136\sigma_e^2 r_{u'u'}^2(\kappa) r_{ee}^2(\kappa)$$
$$+32\sigma_e^6 r_{u'u'}^2(\kappa) + 38\sigma_e^2 r_{ee}^2(\kappa) + 2r_{ee}^2(\kappa - \kappa_2) - 2\sigma_e^2 r_{ee}(\kappa)$$

$$(6.16.33)$$

(6.16.33) differs from zero at least at $\kappa = \kappa_2 \neq 0$ even if the input signal is a white noise. For $\kappa_2 = 0$ the correlation function (6.16.33) differs from zero at least for $\kappa = 0$

$$r_{(e'^2)'e'}(0) \sim 136\sigma_e^2 r_{u'u'}^2(0) r_{e'e'}^2(0) + 32\sigma_e^6 r_{u'u'}^2(0) + 38\sigma_e^2 r_{e'e'}^2(0) + 2r_{e'e'}^2(0) - 2\sigma_e^2 r_{e'e'}^2(0)$$

$$= 136\sigma_e^2 + 32\sigma_e^6 + 38\sigma_e^6 + 2\sigma_e^4 - 2\sigma_e^6 = 204\sigma_e^6 + 2\sigma_e^4 \neq 0$$

∎

Lemma 6.16.17

(Haber, 1993) $r_{u'e'}(\kappa)$ differs from zero for at least one shifting time:

- if the systematic errors are odd functions of both the input signal and the source noise and the odd order moments of the source noise are not zero; or
- if the systematic errors are odd functions of the input signal and even functions of the source noise; or
- if the systematic errors are even functions of the input signal and odd function of the source noise and both the odd order moments of the normalized input signal and those of the source noise are not zero; or
- if the systematic errors are even functions of both the input signal and the source noise and the odd order moments of the normalized input signal are not zero.

Proof.   The linear cross-correlation function is

$$r_{u'e'}(\kappa) \sim E\big\{u'(k-\kappa)\big[f(u'(k))g(e(k)) - \bar{f}(u')\bar{g}(e) + e(k)\big]\big\}$$
$$= E\big\{u'(k-\kappa)f(u'(k))g(e(k))\big\} - E\big\{u'(k-\kappa)\bar{f}(u')\bar{g}(e)\big\} + E\big\{u'(k-\kappa)e(k)\big\}$$
$$= \bar{g}(e)E\big\{u'(k-\kappa)f(u'(k))\big\}$$

$$(6.16.34)$$

because the zero expected value of the normalized input signal occurs as a multiplier in the terms 2 and 3 in (6.16.34) The expression (6.16.34) becomes zero:

- if $g(...)$ is an odd function and the odd order moments of the source noise are zero; or
- if $f(...)$ is an even function and the odd order moments of the normalized input signal are zero.

∎

Lemma 6.16.18

(Haber, 1993) $r_{(u'^2)'e'}(\kappa)$ differs from zero for at least one shifting time:

- if the systematic errors are odd functions of both the input signal and the source

noise and the odd order moments of both the normalized input signal and the source noise are not zero, or
- if the systematic errors are odd functions of the input signal and even functions of the source noise and the odd order moments of the normalized input signal are not zero, or
- if the systematic errors are even functions of the input signal and odd function of the source noise and the odd order moments of the source noise are not zero, or
- if the systematic errors are even functions of both the input signal and the source noise.

Proof. The nonlinear cross-correlation function is

$$r_{\left(u'^2\right)' \varepsilon'}(\kappa) \sim \mathrm{E}\left\{\left[u'^2(k-\kappa) - \mathrm{E}\left\{u'^2(k-\kappa)\right\}\right]\left[f(u'(k))g(e(k)) - \bar{f}(u')\bar{g}(e) + e(k)\right]\right\}$$

$$= \mathrm{E}\left\{u'^2(k-\kappa)f(u'(k))g(e(k))\right\} - \mathrm{E}\left\{u'^2(k-\kappa)\bar{f}(u')\bar{g}(e)\right\}$$

$$+ \mathrm{E}\left\{u'^2(k-\kappa)e(k)\right\} = \bar{g}(e)\left[\mathrm{E}\left\{u'^2(k-\kappa)f(u'(k))\right\} - \bar{f}(u')\right]$$

(6.16.35)

because the expected value of $u'^2(k)$ is one. Expression (6.16.35) becomes zero:
- if $g(\ldots)$ is an odd function and the odd order moments of the source noise are zero; or
- if $f(\ldots)$ is an odd function and the odd order moments of the normalized input signal are zero. ∎

Take an example where both functions $f(\ldots)$ and $g(\ldots)$ are even functions:

$$f(u'(k)) = u'^2(k - \kappa_1)$$
$$g(e(k)) = e^2(k - \kappa_2)$$

Then the cross-correlation function

$$r_{\left(u'^2\right)' \varepsilon'}(\kappa) \sim \sigma_e^2\left[\mathrm{E}\left\{u'^2(k-\kappa)u'^2(k-\kappa)\right\} - 1 \cdot 1\right]$$

$$= \sigma_e^2\left[2r_{u'u'}^2(k-\kappa_1) + 1 - 1\right] = 2\sigma_e^2 r_{u'u'}^2(k-\kappa_1)$$

differs from zero at least for one shifting time even if the input signal is a white noise. ∎

Lemma 6.16.19
(Haber, 1993) $r_{u'(\varepsilon'^2)'}(\kappa)$ differs from zero for at least one shifting time
- if the systematic errors are odd functions of both the input signal and the source noise and the odd order moments of the normalized input signal are not zero or the systematic error is not a static function of the source noise; or
- if the systematic errors are odd functions of the input signal and even functions of the source noise and the odd order moments of either the normalized input

signal or the source noise are not zero; or
- if the systematic errors are even functions of the input signal and odd function of the source noise and the odd order moments of the normalized input signal are not zero; or
- if the systematic errors are even functions of both the input signal and the source noise and the odd order moments of the normalized input signal are not zero.

Proof.   The nonlinear cross-correlation function is

$$r_{u'\left(\varepsilon'^2\right)}'(\kappa) \sim E\left\{u'(k-\kappa)\left[\left[f(u'(k))g(e(k)) - \bar{f}(u')\bar{g}(e) + e(k)\right]^2\right.\right.$$

$$\left.\left.-E\left[\left[f(u'(k))g(e(k)) - \bar{f}(u')\bar{g}(e) + e(k)\right]^2\right]\right]\right\}$$

$$= E\left\{u'(k-\kappa)\left[f^2(u'(k))g^2(e(k)) - \bar{f}^2(u')\bar{g}^2(e) + e^2(k)\right.\right.$$

$$\left.\left.-2\bar{f}(u')\bar{g}(e)f(u'(k))g(e(k)) - 2\bar{f}(u')\bar{g}(e)e(k) + 2f(u'(k))g(e(k))e(k)\right]\right\}$$

$$= E\left\{g^2(e(k))\right\}E\left\{u'(k-\kappa)f^2(u'(k))\right\} - 2\bar{f}(u')\bar{g}^2(e)E\left\{u'(k-\kappa)f(u'(k))\right\}$$

$$+2E\left\{u'(k-\kappa)f(u'(k))\right\}E\left\{g(e(k))e(k)\right\}$$

$$(6.16.36)$$

because the zero expected value of the normalized input signal occurs as a multiplier in the missing terms in (6.16.36). (6.16.36) is zero:
- if $f(...)$ is an even function and the odd order moments of the normalized input signal are zero; or
- if $f(...)$ is an odd and $g(...)$ is an even function and the odd order moments of both the normalized input signal and the source noise are zero; or
- if both $f(...)$ and $g(...)$ are odd functions and the odd order moments of the normalized input signal are zero and $g(...)$ is a dynamic (not static) function of the source noise, i.e., $g(...) = g(e(k-\kappa_1), e(k-\kappa_2),...)$, $\kappa_i > 0$, $i = 1, 2, ...$

Consider the last case. If the odd order moments of the normalized input signal are zero then the first and second terms in (6.16.36) are zero. The remaining third term becomes

$$r_{u'\left(\varepsilon'^2\right)}'(\kappa) \sim 2E\left\{u'(k-\kappa)f(u'(k))\right\}E\left\{g(e(k))e(k)\right\} \qquad (6.16.37)$$

An example for odd functions of both $f(...)$ and $g(...)$ is the linear case

$$f(u'(k)) = u'(k-\kappa_1)$$
$$g(e(k)) = e(k-\kappa_2)$$

In this special case (6.16.37) becomes

$$r_{u'(\varepsilon'^2)}'(\kappa) \sim 2\mathrm{E}\{u'(k-\kappa)u'(k-\kappa_1)\}\mathrm{E}\{e(k-\kappa_2)e(k)\}$$

$$= 2r_{u'u'}(k-\kappa_1)r_{ee}(\kappa_2) \tag{6.16.38}$$

The expression in (6.16.38) differs from zero if $r_{ee}(\kappa_2)$ differs from zero. This occurs for $\kappa_2 = 0$ for a white source noise, which means that when the function $g(...)$ is a static function.   ∎

Lemma 6.16.20
(Haber, 1993) $r_{(u'\varepsilon')\varepsilon'}(\kappa)$ differs from zero for at least one shifting time:
- if the systematic errors are odd functions of both the input signal and the source noise and the odd order moments of the normalized input signal are not zero and either the source noise is a color noise or the systematic error is a static function of the source noise or the systematic error is a function of equally shifted input signal and source noise terms (e.g., $\varepsilon(k) = f\big(u'(k-\kappa_i)\big)$

  $g\big(e(k-\kappa_i)\big) + e(k))$;
- if the systematic errors are odd functions of the input signal and even functions of the source noise and the odd order moments of either the normalized input signal or the source noise are not zero;or
- if the systematic errors are even functions of the input signal and odd function of the source noise and the odd order moments of the normalized input signal are not zero; or
- if the systematic errors are even functions of both the input signal and the source noise and the odd order moments of the normalized input signal are not zero.

Proof.   The nonlinear cross-correlation function

$$r_{(u'\varepsilon')\varepsilon'}(\kappa) \sim \mathrm{E}\{u'(k-\kappa)\varepsilon'(k-\kappa)\varepsilon'(k)\}$$

has similar multiplicative components to

$$r_{u'(\varepsilon'^2)}'(\kappa) \sim \mathrm{E}\{u'(k-\kappa)[\varepsilon'^2(k)-\mathrm{E}\{\varepsilon'^2(k)\}]\} = \mathrm{E}\{u'(k-\kappa)\varepsilon'^2(k)\}$$

only the shifting times between the components can be different. Consequently all statements given in the Proof of Lemma 6.16.19 are also valid here, except those regarding certain shifting times, which means that only the case when the systematic errors are odd functions of both the normalized input signal and the source noise has to be investigated separately. The correlation function $r_{(u'\varepsilon')\varepsilon'}(\kappa)$ has the form:

$$r_{(u'\varepsilon')\varepsilon'}(\kappa) \sim \mathrm{E}\{u'(k-\kappa)\varepsilon'(k-\kappa)\varepsilon'(k)\}$$

$$= \mathrm{E}\{u'(k-\kappa)[f(u'(k-\kappa))g(e(k-\kappa)) - \bar{f}_0(u')\bar{g}_0(e) + e(k-\kappa)]$$

$$\times [f(u'(k))g(e(k)) - \bar{f}_0(u')\bar{g}_0(e) + e(k)]\}$$

$$
\begin{aligned}
&= \mathrm{E}\big\{u'(k-\kappa)f\big(u'(k-\kappa)\big)f\big(u'(k)\big)\big\}\mathrm{E}\big\{g\big(e(k-\kappa)\big)g\big(e(k)\big)\big\} \\
&\quad -\bar{f}_0(u')\bar{g}_0(e)\mathrm{E}\big\{u'(k-\kappa)f\big(u'(k-\kappa)\big)\big\}\mathrm{E}\big\{g\big(e(k-\kappa)\big)\big\} \\
&\quad +\mathrm{E}\big\{u'(k-\kappa)f\big(u'(k-\kappa)\big)\big\}\mathrm{E}\big\{g\big(e(k-\kappa)\big)e(k)\big\} \\
&\quad -\bar{f}_0(u')\bar{g}_0^2(e)\mathrm{E}\big\{u'(k-\kappa)f\big(u'(k)\big)\big\}+\bar{f}_0^2(u')\bar{g}_0^2(e)\mathrm{E}\big\{u'(k)\big\} \\
&\quad -\bar{f}_0(u')\bar{g}_0(e)\mathrm{E}\big\{u'(k-\kappa)\big\}\mathrm{E}\big\{e(k)\big\} \\
&\quad +\mathrm{E}\big\{u'(k-\kappa)f\big(u'(k)\big)\big\}\mathrm{E}\big\{g\big(e(k)\big)e(k-\kappa)\big\} \\
&\quad -\bar{f}_0(u')\bar{g}_0(e)\mathrm{E}\big\{u'(k-\kappa)\big\}\mathrm{E}\big\{e(k-\kappa)\big\}+\mathrm{E}\big\{u'(k-\kappa)\big\}\mathrm{E}\big\{e(k-\kappa)e(k)\big\}
\end{aligned}
$$

$$(6.16.39)$$

If the systematic errors are odd functions of both the normalized input signal and the source noise and all odd order moments of the normalized input signal are zero then (6.16.39) reduces to

$$
\begin{aligned}
r_{(u'\varepsilon')e'}(\kappa) &\sim \mathrm{E}\big\{u'(k-\kappa)f\big(u'(k-\kappa)\big)\big\}\mathrm{E}\big\{g\big(e(k-\kappa)\big)e(k)\big\} \\
&\quad +\mathrm{E}\big\{u'(k-\kappa)f\big(u'(k)\big)\big\}\mathrm{E}\big\{g\big(e(k)\big)e(k-\kappa)\big\}
\end{aligned}
$$

$$(6.16.40)$$

As expected, (6.16.37) and (6.16.40) have the same multiplicative components with various shifting times. An example for odd functions $f(\ldots)$ and $g(\ldots)$ is the linear case

$$
\begin{aligned}
f\big(u'(k)\big) &= u'\big(k-\kappa_1\big) \\
g\big(e(k)\big) &= e\big(k-\kappa_2\big)
\end{aligned}
$$

Expression (6.16.40) now becomes

$$
\begin{aligned}
r_{(u'\varepsilon')e'}(\kappa) &\sim \mathrm{E}\big\{u'(k-\kappa)u'\big(k-\kappa_1-\kappa\big)\big\}\mathrm{E}\big\{e\big(k-\kappa_2-\kappa\big)e(k)\big\} \\
&\quad +\mathrm{E}\big\{u'(k-\kappa)u'\big(k-\kappa_1\big)\big\}\mathrm{E}\big\{e\big(k-\kappa_2\big)e(k-\kappa)\big\} \\
&= r_{u'u'}(\kappa_1)r_{ee}\big(k+\kappa_2\big)+r_{u'u'}\big(\kappa-\kappa_1\big)r_{ee}\big(\kappa-\kappa_2\big)
\end{aligned}
$$

$$(6.16.41)$$

which differs from zero for at least one shifting time if:
- either the input signal is a color noise such that its auto-correlation function exists for all (or at least for $\kappa=\kappa_1$ or $\kappa=\kappa_2-\kappa_1$) shifting times, or
- the systematic error is a static function of the input signal $(\kappa_1=0)$, or
- both the input signal and the source noise are equally shifted in the product terms $(\kappa_1=\kappa_2)$.   ■

Examples of detectable modeling errors by correlation function $r_{(u'\varepsilon')e'}(\kappa)$ test for any input signal even with zero odd order moments are
- $\varepsilon'(k)=u'(k-2)e(k-2)+e(k)$
- $\varepsilon'(k)=u'(k)e(k-3)+e(k)$

- $\varepsilon'(k) = u'(k)e^3(k) + e(k)$
- $\varepsilon'(k) = u'(k-1)e(k-1) + u'(k-2)e(k-2) + e(k)$

The residual

$$\varepsilon'(k) = u'(k-1)e(k-3) + e(k)$$

is undetectable if a white noise input signal is applied, but it can be detected by a color input signal if $r_{u'u'}(\kappa-1)r_{ee}(\kappa-3)$ differs from zero, which happens if $r_{u'u'}(3-1) = r_{u'u'}(2)$ differs from zero.

Lemma 6.16.21

(Haber, 1993) $r_{(u'^2)'(\varepsilon'^2)'}(\kappa)$ differs from zero for at least one shifting time and consequently is able to detect any systematic error in the residuals if the systematic error has product terms of the input signal and the source noise.

Proof.   The nonlinear cross-correlation function under investigation is

$$r_{(u'^2)'(\varepsilon'^2)'}(\kappa) \sim \mathrm{E}\left\{\left[u'^2(k-\kappa) - \mathrm{E}\{u'^2(k)\}\right]\left[\left[f(u'(k))g(e(k)) - \bar{f}(u')\bar{g}(e) + e(k)\right]^2\right.\right.$$

$$-\mathrm{E}\left\{\left[f(u'(k))g(e(k)) - \bar{f}(u')\bar{g}(e) + e(k)\right]^2\right\}\right]\right\}$$

$$= \mathrm{E}\left\{\left[u'^2(k-\kappa) - 1\right]\left[f(u'(k))g(e(k)) - \bar{f}(u')\bar{g}(e) + e(k)\right]^2\right\}$$

$$= \mathrm{E}\left\{\left[u'^2(k-\kappa) - 1\right]\left[f^2(u'(k))g^2(e(k)) + \bar{f}^2(u')\bar{g}^2(e)\right.\right.$$

$$+ e^2(k) - 2f(u')g(e)f(u'(k))g(e(k))$$

$$-2\bar{f}(u')\bar{g}(e)e(k) + 2f(u'(k))g(e(k))e(k)\right]\right\}$$

$$= \mathrm{E}\{g^2(e(k))\}\left[\mathrm{E}\{u'^2(k-\kappa)f^2(u'(k))\} - \mathrm{E}\{f^2(u'(k))\}\right]$$

$$-2\bar{f}(u')\bar{g}^2(e)\left[\mathrm{E}\{u'^2(k-\kappa)f(u'(k))\} + \bar{f}(u')\right] \qquad (6.16.42)$$

$$+ \mathrm{E}\{g(e(k))e(k)\}\left[2\mathrm{E}\{u'^2(k-\kappa)f(u'(k))\} - 2\bar{f}(u')\right]$$

because

$$\mathrm{E}\{u'^2(k)\} = 1$$

and the following terms have no contribution to (6.16.42):

$$\bar{f}^2(u')\bar{g}^2(e)\mathrm{E}\{[u'^2(k-\kappa) - 1]\} = 0$$

$$\mathrm{E}\{[u'^2(k-\kappa) - 1]\}\mathrm{E}\{e^2(k)\} = 0$$

$$\bar{f}(u')\bar{g}(e)\mathrm{E}\big\{\big[u'^2(k-\kappa)-1\big]\big\}\mathrm{E}\{e(k)\} = 0$$

Three different cases will be investigated in turn.

1. *Case when* $f(...)$ *is an odd function:*
Assume all odd order moments of the normalized input signal are zero. Then (6.16.42) reduces to

$$r_{\left(u'^2\right)'\left(\varepsilon'^2\right)'}(\kappa) \sim \mathrm{E}\big\{g^2\big(e(k)\big)\big\}\big[\mathrm{E}\big\{u'^2(k-\kappa)f^2\big(u'(k)\big)\big\} - \mathrm{E}\big\{f^2\big(u'(k)\big)\big\}\big] \qquad (6.16.43)$$

which differs from zero for at least one shifting time. As an example, for an odd $g(...)$ take

$$\varepsilon'(k) = u'(k-\kappa_1)e(k-\kappa_2) + e(k)$$

which means

$$f\big(u'(k)\big) = u'(k-\kappa_1)$$
$$g\big(e(k)\big) = e(k-\kappa_2)$$

The correlation function (6.16.43) becomes in this case

$$r_{\left(u'^2\right)'\left(\varepsilon'^2\right)'}(\kappa) \sim \mathrm{E}\big\{u'^2(k-\kappa)u'^2(k-\kappa_1)\big\} - \mathrm{E}\big\{u'^2(k-\kappa_1)\big\}$$

$$= \sigma_e^2\big[[2r_{u'u'}(k-\kappa_1)+1]-1\cdot 1\big] = 2\sigma_e^2 r_{u'u'}(k-\kappa_1)$$

which differs from zero for at least one shifting time.
  As an example, for an even function $g(...)$ take

$$\varepsilon(k) = u'(k-\kappa_1)e^2(k-\kappa_2) + e(k)$$

which means

$$f\big(u'(k)\big) = u'(k-\kappa_1)$$
$$g\big(e(k)\big) = e^2(k-\kappa_2)$$

The correlation function (6.16.43) becomes in this case

$$r_{\left(u'^2\right)'\left(\varepsilon'^2\right)'}(\kappa) \sim \mathrm{E}\big\{e^4(k-\kappa_2)\big\}\big[\mathrm{E}\big\{u'^2(k-\kappa)u'^2(k-\kappa_1)\big\} - \mathrm{E}\big\{u'^2(k-\kappa_1)\big\}\big]$$

$$= 3\sigma_e^4\big[[2r_{u'u'}(k-\kappa_1)+1]-1\cdot 1\big] = 6\sigma_e^4 r_{u'u'}(k-\kappa_1)$$

which differs from zero for at least one shifting time.

2. *Case when $f(...)$ is an even function and $g(...)$ an odd function:*
The second term in (6.16.42) is zero if all odd order moments of the source noise are zero and (6.16.42) reduces to

$$r_{(u'^2)'(\epsilon'^2)'}(\kappa) \sim E\{g^2(e(k))\}\Big[E\{u'^2(k-\kappa)f^2(u'(k))\} - E\{f^2(u'(k))\}\Big]$$

$$+ E\{g(e(k))e(k)\}\Big[2E\{u'^2(k-\kappa)f(u'(k))\} - 2\bar{f}(u')\Big] \qquad (6.16.44)$$

Expression (6.16.44) detects any modeling error. As an example take

$$\epsilon'(k) = u'^2(k-\kappa_1)e(k-\kappa_2) + e(k)$$

which means

$$f(u'(k)) = u'^2(k-\kappa_1)$$
$$g(e(k)) = e(k-\kappa_2)$$

The correlation function (6.16.44) becomes in this case

$$r_{(u'^2)'(\epsilon'^2)'}(\kappa) \sim E\{e^2(k-\kappa_2)\}\Big[E\{u'^2(k-\kappa)u'^4(k-\kappa_1)\} - E\{u'^4(k-\kappa_1)\}\Big]$$

$$+ E\{e(k-\kappa_2)e(k)\}\Big[2E\{u'^2(k-\kappa)u'^2(k-\kappa_1)\} - 2E\{u'^2(k-\kappa_1)\}\Big]$$

$$= \sigma_e^2\Big[12r_{u'u'}^2(\kappa-\kappa_1)+3-3\Big] + r_{e'e'}(\kappa_2)\Big[2r_{u'u'}^2(\kappa-\kappa_1)+1-2\Big]$$

$$= 12\sigma_e^2 r_{u'u'}^2(\kappa-\kappa_1) + r_{e'e'}(\kappa_2)\Big[2r_{u'u'}^2(\kappa-\kappa_1)-1\Big] \qquad (6.16.45)$$

$$= r_{u'u'}^2(\kappa-\kappa_1)\Big[12\sigma_e^2 + 2r_{e'e'}(\kappa_2)\Big] - r_{e'e'}(\kappa_2)$$

There exist two different cases. If $\kappa_2 = 0$ then (6.16.45) reduces to

$$r_{(u'^2)'(\epsilon'^2)'}(\kappa) \sim r_{u'u'}^2(\kappa-\kappa_1)\Big[12\sigma_e^2 + 2\sigma_e^2\Big] - \sigma_e^2$$

and if $\kappa_2 \neq 0$ then (6.16.45) reduces to

$$r_{(u'^2)'(\epsilon'^2)'}(\kappa) \sim 12\sigma_e^2 r_{u'u'}^2(\kappa-\kappa_1)$$

In both cases (6.16.45) differs from zero for at least one shifting time.

3. *Case when $f(...)$ and $g(...)$ are even functions:*
The last term in (6.16.42) is zero if all odd order moments of the source noise are zero, i.e.,

$$r_{(u'^2)'(\epsilon'^2)'}(\kappa) \sim E\{g^2(e(k))\}\Big[E\{u'^2(k-\kappa)f^2(u'(k))\} - E\{f^2(u'(k))\}\Big]$$

$$-2\bar{f}(u')\bar{g}^2(e)\Big[E\big\{u'^2(k-\kappa)f(u'(k))\big\}+\bar{f}(u')\Big] \tag{6.16.46}$$

Expression (6.16.46) detects the modeling error. As an example take

$$\varepsilon(k)=u'^2(k-\kappa_1)e^2(k-\kappa_2)-E\big\{u'^2(k-\kappa_1)e^2(k-\kappa_2)\big\}+e(k)$$

which means

$$f(u'(k))=u'^2(k-\kappa_1)$$
$$g(e'(k))=e^2(k-\kappa_2)$$

The correlation function (6.16.46) becomes in this case

$$
\begin{aligned}
r_{(u'^2)'(\varepsilon'^2)'}(\kappa) \sim & E\big\{e^4(k-\kappa_2)\big\}\Big[E\big\{u'^2(k-\kappa)u'^4(k-\kappa_1)\big\}-E\big\{u'^4(k-\kappa_1)\big\}\Big] \\
& -2E\big\{u'^2(k-\kappa_1)\big\}E\big\{e^2(k-\kappa_2)\big\} \\
& \times\Big[E\big\{u'^2(k-\kappa)u'^2(k-\kappa_1)\big\}+E\big\{u'^2(k-\kappa_1)\big\}\Big] \\
= & 3\sigma_e^4\big[12r_{u'u'}^2(\kappa-\kappa_1)+3-3\big]-2\sigma_e^4\big[2r_{u'u'}^2(\kappa-\kappa_1)+1+1\big] \\
= & 36\sigma_e^4 r_{u'u'}^2(\kappa-\kappa_1)+4\sigma_e^4
\end{aligned}
$$

which differs from zero for at least one shifting time.   ∎

Theorem 6.16.3 (Haber, 1993) Table 6.16.3 summarizes under what circumstances the normalized correlation functions:
- auto-correlation function of the residuals $r_{e'e'}(\kappa)$, $\kappa\neq0$;
- second-order auto-correlation function of the residuals $r_{(\varepsilon'^2)'\varepsilon'}(\kappa)$;
- cross-correlation function of the input signal and the residuals $r_{u'e'}(\kappa)$;
- cross-correlation function of the square of the input signal and the residual $r_{(u'^2)'\varepsilon'}(\kappa)$;
- cross-correlation function of the input signal and the square of the residual $r_{u'(\varepsilon'^2)'}(\kappa)$;
- cross-correlation function of the product of the residual and the input signal and of the residual $r_{(u'\varepsilon')e'}(\kappa)$;
- cross-correlation function of the square of the input signal and the square of the residuals $r_{(u'^2)'(\varepsilon'^2)'}(\kappa)$,

detect the systematic errors if the residual has product terms of the input signal and the source noise $u^i(k-\kappa_1)e^j(k-\kappa_2)$ (i,j,κ_1,κ_2 integers), as shown in (6.16.14).

Proof. The proof is a summary of the results of the Lemmas 6.16.15 to 6.16.21. ∎

TABLE 6.16.3 Features of different correlation functions when the systematic error in the residuals are product terms of the delayed input signal and the delayed source error (o.o.m.: odd order moments; fct.: function; one-term fct. e.g.,: $u^i(k - \kappa_1)$; two-term fct. e.g.,: $u^i(k - \kappa_1) + u^j(k - \kappa_2)$

Correlation function	κ	$\varepsilon(k) = V^{ue}[u'(k), e(k)] + e(k) = f(u'(k))g(e(k)) - E\{f(u'(k))\}E\{g(e(k))\} + e(k)$			
		$f(u'(k))$ *odd function* $g(e(k))$ *odd function*	$f(u'(k))$ *odd function* $g(e(k))$ *even function*	$f(u'(k))$ *even function* $g(e(k))$ *odd function*	$f(u'(k))$ *even function* $g(e(k))$ *even function*
$r_{\varepsilon'\varepsilon'}(\kappa)$	$\neq 0$	$=0$ if $\begin{cases}\text{o.o.m. of } u' \text{ are zero and} \\ \text{o.o.m. of } e \text{ are zero and} \\ g(\ldots) \text{ is a one - or two -} \\ \text{term fct. and } u' \text{ is white} \\ \text{noise or } f(\ldots) \text{ and} \\ g(\ldots) \text{ are two - term fcts.} \\ \text{with different delays}\end{cases}$ $\neq 0$ otherwise	$=0$ if $\begin{cases}\text{o.o.m. of } u' \text{ are zero and} \\ u' \text{ is white noise and} \\ f(\ldots) \text{ is a one - term fct.}\end{cases}$ $\neq 0$ otherwise	$=0$ if $\begin{cases}\text{o.o.m. of } e \text{ are zero and} \\ g(\ldots) \text{ is a static function}\end{cases}$ $\neq 0$ otherwise	$=0$ if $\begin{cases}\text{o.o.m. of } e \text{ are zero and } f(\ldots) \\ \text{and } g(\ldots) \text{ are two - term} \\ \text{fcts. with different delays}\end{cases}$ $\neq 0$ otherwise
$r_{(\varepsilon'^2)'\varepsilon'}(\kappa)$	$\forall$	$=0$ if o.o.m. of e are zero $\neq 0$ otherwise	$=0$ if $\begin{cases}\text{o.o.m. of } u' \text{ are zero and} \\ \text{o.o.m. of } e \text{ are zero}\end{cases}$ $\neq 0$ otherwise	$=0$ if o.o.m. of e are zero $\neq 0$ otherwise	$\neq 0$
$r_{u'\varepsilon'}(\kappa)$	$\forall$	$=0$ if o.o.m. of e are zero $\neq 0$ otherwise	$\neq 0$	$=0$ if $\begin{cases}\text{o.o.m. of } u' \text{ are zero and} \\ \text{o.o.m. of } e \text{ are zero}\end{cases}$ $\neq 0$ otherwise	$=0$ if o.o.m. of u' are zero $\neq 0$ otherwise
$r_{(u'^2)'\varepsilon'}(\kappa)$	$\forall$	$=0$ if $\begin{cases}\text{o.o.m. of } e \text{ are zero or} \\ \text{o.o.m. of } u' \text{ are zero}\end{cases}$ $\neq 0$ otherwise	$=0$ if o.o.m. of u' are zero $\neq 0$ otherwise	$=0$ if o.o.m. of e are zero $\neq 0$ otherwise	$\neq 0$
$r_{u'(\varepsilon'^2)'}(\kappa)$	$\forall$	$=0$ if $\begin{cases}\text{o.o.m. of } u' \text{ are zero and} \\ g(\ldots) \text{ is a static function}\end{cases}$ $\neq 0$ otherwise	$=0$ if $\begin{cases}\text{o.o.m. of } u' \text{ are zero and} \\ \text{o.o.m. of } e \text{ are zero}\end{cases}$ $\neq 0$ otherwise	$=0$ if o.o.m. of u' are zero $\neq 0$ otherwise	$=0$ if o.o.m. of u' are zero $\neq 0$ otherwise
$r_{(u'\varepsilon')\varepsilon'}(\kappa)$	$\forall$	$=0$ if $\begin{cases}\text{o.o.m. of } u' \text{ are zero and} \\ u' \text{ is white noise and} \\ g(\ldots) \text{ is not a static fct. and} \\ f(\ldots) \text{ and } g(\ldots) \text{ are not equally} \\ \text{shifted fct. - s}\end{cases}$ $\neq 0$ otherwise	$=0$ if $\begin{cases}\text{o.o.m. of } u' \text{ are zero and} \\ \text{o.o.m. of } e \text{ are zero}\end{cases}$ $\neq 0$ otherwise	$=0$ if o.o.m. of u' are zero $\neq 0$ otherwise	$=0$ if o.o.m. of u' are zero $\neq 0$ otherwise
$r_{(u'^2)'(\varepsilon'^2)'}(\kappa)$	$\forall$	$\neq 0$	$\neq 0$	$\neq 0$	$\neq 0$

Example 6.16.2 *Approximation of the noise-free simple Wiener model by simple linear and nonlinear structures (Haber and Zierfuss, 1987) (Continuation of Examples 6.2.1, 6.5.1 and 6.12.1)*

The plot of the normalized auto-correlation function of the residuals in Figure 6.12.1 does not show which model fitting is the best. The cross-correlation functions $r_{(\varepsilon'^2)'\varepsilon'}(\kappa)$, $r_{u'\varepsilon'}(\kappa)$, $r_{(u'^2)'\varepsilon'}(\kappa)$, $r_{(u'\varepsilon')\varepsilon'}(\kappa)$, $r_{(u'^2)'(\varepsilon'^2)'}(\kappa)$ are shown now in Figures 6.16.1 to 6.16.5, respectively. The band $\pm 2/\sqrt{N} = \pm 2/\sqrt{130} = \pm 0.175$ is also marked in all figures. The empirical evaluation of the correlation functions shows the following:

- *second-order auto-correlation function of the residuals:*
  Only values belonging to non-zero shifting times are considered in the test. The bounds are violated with the linear, Hammerstein and parametric Volterra models four times, with the bilinear model three times, and all values lie within the bounds with the general polynomial model;
- *cross-correlation function of the input signal and the residuals:*
  The values are very small for the linear, Hammerstein, and parametric Volterra models. They just achieve the bounds with the bilinear and general polynomial

models but the values are only a little bigger than the bounds. According to this test the linear, Hammerstein, and parametric Volterra models would be the best ones;

- *cross-correlation function of the square of the input signal and the residuals:*
  The values violate the bounds with the linear, Hammerstein, and parametric Volterra models. With the general polynomial model almost all values are very small (less than 20 percent of the bound);
- *cross-correlation function of the product of the residual and the input signal and of the residuals:*
  The values for all models lie within the bounds except around the shifting time equal to zero;
- *cross-correlation function of the square of the input signal and the square of the residuals:*
  All values lie within the bounds. No significant difference can be seen between the various models.

The tests show that the general polynomial model is the best fitting one.

On using only the test based on the cross-correlation function of the square of the input signal and the residuals the correct order of the quality of the approximations can be determined:

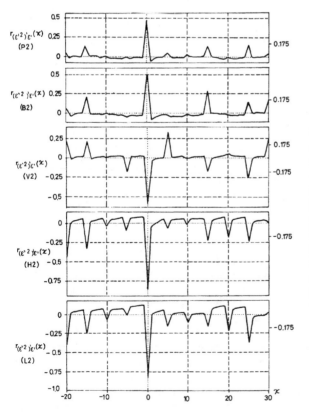

Fig. 6.16.1 The second-order auto-correlation functions of the residuals $r_{\left(\varepsilon'^2\right)\varepsilon'}(\kappa)$ of the estimated linear (L2), Hammerstein (H2), parametric Volterra (V2), bilinear (B2) and general polynomial (P2) models at the identification of the noise-free simple Wiener model

1. *second-order quadratic general polynomial model;*
2. *second-order simple bilinear model;*
3. *second-order quadratic parametric Volterra model;*
4. *second-order quadratic generalized Hammerstein model;*
5. *second-order linear model.*

This order corresponds to the order obtained from the standard deviations of the residuals.   ∎

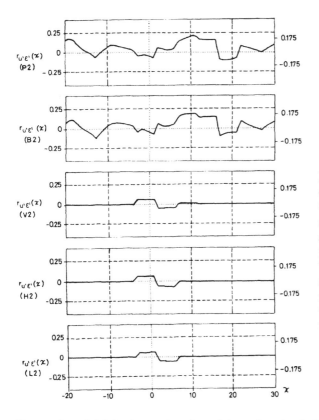

Fig. 6.16.2 The cross-correlation function of the input signal and the residuals $r_{u'\varepsilon'}(\kappa)$ of the estimated linear (L2), Hammerstein (H2), parametric Volterra (V2), bilinear (B2) and general polynomial (P2) models at the identification of the noise-free simple Wiener model

Example 6.16.3 *Approximation of the noisy simple Wiener model by simple linear and nonlinear structures (Haber and Zierfuss, 1987) (Continuation of Examples 6.2.2, 6.5.2 and 6.12.2)*

The plot of the normalized auto-correlation function of the residuals in Figure 6.12.2 does not show which model fitting is the best. The cross-correlation functions $r_{u'\varepsilon'}(\kappa)$,

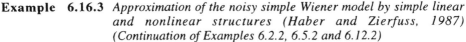

$r_{\left(u'^2\right)'\varepsilon'}(\kappa)$, $r_{(u'\varepsilon')\varepsilon'}(\kappa)$, $r_{\left(u'^2\right)'\left(\varepsilon'^2\right)'}(\kappa)$ are shown now in Figures 6.16.6 to 6.16.9,

respectively. The band $\pm 2/\sqrt{N} = \pm 2/\sqrt{130} = \pm 0.175$ is also marked in all figures. The empirical evaluation of the correlation functions shows the following:

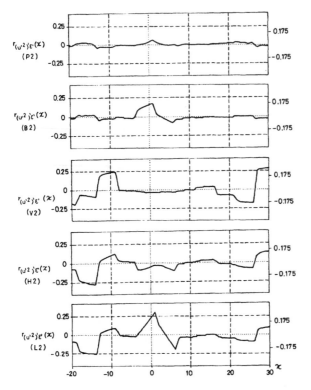

Fig. 6.16.3 The cross-correlation function of the square of the normalized input signal and the residuals $r_{(u'^2)\,\varepsilon'}(\kappa)$ of the estimated linear (L2), Hammerstein (H2), parametric Volterra (V2), bilinear (B2) and general polynomial (P2) models at the identification of the noise-free simple Wiener model

- *cross-correlation function of the input signal and the residuals:*
  The values are very small for the linear, Hammerstein, and parametric Volterra models. They just achieve the bounds with the bilinear and general polynomial models but the values are only a little bigger than the bounds. According to this test the linear, Hammerstein, and parametric Volterra models would be the best ones;
- *cross-correlation function of the square of the input signal and the residuals:*
  The values violate the bounds with the linear and Hammerstein models definitely and only a bit with the parametric Volterra and bilinear models. With the general polynomial model almost all values are very small (less than 30 percent of the bound);
- *cross-correlation function of the product of the residual and the input signal and of the residuals:*
  Most values of all models lie within the bounds except some peaks, which are only a little over the boundary values (with about 10 percent) . The number of places where the outliers leave the bounds are:
  - *linear model:* 4;
  - *generalized Hammerstein model:* 5;
  - *parametric Volterra model:* 3;
  - *bilinear model:* 2;
  - *general polynomial model:* 1.
  The order of the models based on the number of outliers is approximately

correct, but the difference between the correlation functions is not very significant;
- *cross-correlation function of the square of the input signal and the square of the residuals:*
All values lie within the bounds. No significant difference can be seen between the various models.

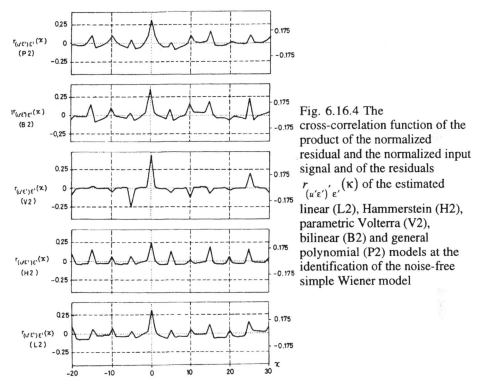

Fig. 6.16.4 The cross-correlation function of the product of the normalized residual and the normalized input signal and of the residuals $r_{(u'\varepsilon')\,\varepsilon'}(\kappa)$ of the estimated linear (L2), Hammerstein (H2), parametric Volterra (V2), bilinear (B2) and general polynomial (P2) models at the identification of the noise-free simple Wiener model

The tests show that the general polynomial model is the best fitting one.

On using only the test based on the cross-correlation function of the square of the input signal and the residuals the correct order of the quality of the approximations can be determined:

1. *second-order quadratic general polynomial model;*
2. *second-order simple bilinear model;*
3. *second-order quadratic parametric Volterra model;*
4. *second-order quadratic generalized Hammerstein model;*
5. *second-order linear model.*

This order corresponds to the order obtained from the standard deviations of the residuals. The same test was also the best one in the noise-free case, as shown in Example 3.16.2.   ∎

By using Tables 6.16.1 to 6.16.3 one can see from the different correlation functions which (wrong) terms include the residuals. Billings and Voon (1986) also presented a list of conclusions based on three correlation functions.

Billings and Voon (1983) recommended three correlation functions, by that a non-proper model fitting can be shown. The following lemma is needed to the proof of their method.

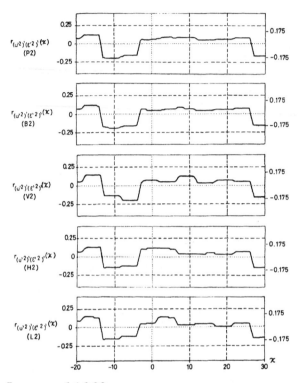

Fig. 6.16.5 The cross-correlation function of the square of the normalized input signal and of the square of the residuals $r_{(u'^2)'(\varepsilon'^2)'}(\kappa)$ of the estimated linear (L2), Hammerstein (H2), parametric Volterra (V2), bilinear (B2), and general polynomial (P2) models at the identification of the noise-free simple Wiener model

Lemma 6.16.22
The cross-correlation function of two normalized signals is equal to zero if and only if the cross-correlation function of the unnormalized signals reduced by their mean values is zero.

Proof. Denote the two signals x_1 and x_2. Then the cross-correlation function of the normalized signals is

$$r_{x_1'x_2'}(\kappa) = E\{x_1'(k-\kappa)x_2'(k)\} = E\left\{\frac{x_1(k-\kappa)-E\{x_1(k)\}}{\sigma_{x_1}} \frac{x_2(k-\kappa)-E\{x_2(k)\}}{\sigma_{x_2}}\right\}$$

$$= \frac{1}{\sigma_{x_1}\sigma_{x_2}}E\left\{\left[x_1(k-\kappa)-E\{x_1(k)\}\right]\left[x_2(k-\kappa)-E\{x_2(k)\}\right]\right\}$$

$$\sim E\left\{\left[x_1(k-\kappa)-E\{x_1(k)\}\right]\left[x_2(k-\kappa)-E\{x_2(k)\}\right]\right\} \qquad \blacksquare$$

Theorem 6.16.4 (Billings and Voon, 1983 with some modifications) The proper model is well fitted to a linear or nonlinear system if and only if the following correlation functions are zero

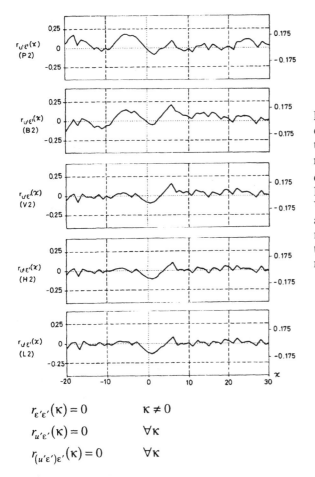

Fig. 6.16.6 The cross-correlation function of the input signal and the residuals $r_{u'\varepsilon'}(\kappa)$ of the estimated linear (L2), Hammerstein (H2), parametric Volterra (V2), bilinear (B2) and general polynomial (P2) models at the identification of the noisy simple Wiener model

$$r_{\varepsilon'\varepsilon'}(\kappa) = 0 \qquad \kappa \neq 0$$

$$r_{u'\varepsilon'}(\kappa) = 0 \qquad \forall \kappa$$

$$r_{(u'\varepsilon')\varepsilon'}(\kappa) = 0 \qquad \forall \kappa$$

If the residuals have terms of type $u(k-i)e(k-j)$ then they can be detected if the input signal is a colored noise (Haber, 1993). In the case of a white noise excitation the systematic error can be detected if $i = 0$ or $i = j$ (Haber, 1993).

Proof. (Haber, 1993) Assume the systematic errors have the form (6.16.2).

1. *Case when* $x(k) = f\big(u'(k-i)\big) - \bar{f}(u')$
   (a) *Assume that* $f(...)$ *is an odd function*
      According to Lemma 6.16.3 $r_{u'\varepsilon'}(\kappa)$ detects the systematic error.
   (b) *Assume that* $f(...)$ *is an even function*
      According to Lemma 6.16.1 $r_{\varepsilon'\varepsilon'}(\kappa)$, $\kappa \neq 0$, detects the systematic error.
2. *Case* $x(k) = g\big(e'(k-j)\big) - \bar{g}(e')$
   (a) *Assume that* $g(...)$ *is an odd function*
      According to Lemma 6.16.8 $r_{\varepsilon'\varepsilon'}(\kappa)$ $\kappa \neq 0$, detects the systematic error.

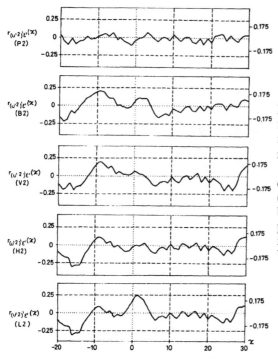

Fig. 6.16.7 The cross-correlation function of the square of the normalized input signal and of the residuals $r_{(u'^2)\,\varepsilon'}(\kappa)$ of the estimated linear (L2), Hammerstein (H2), parametric Volterra (V2), bilinear (B2) and general polynomial (P2) models at the identification of the noisy simple Wiener model

(b) *Assume that $g(...)$ is an even function*
   According to Lemma 6.16.1 $r_{\varepsilon'\varepsilon'}(\kappa)$, $\kappa \neq 0$, detects the systematic error if $j \geq 0$.

3. *Case $x(k) = f(u'(k-i))g(u'(k-j)) - \bar{f}(u')\bar{g}(e')$*
   (a) *Assume that both $f(...)$ and $g(...)$ are odd functions*
      According to Lemma 6.16.20 $r_{(u'\varepsilon')\varepsilon'}(\kappa)$ detects the systematic error with any input signal if $i = 0$ or $i = j$. Without any constraints of the type of the residuals the systematic errors are detected if the input signal is a colored noise.
   (b) *Assume that $f(...)$ is an odd and $g(...)$ is an even function*
      According to Lemma 6.16.17 $r_{u'\varepsilon'}(\kappa)$ detects the systematic error.
   (c) *Assume that $f(...)$ is an even and $g(...)$ is an odd function*
      According to Lemma 6.16.15 $r_{\varepsilon'\varepsilon'}(\kappa)$, $\kappa \neq 0$, detects the systematic error if $j \geq 0$.
   (d) *Assume that both $f(...)$ and $g(...)$ are even functions*
      According to Lemma 6.16.15 $r_{\varepsilon'\varepsilon'}(\kappa)$, $\kappa \neq 0$ detects the systematic error.   ∎

Remarks:
1. Billings and Voon (1983) use the notations $u'(k)$ and $\varepsilon'(k)$ as signals reduced by their expected values, i.e. they are not normalized by their standard deviations.
2. Billings and Voon (1983) recommended to investigate the correlation function

$r_{(u'\varepsilon')\varepsilon'}(\kappa)$ for $\kappa \geq 1$. As is seen from the proof this condition can be dropped.

3. The need of introducing the restrictions in the original form of the Theorem 6.16.4 in the case of a residual type $\varepsilon(k) = f(u(k-i))g(e(k-j)) + e(k)$ with odd $f(\ldots)$ and $g(\ldots)$ functions was explained in the proof of Lemma 6.16.20. There the example $\varepsilon(k) = u(k-i)e(k-j) + e(k)$ was treated in detail.

As is seen from Tables 6.16.1 to 6.16.3 three other correlation functions can be selected for detecting the systematic errors in the residuals. This is summarized in Theorem 6.16.5.

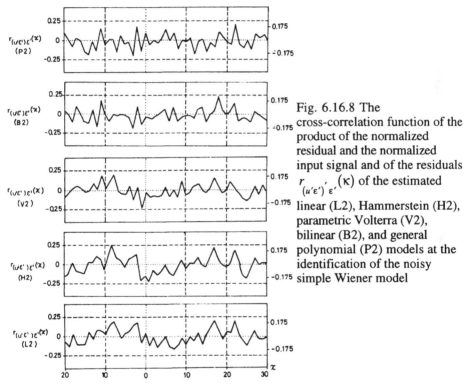

Fig. 6.16.8 The cross-correlation function of the product of the normalized residual and the normalized input signal and of the residuals $r_{(u'\varepsilon')\,\varepsilon'}(\kappa)$ of the estimated linear (L2), Hammerstein (H2), parametric Volterra (V2), bilinear (B2), and general polynomial (P2) models at the identification of the noisy simple Wiener model

Theorem 6.16.5   (Haber, 1993) The proper model is well fitted to a linear or nonlinear system if and only if the following correlation functions are zero:

$$r_{\varepsilon'\varepsilon'}(\kappa) = 0, \qquad \kappa \neq 0, \qquad r_{u'\varepsilon'}(\kappa) = 0, \qquad \forall \kappa$$

$$r_{(u'^2)'(\varepsilon'^2)'}(\kappa) = 0, \qquad \forall \kappa$$

Proof.   Assume the systematic errors have the form (6.16.2)

1. *Case* $x(k) = f(u'(k-i)) - \bar{f}(u')$

    (a) *Assume that* $f(\ldots)$ *is an odd function*

        According to Lemma 6.16.3 $r_{u'\varepsilon'}(\kappa)$ detects the systematic error.

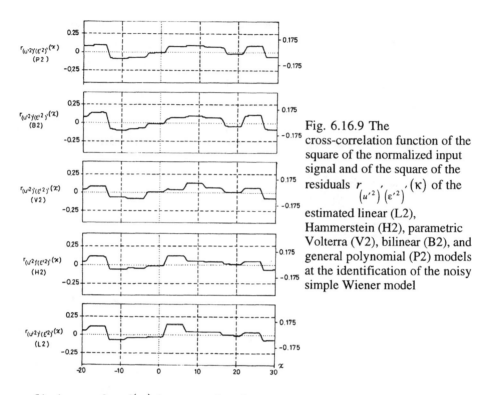

Fig. 6.16.9 The cross-correlation function of the square of the normalized input signal and of the square of the residuals $r_{(u'^2)'(\varepsilon'^2)'}(\kappa)$ of the estimated linear (L2), Hammerstein (H2), parametric Volterra (V2), bilinear (B2), and general polynomial (P2) models at the identification of the noisy simple Wiener model

(b) *Assume that $f(...)$ is an even function*
    According to Lemma 6.16.1 $r_{\varepsilon'\varepsilon'}(\kappa)$, $\kappa \neq 0$, detects the systematic error.

2. *Case $x(k) = g(e'(k-j)) - \bar{g}(e')$*
    (a) *Assume that $g(...)$ is an odd function*
        According to Lemma 6.16.8 $r_{\varepsilon'\varepsilon'}(\kappa)$, $\kappa \neq 0$, detects the systematic error.
    (b) *Assume that $g(...)$ is an even function*
        According to Lemma 6.16.1 $r_{\varepsilon'\varepsilon'}(\kappa)$, $\kappa \neq 0$, detects the systematic error if $j \geq 0$.

3. *Case $x(k) = f(u'(k-i))g(u'(k-j)) - \bar{f}(u')\bar{g}(e')$*
    (a) *Assume that both $f(...)$ and $g(...)$ are odd functions*
        According to Lemma 6.16.21 $r_{(u'^2)'(\varepsilon'^2)'}(\kappa)$ detects the systematic error.
    (b) *Assume that $f(...)$ is an odd and $g(...)$ is an even function*
        According to Lemma 6.16.21 $r_{(u'^2)'(\varepsilon'^2)'}(\kappa)$ detects the systematic error.
    (c) *Assume that $f(...)$ is an even and $g(...)$ is an odd function*
        According to Lemma 6.16.21 $r_{(u'^2)'(\varepsilon'^2)'}(\kappa)$ detects the systematic error if $j \geq 0$.

(d) *Assume that both* $f(\ldots)$ *and* $g(\ldots)$ *are even functions*

According to Lemma 6.16.21 $r_{\left(u'^2\right)\left(\varepsilon'^2\right)}(\kappa)$ detects the systematic error. ■

Leontaritis and Billings (1987) recommended a further model validity test based upon correlation functions. The cross-correlation function of the residuals and any possible model components can be computed and checked for being zero or not. If the correlation function is not zero then the model is not a proper one. As mentioned already, the actual output signal and the actual residuals cannot be possible candidates for building the components that are then correlated with the residuals. Otherwise the residuals would always be correlated with themselves for shifting time zero.

A chi-square test is recommended to check whether a correlation function is zero or not in a shifting time domain. How this test has to be applied was explained in Section 4.9 in detail.

6.17 CHECKING THE ESTIMATED PARAMETERS FOR SIGNIFICANCE

The result of the system identification is a proper model with its estimated parameters. The question is how significant the estimated parameters are. To investigate this problem, we interpret the estimates.

Assumption 6.17.1   Assume that the residuals are stochastic disturbances with zero mean and standard deviation σ_ε. ■

1. *Approximating calculation of the indifference region*

The δ-indifference region of the estimated parameters is defined by the region in which the loss function $J\!\left(\hat{\theta}\right)$ does not increase more than δ,

$$J(\theta) - J\!\left(\hat{\theta}\right) \leq \delta \tag{6.17.1}$$

where

$$J(\theta) = \tfrac{1}{2} \sum_{k=1}^{N} \varepsilon^2(k)$$

Assume the loss function can be differentiated twice with respect to the estimated parameters and it can be approximated by the first three terms of the Taylor series:

$$Q(\theta) - Q\!\left(\hat{\theta}\right) \approx \left.\frac{\partial Q\!\left(\hat{\theta}\right)}{\partial\hat{\theta}^{\mathrm{T}}}\right|_{\theta=\hat{\theta}} \Delta\theta + \tfrac{1}{2}\,\Delta\theta^{\mathrm{T}} H_{\theta\theta}\!\left(\hat{\theta}\right)\Delta\theta$$

where $H_{\theta\theta}$ is the Hessian of the loss function at the estimated parameters. The derivative of the loss function is zero for the estimated parameters,

$$\frac{\partial J\!\left(\hat{\theta}\right)}{\partial\hat{\theta}^{\mathrm{T}}} = 0$$

thus

$$J(\theta) - J\left(\hat{\theta}\right) \approx \tfrac{1}{2} \Delta\theta^T H_{\theta\theta}\left(\hat{\theta}\right) \Delta\theta \qquad\qquad (6.17.2)$$

and

$$\Delta\theta^T H_{\theta\theta}\left(\hat{\theta}\right) \Delta\theta \le 2\delta$$

or

$$\left(\theta - \hat{\theta}\right)^T H_{\theta\theta}\left(\hat{\theta}\right)\left(\theta - \hat{\theta}\right) \le 2\delta$$

determines the δ-indifference region in the parameter space.

2. *Covariance matrix of the estimated parameters*
The covariance matrix of the estimated parameters is

$$\mathrm{cov}\left(\hat{\theta}\right) = \mathrm{E}\left\{\Delta\hat{\theta}\Delta\hat{\theta}^T\right\}$$

It shows the important features of the parameter estimation:
- The elements in the main diagonal are the variances of the estimated parameters:

$$\mathrm{cov}\left(\hat{\theta}\right)[i, i] = \mathrm{var}\left\{\hat{\theta}(i)\right\} = \sigma^2_{\hat{\theta}(i)}$$

- The off-diagonal elements show the correlation between the estimated parameters:

$$\mathrm{cov}\left(\hat{\theta}\right)[i, j] = \mathrm{E}\left\{\Delta\hat{\theta}(i)\,\Delta\hat{\theta}^T(i)\right\} \quad i \ne j$$

The more accurate the parameter estimation is, the smaller the standard deviation of the parameters will be. It is also expedient that the correlation between the parameters is small, because that means that an error in the parameter estimation of a parameter does not influence the estimation of another parameter. Then an uncertainty in an estimated parameter can be considered independently from the uncertainties of the other estimated parameters.
   The covariance matrix of the estimated parameters can be calculated in different ways:
- *Monte Carlo simulation from more identification experiments:*
   If there is a possibility of repeating the identification experiment with the same test signal N_e times then the mean value and the covariance matrix of the estimated parameters can be calculated from the estimated parameters $\hat{\theta}_j$ of the individual experiments as

$$\hat{\theta} = \frac{1}{N_e} \sum_{j=1}^{N_e} \hat{\theta}_j$$

$$\mathrm{cov}\left(\hat{\theta}\right) = \frac{1}{N_e} \sum_{j=1}^{N_e} \left(\hat{\theta}_j - \hat{\theta}\right)\left(\hat{\theta}_j - \hat{\theta}\right)^T$$

- *Analytical calculation from one identification experiment:*
The covariance matrix can be calculated as (Bard, 1974)

$$\text{cov}\left(\hat{\boldsymbol{\theta}}\right) \approx \sigma_\varepsilon^2 \text{E}\left\{\boldsymbol{H_{\theta\theta}^{-1}}\right\} \text{E}\left\{\boldsymbol{H_{\theta\varepsilon}} \boldsymbol{H_{\theta\varepsilon}^T}\right\} \text{E}\left\{\left(\boldsymbol{H_{\theta\theta}^{-1}}\right)^T\right\}$$

where $\boldsymbol{H_{\theta\theta}}$ is the second derivative (the Hessian) of the loss function of the parameter estimation

$$\boldsymbol{H_{\theta\theta}} = \frac{\partial}{\partial\boldsymbol{\theta}}\left[\frac{\partial J\left(\hat{\boldsymbol{\theta}}\right)}{\partial\boldsymbol{\theta}^T}\right] = \frac{\partial^2 J\left(\hat{\boldsymbol{\theta}}\right)}{\partial\boldsymbol{\theta}\partial\boldsymbol{\theta}^T}$$

and $\boldsymbol{H_{\theta\varepsilon}}$ is the mixed partial derivative of the loss function

$$\boldsymbol{H_{\theta\varepsilon}} = \frac{\partial^2 J\left(\hat{\boldsymbol{\theta}}\right)}{\partial\boldsymbol{\varepsilon}\partial\boldsymbol{\theta}^T}$$

If the residual model is linear in the parameters then the convergence matrix of the estimated parameters can be approximated by

$$\text{cov}\left(\hat{\boldsymbol{\theta}}\right) = \sigma_\varepsilon^2 \tilde{\boldsymbol{H}}_{\boldsymbol{\theta\theta}}^{-1}$$

where $\tilde{\boldsymbol{H}}_{\boldsymbol{\theta\theta}}^{-1}$ is the approximating Hessian matrix

$$\tilde{\boldsymbol{H}}_{\boldsymbol{\theta\theta}}^{-1} = \frac{\partial\boldsymbol{\varepsilon}^T}{\partial\hat{\boldsymbol{\theta}}}\frac{\partial\boldsymbol{\varepsilon}}{\partial\hat{\boldsymbol{\theta}}^T} = \sum_{k=1}^{N}\frac{\partial\varepsilon(k)}{\partial\hat{\boldsymbol{\theta}}}\frac{\partial\varepsilon(k)}{\partial\hat{\boldsymbol{\theta}}^T}$$

Proof

$$\tilde{\boldsymbol{H}}_{\boldsymbol{\theta\theta}}^{-1} = \frac{\partial^2 J\left(\hat{\boldsymbol{\theta}}\right)}{\partial\boldsymbol{\theta}\partial\boldsymbol{\theta}^T} = \frac{\partial\boldsymbol{\varepsilon}^T}{\partial\hat{\boldsymbol{\theta}}}\frac{\partial\boldsymbol{\varepsilon}}{\partial\hat{\boldsymbol{\theta}}^T} + \frac{\partial}{\partial\hat{\boldsymbol{\theta}}}\left[\frac{\partial\boldsymbol{\varepsilon}}{\partial\hat{\boldsymbol{\theta}}^T}\right]^T \boldsymbol{\varepsilon} = \frac{\partial\boldsymbol{\varepsilon}^T}{\partial\hat{\boldsymbol{\theta}}}\frac{\partial\boldsymbol{\varepsilon}}{\partial\hat{\boldsymbol{\theta}}^T} \quad\blacksquare$$

Assumption 6.17.2   Assume that Assumption 6.17.1 holds and the residuals have a Gaussian distribution. Further assume a linear or nonlinear model linear-in-parameters. ∎

Then the following properties are valid in addition to 1 and 2. (Bard, 1974; Draper and Smith, 1966; Goodwin and Payne, 1977).

3. *Distribution of the estimated parameters*
The estimated parameters have a Gaussian distribution.

4. *Mean value of the estimated parameters*
The mean value of the estimated parameters can be taken equal to the estimated parameter value.

5. *Standard deviation of the estimated parameters*
The standard deviation of the parameters is

$$\sigma_{\hat{\theta}(i)} = \sqrt{\text{var}\{\hat{\theta}(i)\}} = \sqrt{\text{cov}(\hat{\theta})[i,i]}$$

6. *Confidence limit of the estimated parameters*
Since the distribution of the estimated parameters is normal, the estimated parameters normalized by their estimated standard deviations have a Student (t) distribution. The true parameters lie near the estimated parameters

$$\hat{\theta}(i) - t(N - n_\theta, \alpha)\sigma_{\hat{\theta}(i)} \leq \theta(i) \leq \hat{\theta}(i) + t(N - n_\theta, \alpha)\sigma_{\hat{\theta}(i)}$$

with probability $(1-\alpha)100$ percent. For a large number of data that means, for probability 95 percent $(\alpha = 0.05 = 5 \text{ percent})$,

$$\hat{\theta}(i) - 1.96\sigma_{\hat{\theta}(i)} \leq \theta(i) \leq \hat{\theta}(i) + 1.96\sigma_{\hat{\theta}(i)}$$

and for the probability 90 percent $(\alpha = 0.01 = 10 \text{ percent})$

$$\hat{\theta}(i) - 1.65\sigma_{\hat{\theta}(i)} \leq \theta(i) \leq \hat{\theta}(i) + 1.65\sigma_{\hat{\theta}(i)}$$

7. *Confidence region of all estimated parameters*
The quantity $\left(\theta - \hat{\theta}\right)^T\left[\text{cov}(\hat{\theta})\right]^{-1}\left(\theta - \hat{\theta}\right)$ is Ξ^2-distributed with a degree of freedom equal to the number of parameters. Therefore the inequality

$$\left(\theta - \hat{\theta}\right)^T\left[\text{cov}(\hat{\theta})\right]^{-1}\left(\theta - \hat{\theta}\right) \leq \Xi^2(n_\theta, \alpha)$$

gives a confidence region with confidence $(1-\alpha)100$ percent for the parameters θ that satisfy the inequality. Taking (6.17.1) into account (6.17.2) can be rewritten as

$$J(\theta) - J(\hat{\theta}) \leq \frac{\sigma_\varepsilon^2}{2}\Xi^2(n_\theta, \alpha)$$

8. *Confidence limit for the sum of squares of the residuals*
The sum of squares of the residuals normalized by its variance has a chi-square distribution with $N - n_\theta$ degree of freedom. Therefore the inequality

$$\frac{1}{\sigma_\varepsilon^2}\sum_{k=1}^{N}\varepsilon^2(k) = \frac{2J(\hat{\theta})}{\sigma_\varepsilon^2} \leq \Xi^2(N - n_\theta, \alpha)$$

also gives a confidence region with $(1-\alpha)100$ percent probability for the parameters.

9. *F-test*

Since both $\left(\boldsymbol{\theta}-\hat{\boldsymbol{\theta}}\right)^{\mathrm{T}}\left[\mathrm{cov}\left(\hat{\boldsymbol{\theta}}\right)\right]^{-1}\left(\boldsymbol{\theta}-\hat{\boldsymbol{\theta}}\right)$ and $2J\left(\hat{\boldsymbol{\theta}}\right)/\sigma_\varepsilon^2$ have Ξ^2 distributions with the degree of freedoms n_θ and $N-n_\theta$, respectively, the ratio

$$\frac{\left(\boldsymbol{\theta}-\hat{\boldsymbol{\theta}}\right)^{\mathrm{T}}\left[\mathrm{cov}\left(\hat{\boldsymbol{\theta}}\right)\right]^{-1}\left(\boldsymbol{\theta}-\hat{\boldsymbol{\theta}}\right)/n_\theta}{\left(2J\left(\hat{\boldsymbol{\theta}}\right)/\sigma_\varepsilon^2\right)/\left(N-n_\theta\right)}$$

has an *F*-distribution with degrees of freedom n_θ and $N-n_\theta$. The confidence region of the parameters is determined by the inequality

$$\left(\boldsymbol{\theta}-\hat{\boldsymbol{\theta}}\right)^{\mathrm{T}}\left[\mathrm{cov}\left(\hat{\boldsymbol{\theta}}\right)\right]^{-1}\left(\boldsymbol{\theta}-\hat{\boldsymbol{\theta}}\right)\le\frac{n_\theta}{N-n_\theta}\frac{2J\left(\hat{\boldsymbol{\theta}}\right)}{\sigma_\varepsilon^2}F\left(n_\theta,N-n_\theta,1-\alpha\right)$$

at the probability level $(1-\alpha)100$ percent.

Assumption 6.17.3   Assume that Assumption 6.17.1 holds and the residuals have a Gaussian distribution. Further assume a model nonlinear-in-parameters.   ■

In this case the following properties are valid instead of 3 to 8.

3. *Distribution of the parameters*
The distribution is not Gaussian, therefore the Student, chi-square, and *F*-tests cannot be applied.

4. *Calculation of the indifference region*
The boundary points of the uncertainty region can be drawn by simulation. The loss function has to be calculated in several points of the parameter space. The estimated parameter is that one where the loss function has its global minima. The boundary of the uncertainty region is defined by equation (6.17.1) (Mereau and Prevost, 1979).

6.18 CHECKING THE IDENTIFIED MODEL FOR PHYSICAL INTERPRETATION

The estimated model should be checked whether it is a realistic one or not. It may be that some prior assumptions have to be fulfilled, such as:
- stability;
- aperiodic behavior (no oscillations);
- known steady state relation, etc.

Example 6.18.1 *Extruder*

A thermal process has usually an aperiodic step response. Thus the identified relation between the heating power and the temperature in an extruder should have also this feature.   ■

Example 6.18.2 *Mill*

In its steady state the inlet and outlet flows of a mill are equal to each other, which
means that the steady state relation is linear with unit static gain.   ∎

6.19 REFERENCES

Bard, Y. (1974). *Nonlinear Parameter Estimation*. Academic Press, (New York: USA).
Bendat, J.S. and A.G. Piersol (1971). *Random Data: Analysis and Measurement Procedures*. Wiley
    Interscience, (New York: USA).
Billings, S.A. and W.S.F. Voon (1983). Structure detection and model validity tests in the identification of
    nonlinear systems. *IEE Proc.*, Vol. 130, Pt. D, 4, 193–199.
Billings, S.A. and W.S.F. Voon (1986). Correlation based model validity test for nonlinear models. *Int.
    Journal of Control*, Vol. 44, 1 , pp. 235–244.
Box, G.E.P. and G.M. Jenkins (1976). *Time Series Analysis – Forecasting and Control*. Holden–Day,
    Oakland, (California: USA).
Box, G.E.P. and D.A. Pierce (1970). Distribution of residual auto-correlations in autoregressive-integrated
    moving average time series models. *Journal Amer. Stat. Assoc.*, Vol. 64, pp. 1509.
Cleveland, W.S. and B. Kleiner (1975). A graphical technique for enhancing scatterplots with proving
    statistics. *Technometrics*, Vol. 17, 447–454.
Draper, N.R. and H. Smith (1966). *Applied Regression Analysis*. John Wiley and Sons, (New York: USA).
Durbin, J. and G.S. Watson (1951). Testing for serial correlation in least squares regression III. *Biometrika*,
    Vol. 58, pp. 1–19.
Goodwin, G.C. and R.L. Payne (1977). *Dynamic System Identification. Experiment Design and Data
    Analysis*. Academic Press, (New York: USA).
Haber, R. (1979). Parametric identification of nonlinear dynamic systems based on correlation functions.
    Prepr. 5th IFAC Symposium on Identification and System Parameter Estimation, (Darmstadt: FRG), pp.
    515–522.
Haber, R. and R. Zierfuss (1987). Identification of an electrically heated heat exchanger by several
    nonlinear models using different structure. *Report TUV-IMPA-87/4*, Institute of Machine- and
    Processautomation, Technical university of Vienna, (Vienna: Austria).
Haber, R. (1988). Parametric identification of nonlinear dynamic systems based on nonlinear
    crosscorrelation functions. *IEE Proceedings*, Vol. 135, Pt. D, 6, pp. 405–420.
Haber, R. and R. Zierfuss (1991). Identification of nonlinear models between the reflux and the top
    temperature of a distillation column. *Prepr. 9th IFAC Symposium on Identification and System
    Parameter Estimation*, (Budapest: Hungary), pp. 486–491.
Haber, R. (1993). Model validity tests based on correlation functions. *Report FHK-AV-PLT-93/1, Cologne
    Institute of Technology (Fachhochschule Köln), Department of Process Engineering, Laboratory for
    Process Control*, (Köln: Germany).
Haber, R. and R. Zierfuss (1987). Simulation of different model validity tests based on correlation functions
    (in German). *Report, TUV-IMPA-87/2, Institute of Machine and Process Automation, Technical
    University of Vienna*, (Vienna: Austria).
Hoffmeyer–Zlotnik, H.J., J. Wernstedt and A. Kurz (1978a). Applications of methods of statistic
    identification for developing models of the catchment area of a river in a high and low level range.
    Prepr. 7th IFAC World Congress, (Helsinki: Finland), pp. 1453–1459.
Hoffmeyer–Zlotnik, H.J., J. Wernstedt, A. Becker and P. Braun (1978b). Modeling of hydrologic systems by
    means of statistical methods (in German). *Messen, Steuern, Regeln*, Vol. 7, pp. 385–389.
Hoffmeyer–Zlotnik, H.J., P. Otto, A. Seifert, H. Heym and H. Luckers (1979). Application of experimental
    models for automatic control and prediction of water flows in the mountain of medium height (in
    German). *24. Internat. Wissenschaft. Kolloquium, Technical University of Ilmenau*, (Ilmenau: GDR), pp.
    173–177.
Leontaritis, I.J. and S.A. Billings (1987). Model selection and validation methods for nonlinear systems, *Int.
    Journal of Control*, Vol. 45, 1, 311–341.
Mereau, P. and G. Prevost (1979). Calculation of uncertainty intervals in non-linear estimation. *Prepr. of 5th
    IFAC Symposium on Identification and System Parameter Estimation*, (Darmstadt: FRG), 365–372.
Rajbman, N.S. and V.M. Chadeev (1980). *Identification of Industrial Processes*. North Holland Publ. Co.,
    (Amsterdam: The Netherlands), pp. 435.
Tuschák, R., T. Bézi, G. Tevesz, J. Hetthéssy and R. Haber (1982). Practical experiences on the setup and
    identification of a distillation pilot plant. *Prepr. of 6th IFAC Symposium on Identification and System
    Parameter Estimation*. (Washington: USA), pp. 663–668.

7. Case Studies on Identification of Real Processes

7.1 ELECTRICALLY EXCITED BIOLOGICAL MEMBRANE

7.1.1 Description of the process
The electrical processes in a biological membrane are the bases of information processing and transmission in an organism. Therefore the modeling of an electrically excited membrane is of great importance. After several theoretical efforts Hodgkin *et al.*, (1957) carried out measurements on the giant axon of *Loligo*. The current fed to the cells was changed in such a way that the potential on the electrodes in the fibre was kept constant. By means of experiments repeated at different potential levels the step responses of the current were recorded. For theoretical considerations the current was divided into four components:
1. capacitive;
2. leakage;
3. sodium;
4. potassium.

The values of the two latter ones always changed with time and were also a function of the potential. Hodgkin and Huxley (1952) presented the time functions of the conductivity of the Na^+ and K^+ ions at switching on the potentials of different levels.

Further on we deal only with the modeling of the conductivity of the K^+ ions. The measured $K^+(t)$ time functions are given in Figure 7.1.1 at different excitation (potential) levels.

Hodgkin and Huxley gave a physical interpretation to the process that is briefly presented here in the interpretation of Fedina (1974). The conductivity of K^+ ions $y(t)$ – hereinafter conductivity – can be described by the equations

$$y(t) = \delta\gamma^4(t)$$
$$\frac{d\gamma(t)}{dt} = \alpha(U)(1 - \gamma(t)) - \beta(U)\gamma(t)$$

where δ is a constant, and $\alpha(U)$ and $\beta(U)$ are transfer constants of the ion stream from outside to the inside and *vice versa*, respectively. The K^+ ions cross the membrane only if four molecules are in a given region and γ and $(1 - \gamma)$ are the portions of these molecules in and outside the membrane, respectively. The step response of $\gamma(t)$ to an excitation $U \cdot 1(t)$ is

$$\gamma(t) = \gamma_\infty - (\gamma_\infty - \gamma_0)\exp(-t/T)$$

where

$$\gamma_0 = \frac{\alpha(0)}{\alpha(0) + \beta(0)}$$
$$\gamma(\infty) = \gamma_\infty = \frac{\alpha(U)}{\alpha(U) + \beta(U)}$$

$$T = \frac{1}{\alpha(U) + \beta(U)}$$

The transfer rate constants are

$$\alpha(U) = \frac{0.01(U + 10)}{\exp(1 + 0.1U) - 1}$$

and

$$\beta(U) = 0.125\exp(0.0125U)$$

Based on the above, the conductivity response is described by the equation

$$y(t) = \left[y_\infty^{1/4} - \left(y_\infty^{1/4} - y_0^{1/4} \right)\exp(-t/T) \right]^4$$

It can be interpreted as the output signal of a first-order lag term of the fourth power where the time constants depend on the potential level.

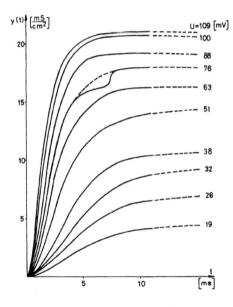
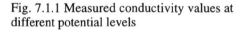

Fig. 7.1.1 Measured conductivity values at different potential levels

7.1.2 Conclusions drawn from the measured records

The measured conductivity values are plotted in Figure 7.1.1 at different constant voltage values. The records can be considered as step responses, where the magnitude of the excitation is equal to the voltage value. In Figure 7.1.2 the measured conductivity values are normalized to their steady state values at different potential levels.

The following qualitative conclusions can be drawn from Figures 7.1.1 and 7.1.2 (Haber and Wernstedt, 1979):

    1. The forms of the conductivity functions of different excitation levels are similar to the step responses of the linear systems;

2. The relation between the steady state value of the measured output signal and the input (excitation) signal is nonlinear;
3. The measured functions being normalized to their steady state values are similar to the linear step responses. The normalized step responses belonging to different levels do not coincide, i.e., different time constants describe them.

On the basis of the above mentioned conclusions the model should be:
- nonlinear;
- dynamic;
- separable into a static nonlinear and a dynamic quasi-linear part;
- the parameters of the quasi-linear section should be signal dependent.

There are three models satisfying the postulates, namely: the quasi-linear, the Hammerstein and the Wiener models, each of them with signal dependent parameters. It is obvious that the quasi-linear and Hammerstein models are equivalent; the nonlinear static section represents the signal dependent gain. The Wiener and quasi-linear models describe different physical relations and are not equivalent.

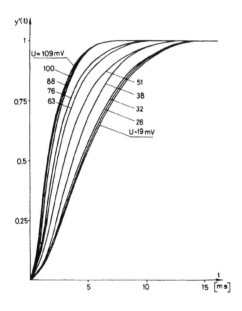

Fig. 7.1.2 Measured conductivity values normalized to their steady state values at different potential levels

7.1.3 Modeling by the simple Wiener model with signal dependent time constant
The structure of the model is illustrated in Figure 7.1.3a. The notation $T(u, v)$ means that the quasi-linear term can be described by a first-order differential equation and the coefficient of the derivative of v — hereinafter, time constant — may depend on the input u or output v of the quasi-linear part. The nonlinear static part of the model $Y = Y(V)$ can be obtained from the potential levels U and the corresponding steady state conductivity values $Y(U)$. The measured points as well as the analytic relation are plotted in Figure 7.1.3b. The $Y = Y(V) = Y(U)$ relation was determined analytically by static regression and it resulted in the following polynomial

$$Y\left[\mathrm{mS/cm^2}\right] = 7.42\cdot10^{-2}U[\mathrm{mV}] + 1.39\cdot10^{-2}U^2[\mathrm{mV}]$$
$$-3.05\cdot10^{-4}U^3[\mathrm{mV}] + 2.56\cdot10^{-6}U^4[\mathrm{mV}]$$
$$-7.68\cdot10^{-9}U^5[\mathrm{mV}]$$

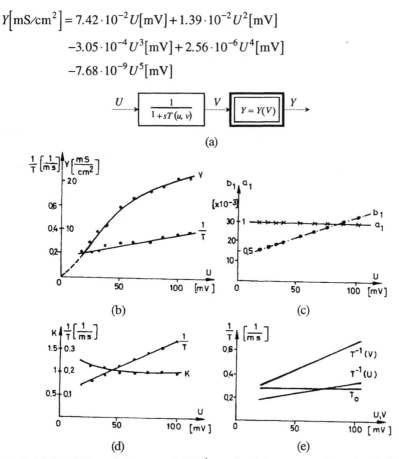

Fig. 7.1.3 Modeling of the potential/ K^+ conductivity process by a simple Wiener model with signal dependent parameters: (a) scheme of the model; (b) steady state characteristics $Y(U)$ and grapho-analytically estimated time constants of the quasi-linear transfer functions at different potential levels; (c) estimated parameters of the quasi-linear pulse transfer functions at different potential levels obtained by the LS method; (d) estimated parameters of the quasi-linear transfer functions at different potential levels obtained by the LS method; (e) different estimated potential level/time constant functions obtained by nonlinear identification of the total record

Since the equation is single-valued the output signals of the linear term can be easily calculated by simply transforming the measured conductivity values by the inverse steady state characteristic $V = Y^{-1}(U)$. They can be considered subsequently as the measured output signals of the linear term of the simple Wiener model. Figure 7.1.4 presents the above signals. The step responses of Figure 7.1.4 do not start with zero tangents – contrary to the measured conductivity values in Figure 7.1.1. The step responses of the linear part can be modeled well by first-order lags with unity gain and signal dependent time constants

$$G(s, u, v) = \frac{1}{1 + sT(u, v)}$$

(In some references it is usual to use the differential operator $p = d/dt$ instead of the Laplace variable s, which is a more exact approach. Because we concentrate on the applicable block-oriented model structures we therefore use the formal meaning and structure of the regular transfer functions.) The individual quasi-linear step responses can be identified either by a grapho-analytical method (Figure 7.1.3b) or by an LS method. The estimated parameters of the linear pulse transfer functions

$$G(q^{-1}) = \frac{b_1 q^{-1}}{1 + a_1 q^{-1}}$$

are given in Figure 7.1.3c and the corresponding gain and inverse of the time constants in Figure 7.1.3d as a function of the potential level. (The sampling time was $\Delta T = 0.1$ [ms].) The reciprocal values of the estimated time constants lie on a straight line, their equation can be determined by static regression. The result of the regression is

$$T^{-1}[1/ms] = 0.168 + 0.00193\, U\,[mV]$$

if the grapho-analytical evaluation is used and

$$T^{-1}[1/ms] = 0.12 + 0.0022\, U\,[mV]$$

if LS identification is applied.

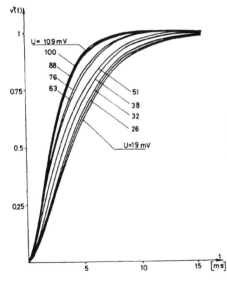

Fig. 7.1.4 Measured conductivity values transformed through the inverse steady state characteristic at different potential levels

The identification procedure presented could be performed only because the input signals were steps. If they had had other forms, linear models could not be estimated at different potential levels and the signal dependencies of the parameters could not be regressed. In this case nonlinear identification is needed. To collect all information from the step responses, attach them one after another, as seen in Figure 7.1.5. The long records contain 10 input steps $U \cdot 1(t)$ and responses $v(t)$, the latter ones are calculated from the measured output signals by transforming them through the inverse steady state characteristics. The static transformation is not an assumption of the method, yet it reduces the number of unknown parameters because now the model between $u(t)$ and $v(t)$ has a unity gain. Since all step responses start from zero and end at their steady state values, those parts of the total record where the input–output data pairs turn back to zero before the next step are not measured data, they should be omitted from the regression. They are needed, however, for filling the memory vectors; before the next step there should be more zero values than the model order.

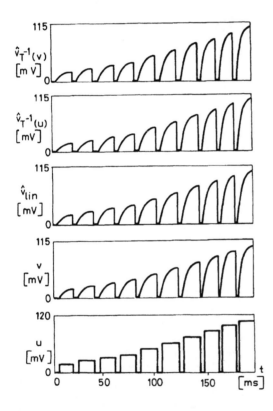

Fig. 7.1.5 Potential level u, calculated v from the measured conductivity values and inverse steady state characteristic and output signal $\hat{v}$ of the linear dynamic term of the simple Wiener model computed on the basis of the identified nonlinear models

Based on the previous investigations, three different models were identified. The type of models, their difference equations using the bilinear discretization, and the estimated parameters are as follows:

1. *First-order model with constant parameters (model K_0, T_0):*

$$y(k) = -a_1 y(k-1) + b_0 \left[u(k) + u(k-1) \right]$$

$$a_1 = \frac{\Delta T - 2T_0}{\Delta T + 2T_0} = -0.8454$$

$$b_0 = K_0 \frac{\Delta T}{\Delta T + 2T_0} = 0.08189$$

$$K_0 = 1.059$$

$$T_0^{-1} = 0.279 [1/\text{ms}]$$

2. *First-order model with potential level dependent time constant*
   (model $K_0 = 1, \overline{T}_1^{-1}(u)$)

$$y(k) = -a_1 y(k-1) + b_0 [u(k) + u(k-1)]$$
$$+ c_0 [u(k)/50.0] \{ [u(k) + u(k-1)] - [y(k) + y(k-1)] \}$$

$$a_1 = \frac{\Delta T - 2T_0}{\Delta T + 2T_0} = -0.9216 (\pm 0.009491)$$

$$b_0 = K_0 \frac{\Delta T}{\Delta T + 2T_0} = 0.04402 (\pm 0.004626)$$

$$c_0 = 50.0 \frac{T_0}{T_1} \frac{T_0}{\Delta T + 2T_0} = 0.02508 (\pm 0.002931)$$

(The scaling factor (50.0) was used to improve the numerical conditions of the parameter estimation.) The constant part of the time constant could be calculated from both a_1 and b_0, therefore the mean value of the computed values was accepted

$$T_0^{-1} = (0.136 + 0.153)/2 = 0.144 [1/\text{ms}]$$

$$\overline{T}_1^{-1} = 0.00174 [1/\text{ms/mV}]$$

Finally, the model parameters are as follows:

$$K_0 = 1$$

$$T^{-1} [1/\text{ms}] = 0.144 + 0.00174 \, U [\text{mV}]$$

3. *First-order model with conductivity dependent time constant (model*
   $K_0 = 1, \overline{T}_1^{-1}(v)$)

$$y(k) = -a_1 y(k-1) + b_0 [u(k) + u(k-1)]$$
$$+ c_0 [y(k)/50.0] \{ [u(k) + u(k-1)] - [y(k) - y(k-1)] \}$$

$$a_1 = \frac{\Delta T - 2T_0}{\Delta T + 2T_0} = -0.8821 (\pm 0.002981)$$

$$b_0 = K_0 \frac{\Delta T}{\Delta T + 2T_0} = 0.0592 (\pm 0.001485)$$

$$c_0 = 50.0 \frac{T_0}{T_1} \frac{\Delta T}{\Delta T + 2T_0} = 0.05894\,(\pm 0.002942)$$

The constant part of the time constant can be calculated from both a_1 and b_0, therefore the mean value of the computed values is accepted

$$T_0^{-1} = (0.209 + 0.210)/2 = 0.2095\,[1/\text{ms}]$$

Finally, the model parameters are as follows:

$$K_0 = 1$$

$$T^{-1}\,[1/\text{ms}] = 0.209 + 0.004176\,U\,[\text{mV}]$$

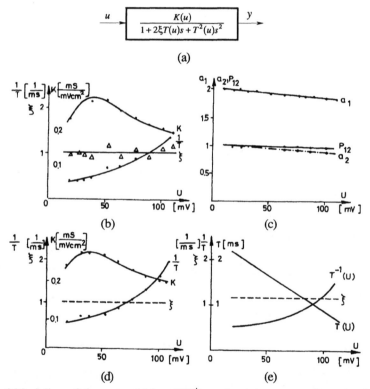

(a)

(b)

(c)

(d)

(e)

Fig. 7.1.6 Modeling of the potential level/ K^+ conductivity process by a simple quasi-linear model with signal dependent parameters: (a) scheme of the model; (b) gain, time constant and damping factor as a function of the potential level obtained by the grapho-analytical evaluation of the step responses; (c) estimated parameters of the quasi-linear pulse transfer functions at different potential levels obtained by the LS method (p_{12} is a double pole), (d) estimated parameters of the quasi-linear transfer functions at different potential levels obtained by the LS method; (e) different estimated potential level/time constant functions obtained by nonlinear identification of the total record

The measured input $u(t)$ and output $v(t)$ signals and the three model outputs $\hat{v}(t)$ are seen in Figure 7.1.5. The models with signal dependent parameters fit better than the linear first-order model (with constant parameters). The parameters of the model $K_0, \overline{T}_1^{-1}(u)$ are almost the same as obtained by identifying the individual step responses separately and by performing a regression for the parameters as a function of the potential level. The third model $K_0, \overline{T}_1^{-1}(v)$ cannot be identified by linear methods, this model can be created only by nonlinear identification, since the signal $v(t)$, the parameters depend on, is never constant. The output signals $v(t)$ in Figure 7.1.5 are not the conductivity values but their transformation through the inverse steady state characteristic. Figure 7.1.8 shows the measured conductivity values and the step responses of the simple Wiener models with signal dependent parameters. The assumption of that the time constant depends on the conductivity value seems to be better than the assumption that it depends on the potential level.

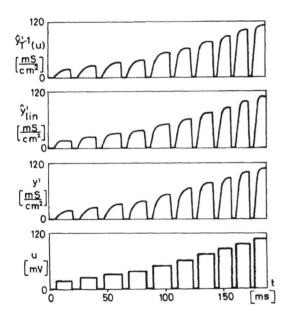

Fig. 7.1.7 Potential levels u, measured conductivity values normalized to their steady state values y', and computed output signals $\hat{y}'$ based on the identified nonlinear models with unit gain

7.1.4 Modeling by a quasi-linear second-order lag with signal dependent gain and time constant

Since the measured conductivity responses start from a zero tangent (Figure 7.1.1) it is worth trying a second-order lag approximation,

$$G(s, u) = \frac{K(u)}{1 + 2\xi(u)T(u)s + T^2(u)s^2}$$

First the step responses were evaluated by the grapho-analytical method and the signal dependence of the parameters were estimated by static regression. The results were (Haber and Wernstedt, 1979):

$$K[\text{mS/mVcm}^2] = 7.42 \cdot 10^{-2} + 1.39 \cdot 10^{-2} U[\text{mV}] - 3.05 \cdot 10^{-4} U^2[\text{mV}]$$
$$+ 2.56 \cdot 10^{-6} U^3[\text{mV}] - 7.68 \cdot 10^{-9} U^4[\text{mV}]$$

$$T^{-1}[1/\text{ms}] = 0.327 + 0.001114 U[\text{mV}] + 0.0000694 U^2[\text{mV}]$$

$$\xi \approx 1$$

Both the regression lines and the separately estimated parameters are seen in Figure 7.1.6b.

The data were sampled by $\Delta T = 0.1$ [ms] and the step responses were identified by an LS algorithm as well (Haber and Keviczky, 1979). The results of the discrete time identification are seen in Figure 7.1.6c and the parameters of the corresponding continuous models are plotted in Figure 7.1.6d. The discrete time identification proved the correctness of the grapho-analytical evaluation – since Figure 7.1.6b and Figure 7.1.6d almost cover each other.

It was possible to estimate the parameters by one regression if the nonlinear identification method was used for the whole record of the experiment. The whole record consists of the chain of the individual step responses – as in the case of fitting the simple Wiener model.

To reduce the number of the unknown parameters and to use the *a priori* knowledge, the measured conductivity responses were normalized with their steady state values. It means that the gain of the identified (normalized) model is unity. Based on the previous investigations, two different models were identified.

Although the reciprocal value of the time constant is a quadratic function of the potential level, the time constant itself is approximately a linear function of the same potential level. This fact can be verified by simply drawing the estimated time constants against the potential level. The advantage of using the linear dependence is that requires less number of parameters.

The type of models, their difference equations using the bilinear discretization and the estimated parameters are as follows:

1. *Second-order lag term with constant parameters (model K_0, T_0, ξ_0)*

$$y(k) = -a_1 y(k-1) - a_2 y(k-2) + b_0[u(k) + 2u(k-1) + u(k-2)]$$
$$N_0 = 4T_0^2 - 4\Delta T\, T_0\xi_0 + \Delta T^2$$
$$D_0 = 4T_0^2 + 4\Delta T\, T_0\xi_0 + \Delta T^2$$
$$a_1 = \frac{2\Delta T^2 - 8T_0^2}{D_0} = -1.419$$
$$a_2 = \frac{N_0}{D_0} = 0.5394$$
$$b_0 = K_0 \frac{\Delta T^2}{D_0} = 0.02941 [\text{mS/mVcm}^2]$$

The estimated model parameters are

$$K_0 = 0.977 \left[\text{mS/mVcm}^2 \right]$$
$$T_0 = 1.487 [\text{ms}]$$
$$\xi_0 = 0.772$$

2. *Second-order lag term with signal dependent time constant (model $K_0, T_1(u), \xi_0$):*

$$y(k) = -a_1 y(k-1) - a_2 y(k-2) + b_0 \left[u(k) + 2u(k-1) + u(k-2) \right]$$
$$+ c_0 x(k) \left[y(k) - y(k-2) \right]$$
$$+ d_0 x(k) \left[y(k) + 2y(k-1) + y(k-2) \right] + d_1 x^2(k) \left[y(k) + 2y(k-1) + y(k-2) \right]$$
$$N_0 = 4T_0^2 - 4\Delta T T_0 \xi_0 + \Delta T^2$$
$$D_0 = 4T_0^2 + 4\Delta T T_0 \xi_0 + \Delta T^2$$
$$a_1 = \frac{2\Delta T^2 - 8T_0^2}{D_0} = -1.516 \, (\pm 0.02374)$$
$$a_2 = \frac{N_0}{D_0} = 0.5637 \, (\pm 0.02264)$$
$$b_0 = K_0 \frac{\Delta T^2}{D_0} = 0.01173 \, (\pm 0.00509) \left[\text{mS/mVcm}^2 \right]$$
$$c_0 = 4\Delta T \xi_0 \frac{T_1}{D_0} = 0.07100 \, (\pm 0.005203)$$
$$d_0 = 8 \frac{T_0 T_1}{D_0} 50$$
$$d_1 = 4 \frac{T_1^2}{D_0} 50^2$$

The model parameters were calculated from the estimated values a_1, a_2, b_0 and c_0:

$$K_0 = 0.9835 \left[\text{mS/mVcm}^2 \right]$$
$$T_0 [\text{ms}] = 2.41 - 0.0157 \, U [\text{mV}]$$
$$\xi_0 = 1.1384$$

(d_0 and d_1 were used only for comparing them with the values obtained from the first four parameters).

The parameters of the nonlinear model are shown in Figure 7.1.6e. Here also the reciprocal values of the time constants are drawn. The curve is similar to those obtained by the evaluation of the individual step responses. Figure 7.1.7 shows the measured potential levels, normalized conductivity values and model outputs. The priority of the nonlinear model over the linear one can be seen well.

Finally, the conductivity values calculated by means of the nonlinear model $K(u), T_1(u), \xi_0$ is shown in Figure 7.1.8.

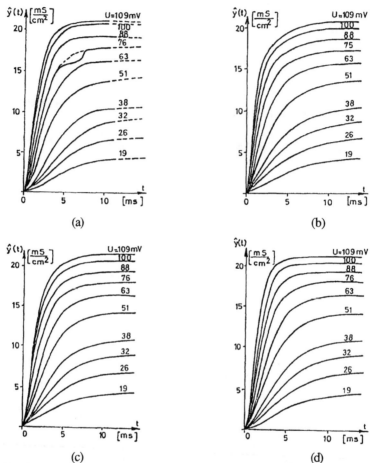

Fig. 7.1.8 Comparison of the identified nonlinear models of the potential
level/conductivity process: (a) measured conductivity values;

(b) simple Wiener model with potential dependent time constant $\left(K_0, T_1^{-1}(u)\right)$;

(c) simple Wiener model with conductivity dependent time constant $\left(K_0, T_1^{-1}(v)\right)$

(d) quasi-linear model with potential dependent time constant $\left(K_0, T_1(u), \xi_0\right)$

7.1.5 Conclusions

The first five remarks concern the concrete case, the electrically excited biological
membrane:

1. It was successful in elaborating simple, clear, nonlinear dynamic models of
    the potential level/conductivity model.
2. The coincidence of the model outputs with the measured values is worse
    than that of the sophisticated Hodgkin–Huxley model.
3. The measured step responses show that the time constant is inversely

proportional to the potential level. This fact was proved by many identification methods.

4.   The simple Wiener model with potential level dependent time constants has been identified by Hoyt (1963) and Tischmeyer (1974) by linear methods. Their methods were unable to identify the dependence on the conductivity value that showed a better model.

5.   The second-order model has a physical interpretation. Antomonov and Kotova (1968) have shown that

$$\frac{dg_K(t)}{dt} = g_{Na}(t)$$

$$\frac{dg_{Na}(t)}{dt} = Ku(t) - a_1 g_{Na}(t) - a_0 g_K(t)$$

and consequently

$$\frac{d^2 g_K(t)}{dt^2} + a_1 \frac{dg_K(t)}{dt} + a_0 g_K(t) = Ku(t)$$

The next five remarks concern the nonlinear identification technique applied:

6.   Separate tests performed in different working points can be evaluated by two different methods:
   • Linear identification of the individual records and analytical regression of the estimated parameters as a function of the working point (in this case the potential level);
   • The individual records have to be chained one after the other and the total record covering several working points can be identified by a nonlinear model.

7.   The chaining of records is an easy task; however, the regression procedure needs some attention. If the final values of a record coincides with the first values of the next record chained to the first one, then the regression can be performed as if the two records were measured at once. (This case would happen, e.g., if the step responses were to contain not only their upward section, but, after having stopped the excitation, their return to zero). If the final and initial values of the successive records do not coincide, then the memory vector of the delayed input and output values and their functions have to be kept updated for the whole data record, but the regression must not be executed in practice sofar as the memory vector consists only of terms which concern to their own individual record (working point). It means that at every working point change some data have to be omitted from the regression and this number is equal to the sum of the order of the equation and the (relative) delay time.

8.   Nonlinear identification methods allow the identification of such models as cannot be identified by linear techniques, at least not easily. An example of this is an output signal dependent parameter, e.g., a time constant. These signal dependencies can be identified from one record by applying the nonlinear difference equation of the proper model. If only linear techniques

were used – either grapho-analytical or LS – several step responses with small amplitudes had to be performed in different working points.

9.   The above case study demonstrates that *a priori* knowledge about the structure and parameters of the process may reduce the number of unknown parameters. In our case the steady state character was estimated from the steady state values of the step responses, and therefore the dynamic nonlinear models identified can have a unity gain.

10. The chained records consist of several step responses. This signal is not a good exciting signal, and if possible, better exciting signals should be applied for identification. On the other hand, the step responses may be the best tests for structure identification.

7.2 FERMENTATION PROCESS

7.2.1 Description of the process
A laboratory scale BIOTEC fermentor was installed at the Department of Agricultural Chemical Technology, Technical University of Budapest. In the experiment the microbe *Hansenula anomala* was cultivated from ethanol. The aim of the control of the fermentation process was to minimize the respiratory coefficient, concerning the temperature and pH, in chemostat continuous cultivation. The respiratory coefficient is the quotient of the CO_2 produced and O_2 used during a unit of time. It shows which part of the carbon source will be built into the cell and how great portion will be used as the energy source. The smaller the value the respiratory coefficient has the more useful the components of the ethanol become to the cell. The control variables were the temperature ϑ and the pH, and the aim of the experiment was to minimize the respiratory coefficient (RQ).

7.2.2 Process identification and computation of the optimal control signals
The temperature and pH were changed stepwise according to an orthogonal factor plan to evaluate the stationary values of the responses and to set the control variables according to the dual extremum control (Nyeste and Szigeti, 1982). The records of the temperature, pH, O_2, CO_2 and RQ are drawn in Figure 7.2.1. The respiratory coefficient was calculated from the measured O_2 and CO_2 content of the gas outlet. The data were sampled by $\Delta T = 1$ [h] and the parameters of a two variable generalized Hammerstein model were estimated by the maximum likelihood method assuming linear color noise filter at the output of the process. The parameters of the best model are as follows (Boros, 1982):

$$y(k) = -0.2438y(k-1) - 0.1216y(k-2) + 20.4867$$
$$+0.5062u_1(k-1) - 1.9232u_1(k-2) - 0.1753u_2(k-1) + 0.1288u_2(k-2)$$
$$+0.009592u_1^2(k-1) + 0.03484u_1^2(k-2)$$
$$+0.005717u_1(k-1)u_2(k-1) - 0.005898u_1(k-2)u_2(k-2)$$
$$+0.002551u_2^2(k-1) + 0.002759u_2^2(k-2) + e(k) + 0.5471e(k-1) + 0.195e(k-2)$$

Here $u_1(k)$ is the temperature, $u_2(k)$ the pH and $y(k)$ the respiratory coefficient (RQ), respectively. The steady state relation of the noise-free model is

$$Y = 15.0 - 1.038U_1 - 0.034U_2 + 0.01846U_1^2 - 0.00013U_1U_2 + 0.00389U_2^2$$

The optimal values of the control variables can be obtained by derivation of the above equation and setting the partial derivatives equal to zero:

$$\frac{dY}{dU_1} = -1.038 + 0.03692U_1 - 0.00013U_2 = 0$$

$$\frac{dY}{dU_2} = -0.034 - 0.00013U_1 + 0.00778U_2 = 0$$

and hence

$$U_1^o = 28.13 \ [°C]$$
$$U_2^o = 4.84$$

and the minimum value of the respiratory coefficient is

$$Y^o = 0.335$$

The optimal control variables coincide well with the results of the dual extremum control using only the steady state model, as is seen in the last section of Figure 7.2.1. In the time interval between 110 and 120 [h] the average value of the respiratory coefficient is equal to the optimal extremum value obtained from the nonlinear dynamic model identification.

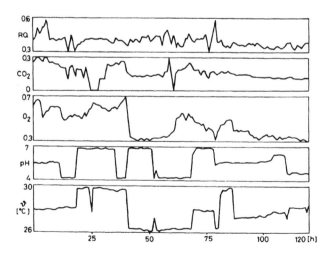

Fig. 7.2.1 Measured records of the temperature ϑ, pH, O_2, CO_2 and the respiratory coefficient RQ at the fermentation process

7.2.3 Conclusions

1. The generalized Hammerstein model was successfully used for the input–output equivalent identification of a complicated process.
2. By means of the generalized Hammerstein model correct control variables could be

computed which ensure the steady state minimum value of the output signal.
3. The optimal control variables and extremum value of the output signal coincides
   with those obtained via evaluation of the stationary behavior of the process.
4. The generalized Hammerstein model can be used also in those situations when the
   control variables are not changed stepwise, i.e., if static modeling techniques fail.

7.3 ELECTRICALLY HEATED HEAT EXCHANGER

7.3.1 Description of the plant

An electrically heated heat exchanger was installed at the Department of Automation,
Technical University of Budapest. The plant is used as a temperature control laboratory
system. Its scheme can be seen in Figure 7.3.1. The copper cylinder is 100 [mm]
diameter and 2.75 $[\ell]$ in volume. The temperature of the inlet water was constant at
$\vartheta_0 = 16$ [°C] during the experiments. The temperature of the outlet water ϑ varied
with the change in the heating power P and the flow of water Q. The purpose of the
identification was to establish a relation between the heating power and the temperature
difference

$$\Delta\vartheta(t) = \vartheta(t) - \vartheta_0 \tag{7.3.1}$$

at different water loads. The heating power could only be set at a minimum or
maximum value, therefore the water load dependence of the power/water temperature
difference model was only of interest.

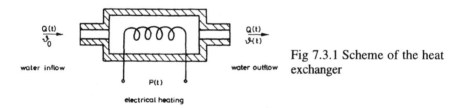

Fig 7.3.1 Scheme of the heat exchanger

Figure 7.3.2 shows the 13 temperature difference step responses for switching on and off
the heating power at different water flow values. During the detection of the step
responses the water flow was kept constant, a fact that makes a grapho-analytical
evaluation of the measurements also possible.
    The heat balance of the plant is

$$Q(t)c_p\vartheta_0 + P(t) = Q(t)c_p\vartheta(t) + \alpha\rho Vc_p\frac{d\vartheta(t)}{dt} \tag{7.3.2}$$

where:
   • c_p is the specific heat of the water,
   • ρ is the density of the water,
   • α is an uncertainty factor,

which takes the disregarded time lag effects of the heat exchange into account. (If α is
larger than one then the change of the temperature is slower, as expected from the
theoretical heat balance, and if α is less than one then the change of the temperature is

faster as expected.)

If the water flow is constant,

$$Q(t) = Q$$

during a transient caused by a small change in the heating power, then by means of the Laplace transformation a first-order linear model can be derived between the heating power and the temperature difference:

$$\frac{\Delta\vartheta(s)}{P(s)} = \frac{K}{1+sT} \qquad (7.3.4)$$

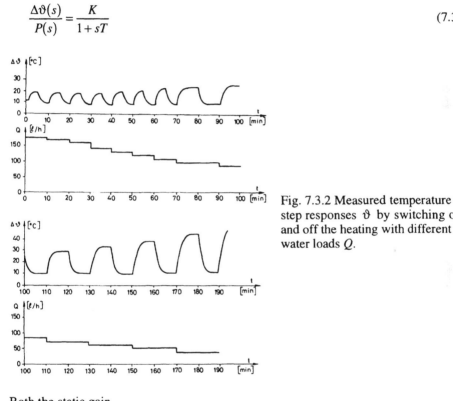

Fig. 7.3.2 Measured temperature step responses ϑ by switching on and off the heating with different water loads Q.

Both the static gain

$$K = \frac{1}{c_p Q} \qquad (7.3.5)$$

and the time constant

$$T = \frac{\alpha \rho V}{Q} \qquad (7.3.6)$$

depend on the water flow. The uncertainty factor α appears only in the time constant. Substituting the known parameters into (7.3.5) and (7.3.6) we obtain

$$K\left[°C/kW\right] = \frac{1}{4.18[kWs/kcal]Q[\ell/3600s]} = \frac{861.25}{Q[\ell/h]} \qquad (7.3.7)$$

and

$$T[\min] = \alpha\frac{1[kg/\ell]2.75[\ell]}{Q[\ell/3600s]}\frac{1}{60}\left[\frac{\min}{s}\right] = \frac{165\alpha}{Q[\ell/h]} \qquad (7.3.8)$$

In the following, different identification techniques such as:
- a grapho-analytical method based on the individual step responses,
- a discrete time identification from the whole measurement record in one step, and
- a stepwise regression

will be presented, and different model types such as:
- a linear model,
- a general nonlinear model linear in the parameters,
- a quasi-linear continuous time model with signal dependent parameters,
- a quasi-linear discrete time model with signal dependent parameters

will be fitted to the measurement data.

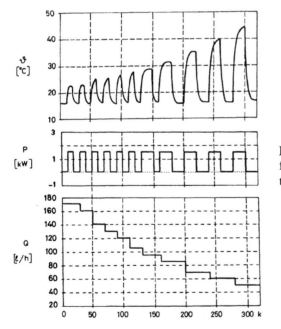

Fig. 7.3.3 Sampled measured water flow Q, heating power P, and temperature ϑ records

The water flow range was selected to be $50[\ell/h] \leq Q \leq 170[\ell/h]$ for the identification. Figure 7.3.3 shows the plot of the sampled water flow, heating power and temperature. The measured signals were sampled with $\Delta T = 0.5$ [min], which resulted in $N = 320$ data pairs. In order to improve the numerical conditions of the parameter estimation, the normalized values

$$Q'(k) = 0.005Q(k) \tag{7.3.9}$$
$$P'(k) = 0.6667P(k) \tag{7.3.10}$$
$$\Delta\vartheta'(k) = 0.02\Delta\vartheta(k) \tag{7.3.11}$$

were used when the parameters were estimated from the whole measurement record. As the measured temperature has almost no noise the simple least squares method can be used.

7.3.2 Tuning the uncertainty factor in the theoretically derived model
In order to simulate the theoretically derived signal dependent model in a simple way the continuous time model was transformed to a difference equation by means of the Euler transformation

$$\Delta\vartheta(k) - \Delta\vartheta(k-1) = -\frac{\Delta T}{T}\Delta\vartheta(k-1) + K\frac{\Delta T}{T}P(k-1) \tag{7.3.12}$$

Substituting (7.3.5) and (7.3.6) into (7.3.12) one obtains

$$\Delta\vartheta(k) - \Delta\vartheta(k-1) = -\frac{\Delta T \, Q(k-1)}{\alpha\rho V}\Delta\vartheta(k-1) + \frac{\Delta T}{\alpha\rho V c_p}P(k-1) \tag{7.3.13}$$

In (7.3.13) the water flow Q was replaced by its actual value $Q(k)$ to obtain a general valid model that can also be used for varying Q values. Since any change in the process can cause a change in the output signal after at least one sampling time, $Q(k-1)$ was substituted instead of $Q(k)$. Using the numerical values of (7.3.7) and (7.3.8) the resulting difference equation is

$$\Delta\vartheta(k) - \Delta\vartheta(k-1) = -0.0030303\frac{1}{\alpha}Q(k-1)\Delta\vartheta(k-1) + 2.61\frac{1}{\alpha}P(k-1) \tag{7.3.14}$$

Equation (7.3.14) was simulated for three different uncertainty factors. All step responses belonging to different water flows are collected in one time function $\hat{\vartheta}(k)$ in Figures 7.3.4 to 7.3.6, corresponding to the three α values 0.8, 1.0, 1.2, respectively. The corresponding mean square errors (MSE) are:
- $\alpha = 0.8$ MSE = 1.704
- $\alpha = 1.0$ MSE = 1.943
- $\alpha = 1.2$ MSE = 2.458

which means that the best model seems to be the one with $\alpha = 0.8$. The plots show the simulated outlet temperature based on the derived theoretical model and the modeling error defined as the difference between the measured and simulated outlet temperatures $\left(\vartheta(k) - \hat{\vartheta}(k)\right)$. All model outputs, and especially that for $\alpha = 0.8$, fit well to the measured temperature record.

Since all variables are measurable in (7.3.14) and only the uncertainty factor is unknown it can be estimated from the equation

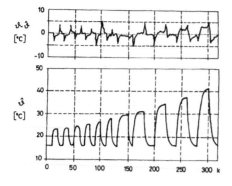

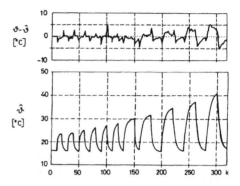

Fig. 7.3.4 Computed temperature $\hat{\vartheta}(k)$ based on the theoretically derived model with uncertainty factor $\alpha = 0.8$ and modeling error $\left(\vartheta(k) - \hat{\vartheta}(k)\right)$

Fig. 7.3.5 Computed temperature $\hat{\vartheta}(k)$ based on the theoretically derived model with uncertainty factor $\alpha = 1.0$ and modeling error $\left(\vartheta(k) - \hat{\vartheta}(k)\right)$

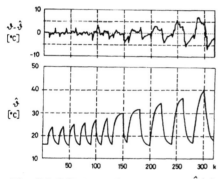

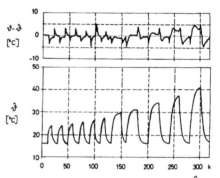

Fig. 7.3.6 Computed temperature $\hat{\vartheta}(k)$ based on the theoretically derived model with uncertainty factor $\alpha = 1.2$ and modeling error $\left(\vartheta(k) - \hat{\vartheta}(k)\right)$

Fig. 7.3.7 Computed temperature $\hat{\vartheta}(k)$ based on the theoretically derived model with estimated uncertainty factor $\hat{\alpha} = 0.904$ and modeling error $\left(\vartheta(k) - \hat{\vartheta}(k)\right)$

$$\Delta\vartheta(k) - \Delta\vartheta(k-1) = \frac{1}{\alpha}\left[-0.0030303Q(k-1)\Delta\vartheta(k-1) + 2.61P(k-1)\right] \quad (7.3.15)$$

linear in $1/\alpha$, by the least squares method. To improve the numerical condition of the parameter estimation replace the measured values by its normalized ones according to (7.3.9) to (7.3.11)

$$\Delta\vartheta'(k) - \Delta\vartheta'(k-1) = \frac{1}{\alpha}\left[-0.60606Q'(k-1)\Delta\vartheta'(k-1) + 0.0783P'(k-1)\right] \quad (7.3.16)$$

The estimated uncertainty factor is $\hat{\alpha} = (0.904 \pm 0.0312)$. Figure (7.3.7) shows the simulated outlet temperature based on the identified model with $\hat{\alpha} = 0.904$ and the modeling error.

7.3.3 Process identification by the grapho-analytical method

The 11 temperature step responses – in both directions – show first-order features. The grapho-analytically evaluated static gains and time constants are summed up in Table 7.3.1. The time constants T_+ valid for increasing of the heat flow hardly differ from the time constants T_- determined from the step responses caused by decreasing the heating power. Therefore the average value of the two time constants is interpreted as the time constant of the process at each water flow

$$T = 0.5(T_+ + T_-)$$

Figure 7.3.8 shows the identified static gains together with the theoretically determined function (7.3.7). The coincidence is excellent. Figure 7.3.9 shows the theoretically derived water flow dependence of the time constant for three different uncertainty factors $\alpha = 0.8$, 1.0 and 1.2. As is seen, the experimentally determined time constants fall approximately inside the range determined by the uncertainty factors $(0.7 \le \alpha \le 1.2)$.

Both the static gain and time constant are nonlinear functions of the water flow. It is of interest to plot the reciprocal values of the grapho-analytically determined parameters and check whether they can be approximated by a linear function of the water flow. The reciprocal value of the static gain is drawn in Figure 7.3.10 and that of the time constant in Figure 7.3.11. As is seen, the static gain can be well approximated by a line that goes through the origin. The parameters can be determined by static regression and are as follows:

- *linear function crossing the origin:*

$$K^{-1} = \overline{K}_1^{-1}Q = (0.001191 \pm 0.00002036)Q \tag{7.3.17}$$

- *linear function not crossing the origin:*

$$K^{-1} = \overline{K}_0^{-1} + \overline{K}_1^{-1}Q = (-0.01548 \pm 0.005227) + (0.001318 \pm 0.00004574)Q$$

TABLE 7.3.1 The grapho-analytically identified static gains and time constants of the 11 temperature step responses. T_+ refers to positive jumps in the heating power and T_- to jumps in the opposite direction. The table also contains the coefficients of the equivalent pulse transfer functions

No	Q [ℓ/h]	$\Delta\vartheta$ [°C]	K [°C/kW]	1/K [kW/°C]	T_+ [min]	T_- [min]	T [min]	1/T [1/min]	a_1	b_1 [°C/kW]	$1/b_1$ [kW/°C]
1	170	7.20	4.80	0.208	1.09	1.20	1.15	0.87	-0.647	1.694	0.590
2	160	7.32	4.88	0.205	1.09	1.46	1.28	0.78	-0.677	1.576	0.634
3	140	9.31	6.21	0.161	1.09	1.13	1.11	0.90	-0.637	2.254	0.444
4	130	9.78	6.52	0.153	1.31	1.39	1.35	0.74	-0.690	2.021	0.495
5	120	11.01	7.34	0.136	1.64	1.46	1.55	0.65	-0.724	2.026	0.494
6	105	11.85	7.90	0.127	1.39	1.17	1.28	0.78	-0.677	2.552	0.392
7	95	12.87	8.58	0.117	1.46	0.77	1.11	0.90	-0.637	3.115	0.321
8	85	15.75	10.50	0.095	1.24	1.64	1.44	0.69	-0.707	3.077	0.325
9	70	19.60	13.07	0.077	2.00	1.82	1.91	0.52	-0.770	3.006	0.333
10	60	25.06	16.71	0.060	2.19	2.00	2.10	0.48	-0.788	3.543	0.282
11	50	28.15	18.77	0.053	2.37	2.00	2.19	0.46	-0.796	3.829	0.261

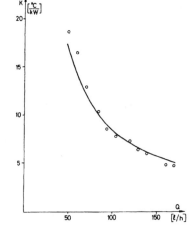

Fig. 7.3.8 Theoretically derived water
flow/static gain characteristics
(continuous line). The grapho-analytically
identified static gains are indicated by
circles

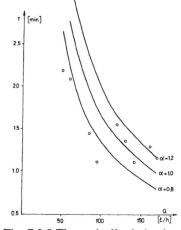

Fig. 7.3.9 Theoretically derived water
flow/time constant characteristics for three
different uncertainty factors
($\alpha = 0.8, 1.0, 1.2$) (continuous line). The
grapho-analytically identified time
constants are indicated by circles

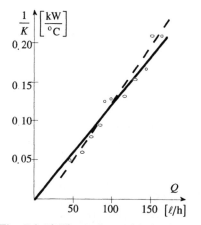

Fig. 7.3.10 The reciprocal values of the
grapho-analytically identified static gains
vs the water flow (marked by circles).
Lines fitted to the values:
• one through the origin (continuous line)
• not through the origin (dashed line)

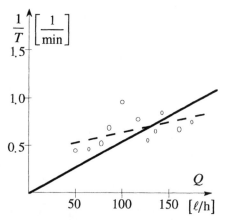

Fig. 7.3.11 The reciprocal values of the
grapho-analytically identified time
constants vs the water flow (marked by
circles). Lines fitted to the values:
• one through the origin (continuous line)
• not through the origin (dashed line)

The reciprocal values of the time constants can also be approximated by a line but the
fit is not as good as in the case of the gain and the line does not go through the origin.
The estimated parameters for both cases are:
    • *linear function through the origin:*

$$T^{-1} = \overline{T}_1^{-1} Q = (0.006175 \pm 0.000437) Q$$

- *linear function not through the origin:*

$$T^{-1} = \overline{T}_0^{-1} + \overline{T}_1^{-1} Q = (0.3694 \pm 0.009851) + (0.003128 \pm 0.0008619) Q \quad (7.3.18)$$

The large parameter/(standard deviation) ratio of the first term in (7.3.18) shows also that the inverse time constant function must have both constant and linear terms.

A global valid nonlinear discrete time model can be obtained if we substitute (7.3.17) and (7.3.18) into the first-order linear difference equation (7.3.12)

$$\Delta\vartheta(k) = \left[1 - \Delta T \, \overline{T}_0^{-1}\right] \Delta\vartheta(k-1) + \Delta T \, \overline{T}_1^{-1} \left(1/\overline{K}_1^{-1}\right) P(k-1)$$
$$+ \Delta T \, \overline{T}_0^{-1} \left(1/\overline{K}_1^{-1}\right) \left[P(k-1)/Q(k-1)\right] - \Delta T \, \overline{T}_1^{-1} Q(k-1) \Delta\vartheta(k-1) \quad (7.3.19)$$

and numerically

$$\Delta\vartheta(k) = 0.8153 \Delta\vartheta(k-1) + 1.3132 P(k-1)$$
$$+ 155.08\left[P(k-1)/Q(k-1)\right] - 0.001564 Q(k-1) \Delta\vartheta(k-1) \quad (7.3.20)$$

Figure 7.3.12 shows the simulated outlet temperature based on the identified model and the modeling error. The mean square error is $MSE = 1.682$.

The results of the grapho-analytical identification can also be used for determining the structure of the model and the unknown parameters can be estimated by the least squares method. The advantage of the continuous time water flow dependent model lies in its physical interpretation. While deriving the difference equation (7.3.19) the Euler discretization was applied. Since both the water flow and the heating power were constant during each temperature transient, the discretization by the assumption of a zero order holding device would be the correct one. The discrete time equivalent of (7.3.4) is

$$\Delta\vartheta(k) = -a_1 \Delta\vartheta(k-1) + b_1 P(k-1) \quad (7.3.21)$$

where

$$a_1 = -\exp(-\Delta T/T)$$

and

$$b_1 = K(1 + a_1) = K\left[1 - \exp(-\Delta T/T)\right] \quad (7.3.23)$$

The parameters a_i and b_i are drawn against the water flow in Figures 7.3.13 and 7.3.14. Furthermore the reciprocal values of the parameters b_i are also drawn in Figure 7.3.15. As is seen, all three parameters $(a_i, b_i, 1/b_i)$ can be well approximated by linear functions of the water flow. Static regression resulted in the following linear equations:

$$\hat{a}_1 = \hat{a}_{10} + \hat{a}_{11} Q = (-0.8266 \pm 0.03512) + (0.001133 \pm 0.0003073) Q \quad (7.3.24)$$
$$\hat{b}_1 = \hat{b}_{10} + \hat{b}_{11} Q = (4.556 \pm 0.2079) + (-0.01808 \pm 0.01819) Q \quad (7.3.25)$$

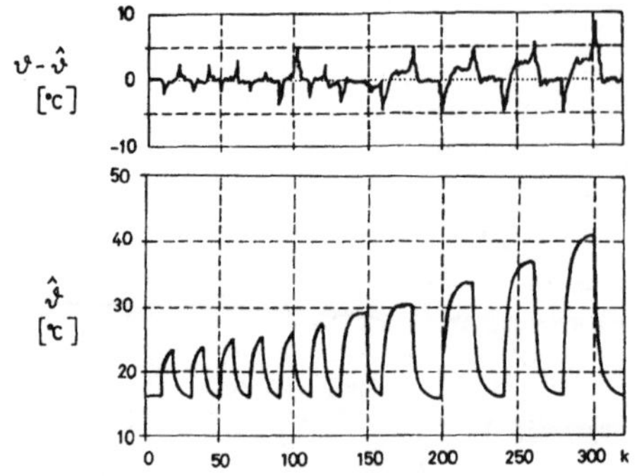

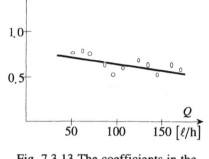

Fig. 7.3.12 Computed temperature $\hat{\vartheta}(k)$ and the modeling error $\left(\vartheta(k) - \hat{\vartheta}(k)\right)$ by fitting a first-order continuous time quasi-linear model with water flow dependent gain and time constant from the grapho-analytically determined parameters

Fig. 7.3.13 The coefficients in the denominators of the grapho-analytically identified first-order pulse transfer functions vs the water flow (indicated by circles). A straight line is fitted to the values

$$\hat{b}_1^{-1} = \hat{\bar{b}}_{10}^{-1} + \hat{\bar{b}}_{11}^{-1}Q = (0.09496 \pm 0.03784) + (0.002976 \pm 0.0003311)Q$$

As is seen from Figure 7.3.15 the reciprocal values of the parameters $\hat{b}_i$ can be approximated also by a straight line through the origin

$$\hat{b}_1^{-1} = \hat{\bar{b}}_{11}^{-1}Q = (0.003759 \pm 0.0001367)Q \qquad (7.3.26)$$

This approximation is a somewhat worse than that of (7.3.25) but has one parameter fewer. A model having fewer parameters is generally to be preferred if the fit is hardly worse.

Two discrete time models having signal dependent parameters were set up:
1. *Both a_i and b_i are linear functions of the water flow*
Having substituted (7.3.24) and (7.3.25) into (7.3.21) we obtain

$$\Delta\vartheta(k) = -a_{10}\Delta\vartheta(k-1) - a_{11}Q(k-1)\Delta\vartheta(k-1) + b_{10}P(k-1) + b_{11}P(k-1)Q(k-1)$$
$$(7.3.27)$$

or, numerically,

$$\Delta\vartheta(k) = 0.8266\Delta\vartheta(k-1) - 0.001133Q(k-1)\Delta\vartheta(k-1)$$
$$+4.56P(k-1) - 0.01808P(k-1)Q(k-1) \qquad (7.3.28)$$

To get a globally valid nonlinear *lag* type model Q was replaced by $Q(k-1)$ in (7.3.24) and (7.3.25).

2. *Both a_i and $1/b_i$ are linear functions of the water flow*
Having substituted (7.3.24) and (7.3.26) into (7.3.21) we obtain

$$\Delta\vartheta(k) = -a_{10}\Delta\vartheta(k-1) - a_{11}Q(k-1)\Delta\vartheta(k-1) + \left(1/\overline{b}_{11}^{-1}\right)\left[P(k-1)/Q(k-1)\right]$$

$$(7.3.29)$$

or, numerically,

$$\Delta\vartheta(k) = 0.8266\Delta\vartheta(k-1) - 0.001133Q(k-1)\Delta\vartheta(k-1) + 266.03\left[P(k-1)/Q(k-1)\right]$$

$$(7.3.30)$$

To obtain a globally valid nonlinear *lag* type model Q was replaced by $Q(k-1)$ in (7.3.24) and (7.3.26).   ∎

The simulated outlet temperatures based on the identified model and the modeling errors are plotted

- in Figure 7.3.16 for the linear functions $a(Q)$ and $b(Q)$ and

- in Figure 7.3.17 for the linear functions $a(Q)$ and $b^{-1}(Q)$

The mean square errors are
- MSE = 1.757 for the linear functions $a(Q)$ and $b(Q)$, and
- MSE = 1.848 for the linear functions $a(Q)$ and $b^{-1}(Q)$

Both model outputs fit well to the measured temperatures.

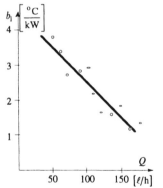
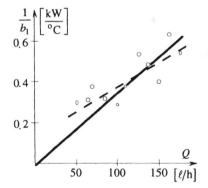

Fig. 7.3.14 The coefficients in the numerators of the grapho-analytically identified first-order pulse transfer functions vs the water flow (indicated by circles). A straight line is fitted to the values

Fig. 7.3.15 The reciprocal values of the coefficients in the numerators of the grapho-analytically identified first-order pulse transfer functions vs the water flow (indicated by circles). Two lines are fitted to the values:
• one through the origin (continuous line)
• not through the origin (dashed line)

7.3.4 Discrete time identification of a linear model
The parameter estimation was performed from the $2 \cdot 11 = 22$ temperature step responses in one step. That means that the whole heating power record in Figure 7.3.3 was

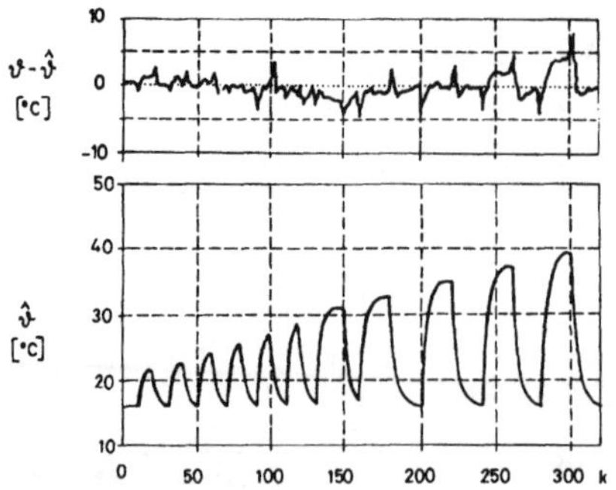

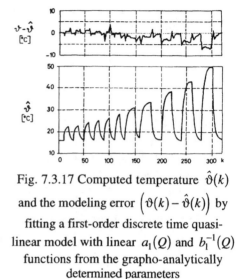

Fig. 7.3.16 Computed temperature $\hat{\vartheta}(k)$ and the modeling error $\left(\vartheta(k) - \hat{\vartheta}(k)\right)$ by fitting a first-order discrete time quasi-linear model with linear $a_1(Q)$ and $b_1(Q)$ functions from the grapho-analytically determined parameters

Fig. 7.3.17 Computed temperature $\hat{\vartheta}(k)$ and the modeling error $\left(\vartheta(k) - \hat{\vartheta}(k)\right)$ by fitting a first-order discrete time quasi-linear model with linear $a_1(Q)$ and $b_1^{-1}(Q)$ functions from the grapho-analytically determined parameters

considered as the input signal and the whole temperature record as the output signal. Instead of the measured values itself their normalized values (7.3.10) and (7.3.11) were used during the LS parameter estimation. The estimated model equation between the normalized variables is

$$\Delta\hat{\vartheta}'(k) = (0.8554 \pm 0.008941)\Delta\hat{\vartheta}'(k-1) + (0.05472 \pm 0.002835)P'(k-1) \qquad (7.3.31)$$

The standard deviation of the residuals is $\sigma_{\varepsilon'} = 0.02281$. Using the normalized equations (7.3.10) and (7.3.11), Equation (7.3.31) can be rewritten as

$$\Delta\hat{\vartheta}(k) = 0.8554\Delta\hat{\vartheta}(k-1) + 1.8241P(k-1) \qquad (7.3.32)$$

Figure 7.3.18 shows the simulated outlet temperature based on the identified model and the modeling error. As expected, the linear model is bad.

7.3.5 Discrete time identification of a two-variable general quadratic model being linear in the parameters

The simplest nonlinear model is the quadratic one. We have tried to fit a *formal* quadratic model to the measured temperature record. *Formal* means that the model components are only linear or quadratic functions of the measurable signals; however, the model itself (e.g., its steady state characteristics) may be of higher degree than quadratic. Since the model parameters depend on the water flow, as well, the water flow was assumed as a second input signal (or a measurable state variable). Former experiences have shown that the model components that include the output signal as multiplicative terms can improve a model considerably. Therefore these kinds of nonlinear terms were considered in the model. The structure of the above model is extended parametric Volterra model. As the heating power is a binary signal its square was not included in the model because it would not deliver any new information. The identified model of the normalized signals with the estimated parameters is as follows

$$\Delta\hat{\vartheta}'(k) = (0.9149 \pm 0.02151)\Delta\hat{\vartheta}'(k-1) + (0.01055 \pm 0.009448)$$
$$+ (0.1157 \pm 0.006227)P'(k-1) + (-0.0804 \pm 0.03637)Q'(k-1)$$
$$+ (0.08395 \pm 0.03138)Q'^2(k-1) + (-0.08201 \pm 0.01142)P'(k-1)Q'(k-1)$$
$$+ (0.3006 \pm 0.05553)Q'(k-1)\Delta\hat{\vartheta}'(k-1)$$

$$(7.3.33)$$

The standard deviation of the residuals is $\sigma_{\varepsilon'} = 0.01598$. Although seven parameters are estimated the parameter/standard deviation ratio is large; e.g., the black box model is a good input–output equivalent model of the process. The global valid nonlinear model between the measured variables can be obtained by using the normalizing equations (7.3.9) to (7.3.11)

$$\Delta\hat{\vartheta}(k) = 0.9149\Delta\hat{\vartheta}(k-1) + 0.5275 + 3.857P(k-1)$$
$$-0.0201Q(k-1) + 0.0001049Q^2(k-1)$$

$$(7.3.34)$$

$$-0.01367P(k-1)Q(k-1) + 0.001503Q(k-1)\Delta\hat{\vartheta}(k-1)$$

Figure 7.3.19 shows the simulated outlet temperature based on the identified model and modeling error. As is seen, the model fit is surprisingly good.

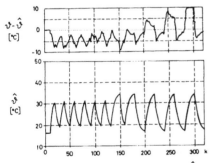

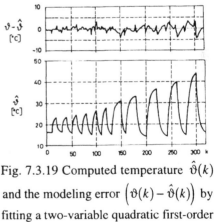

Fig. 7.3.18 Computed temperature $\hat{\vartheta}(k)$ and the modeling error $\left(\vartheta(k) - \hat{\vartheta}(k)\right)$ by fitting a first-order discrete time quasi-linear model to the whole measurement record by discrete time identification

Fig. 7.3.19 Computed temperature $\hat{\vartheta}(k)$ and the modeling error $\left(\vartheta(k) - \hat{\vartheta}(k)\right)$ by fitting a two-variable quadratic first-order extended parametric Volterra model to the whole measurement record

7.3.6 Discrete time identification by means of stepwise regression

Stepwise regression is a systematic search for the proper structure if the possible model components are chosen correctly. We have selected the constant term, the linear, and quadratic functions of the normalized heating power, water flow, and temperature as the possible components. The order of the model was restricted to one, therefore only variables delayed by one sampling unit were considered. Besides the squares of the variables their cross-products were also taken into account. As mentioned before the heating power is a binary signal thus its square was not included in the model to avoid that the constant, linear and quadratic functions of the heating power would be linearly dependent. The possible model components are as follows:

1. $P'(k-1)$;
2. $Q'(k-1)$;
3. $\Delta\vartheta'(k-1)$;
4. $Q'^2(k-1)$;
5. $\Delta\vartheta'^2(k-1)$;
6. $P'(k-1)Q'(k-1)$;
7. $P'(k-1)\Delta\vartheta'(k-1)$;
8. $Q'(k-1)\Delta\vartheta'(k-1)$;
9. 1.

If all possible model structures were tried out then the least squares parameter estimation should be executed $2^9 - 1 = 511$-times. By means of the stepwise regression the number of the regressions could be decreased to 47. Table 7.3.2 summarizes the whole selection procedure. Both the F to enter and F to remove thresholds were chosen 4, which resulted in a five-component model. The possible components came into the model in the following sequence:

1. $\Delta\vartheta'(k-1)$;
2. $P'(k-1)$;
3. $Q'(k-1)\Delta\vartheta'(k-1)$;
4. $P'(k-1)Q'(k-1)$;
5. $\Delta\vartheta'^2(k-1)$.

The five-term model could be reduced by eliminating the term $Q'(k-1)\Delta\vartheta'(k-1)$. The final model with the estimated parameters and with their standard deviations is as follows

$$\Delta\hat\vartheta'(k) = (0.6612 \pm 0.01735)\Delta\hat\vartheta'(k-1) + (0.1202 \pm 0.0052)P'(k-1)$$
$$+(-0.0944 \pm 0.00864)P'(k-1)Q'(k-1) \qquad (7.3.35)$$
$$+(0.3328 \pm 0.03787)\Delta\hat\vartheta'^2(k-1)$$

The standard deviation of the residuals is $\sigma_{\varepsilon'} = 0.01616$. The nonlinear model between the measured variables can be obtained by using the normalized equations (7.3.9) to (7.3.11)

$$\Delta\hat\vartheta(k) = 0.6612\Delta\hat\vartheta(k-1) + 4.007P(k-1)$$
$$-0.0157341P(k-1)Q(k-1) + 0.006656\Delta\hat\vartheta^2(k-1) \qquad (7.3.36)$$

Figure 7.3.20 shows the simulated outlet temperature based on the identified model and the modeling error. As is seen, the model fit is very good.

TABLE 7.3.2 Selection of the best model structure from all possible first-order quadratic models being linear in the parameters by means of the stepwise regression based on the F-test

No	No of comp	Components	σ_ε	F to enter	Comp	σ_ε	F value
						F to remove	
1	1	1	0.1245				
2	1	2	0.1897				
3	1	3	0.0360				
4	1	4	0.2080				
5	1	5	0.0774				
6	1	6	0.1686				
7	1	7	0.07665				
8	1	8	0.08454				
9	1	9	0.1614				
10	2	3, 1	0.02281	474.1	1	0.0360	474.1
11	2	3, 2	0.03361		3	0.1245	9156.0
12	2	3, 4	0.03363				
13	2	3, 5	0.03365				
14	2	3, 6	0.02887				
15	2	3, 7	0.02540				
16	2	3, 8	0.03362				
17	2	3, 9	0.03356				
18	3	3, 1, 2	0.02065				
19	3	3, 1, 4	0.02070				
20	3	3, 1, 5	0.01894				
21	3	3, 1, 6	0.01800				
22	3	3, 1, 7	0.02052		1	0.03362	853.0
23	3	3, 1, 8	0.01750	221.55	3	0.08043	6379.0
24	3	3, 1, 9	0.02163		8	0.02281	221.6
25	2	1, 8	0.08043				
26	4	3, 1, 8, 2	0.01726				
27	4	3, 1, 8, 4	0.01737		1	0.02338	315.4
28	4	3, 1, 8, 5	0.01740		3	0.04988	2558.0
29	4	3, 1, 8, 6	0.01654	37.78	6	0.01750	37.8
30	4	3, 1, 8. 7	0.01691		8	0.01800	58.3
31	4	3, 1, 8, 9	0.01721				
(32)	3	1, 6, 8	0.04988				
(33)	3	3, 6, 8	0.02338				
34	5	3, 1, 8, 6, 2	0.01642		1	0.02338	350.1
35	5	3, 1, 8, 6, 4	0.01654		3	0.02157	251.1
36	5	3, 1, 8, 6, 5	0.01609	17.87	5	0.01654	17.9
37	5	3, 1, 8, 6, 7	0.01631		6	0.01740	53.4
38	5	3, 1, 8, 6, 9	0.01625		8	0.01616	2.7
(39)	4	1, 5, 6, 8	0.02157				
(40)	4	3, 5, 6, 8	0.02338		1	0.02642	528.6
(41)	4	1, 3, 5, 6	0.01616		3	0.03821	1451.0
(42)	3	3, 5, 6	0.02642		5	0.01800	76.1
(43)	3	1, 5. 6	0.03821		6	0.01894	118.1
44	5	1, 3, 5, 6, 2	0.01608				
45	5	1, 3, 5, 6, 4	0.01614				
46	5	1, 3, 5, 6, 7	0.01617				
47	5	1, 3, 5, 6, 9	0.01606	3.94			

7.3.7 Discrete time identification of the parameters of a continuous time quasi-linear model with signal dependent parameters

In Section 7.3.3 a continuous time first-order quasi-linear model with signal dependent static gain and time constant was set up based on the grapho-analytical identification. If we keep the structure of (7.3.19) the unknown parameters of the difference equation can be estimated by the least squares method. Instead of using (7.3.19) we estimated the parameters of the model equation between the normalized variables, which can be obtained by substituting (7.3.9) to (7.3.11) into (7.3.19). The structure of the difference equation does not change and the LS parameter estimation leads to the model

$$\Delta\hat{\vartheta}'(k) = (0.8858 \pm 0.02026)\Delta\hat{\vartheta}'(k-1) + (0.02759 \pm 0.006128)P'(k-1)$$

$$+(0.01938 \pm 0.002482)\left[P'(k-1)/Q'(k-1)\right] \qquad (7.3.37)$$

$$+(-0.2661 \pm 0.05163)Q'(k-1)\Delta\hat{\vartheta}'(k-1)$$

The standard deviation of the residuals is $\sigma_{\varepsilon'} = 0.01604$. Using the normalized equations (7.3.9) to (7.3.11) the model equation between the measured signals is

$$\Delta\hat{\vartheta}(k) = 0.8858\Delta\hat{\vartheta}(k-1) + 0.9197P(k-1)$$

$$+129.207\left[P(k-1)/Q(k-1)\right] - 0.0013305Q(k-1)\Delta\hat{\vartheta}(k-1) \qquad (7.3.38)$$

Comparing (7.3.38) with (7.3.20) shows that the one-step discrete time identification leads to similar results as the two-step identification – grapho-analytical evaluation of the individual step responses and static regression of the signal dependence of the estimated parameters.

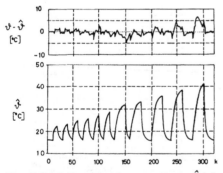

Fig. 7.3.20 Computed temperature $\hat{\vartheta}(k)$ and the modeling error $\left(\vartheta(k) - \hat{\vartheta}(k)\right)$ by fitting a two-variable quadratic first-order general model to the whole measurement record by the stepwise regression method

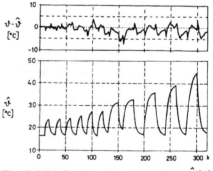

Fig. 7.3.21 Computed temperature $\hat{\vartheta}(k)$ and the modeling error $\left(\vartheta(k) - \hat{\vartheta}(k)\right)$ by fitting a first-order continuous time quasi-linear model with water flow dependent gain and time constant by least squares parameter estimation from the whole measurement record

Figure 7.3.21 shows the simulated outlet temperature based on the identified model and the modeling error. Here we can see that the computed temperature is similar to the one in Figure 7.3.12 computed on the basis of the model (7.3.20) by the two-step identification.

7.3.8 Discrete time identification of the parameters of a discrete time quasi-linear model with signal dependent parameters

In Section 7.3.3 two discrete time first-order quasi-linear models with signal dependent parameters were set up based on the grapho-analytical identification. If we keep the structures of (7.3.27) and (7.3.29) the unknown parameters of the difference equation can also be immediately estimated by the least squares method.

1. *Both a_i and b_i are linear functions of the water flow*

The structure of the model equation is given in (7.3.27). The transformation of (7.3.27) into a model of the normalized signals does not change the structure. Least squares parameter estimation leads to the model

$$\Delta\hat{\vartheta}'(k) = (0.9288 \pm 0.01836)\Delta\hat{\vartheta}'(k-1) + (0.1063 \pm 0.005973)P'(k-1)$$
$$+(-0.06785 \pm 0.0109)P'(k-1)Q'(k-1) \qquad (7.3.39)$$
$$+(-0.3678 \pm 0.04755)Q'(k-1)\Delta\hat{\vartheta}'(k-1)$$

The standard deviation of the residuals is $\sigma_{\varepsilon'} = 0.01654$. Using the normalized equations (7.3.9) to (7.3.11) the model equation between the measured signals is

$$\Delta\hat{\vartheta}(k) = 0.9288\Delta\hat{\vartheta}(k-1) + 3.5451P(k-1)$$
$$-0.01131P(k-1)Q(k-1) - 0.001839Q(k-1)\Delta\hat{\vartheta}(k-1) \qquad (7.3.40)$$

Comparing (7.3.40) with (7.3.28) shows that the one-step discrete time identification leads to similar results as the two-step identification.

Figure 7.3.22 shows the simulated outlet temperature based on the identified model and the modeling error. As expected the computed temperature is similar to that in Figure 7.3.16 computed on the basis of the model (7.3.28) by the two-step identification.

2. *Both a_i and $1/b_i$ are linear functions of the water flow*

The structure of the model equation is given in (7.3.29). The transformation of (7.3.29) into a model of the normalized signals does not change the structure. Least squares parameter estimation leads to the model

$$\Delta\hat{\vartheta}'(k) = (0.8159 \pm 0.0134)\Delta\hat{\vartheta}'(k-1)$$
$$+(0.02977 \pm 0.0009427)\left[P'(k-1)/Q'(k-1)\right] \qquad (7.3.41)$$
$$+(-0.07891 \pm 0.03152)Q'(k-1)\Delta\hat{\vartheta}'(k-1)$$

The standard deviation of the residuals is $\sigma_{\varepsilon'} = 0.01652$. Using the normalized equations (7.3.9) to (7.3.11) the model equation between the measured signals is

$$\Delta\hat{\vartheta}(k) = 0.8159\Delta\hat{\vartheta}(k-1) + 198.477\big[P(k-1)/Q(k-1)\big]$$
$$-0.0003945Q(k-1)\Delta\hat{\vartheta}(k-1)$$

(7.3.42)

Comparing (7.3.42) with (7.3.30) shows that the one-step discrete time identification leads to almost the same results as the two-step identification.

Figure 7.3.23 shows the simulated outlet temperature based on the identified model and the modeling error. As expected, the computed temperature is similar to that in Figure 7.3.17 computed on the basis of the model (7.3.30) by the two-step identification.

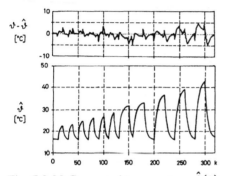 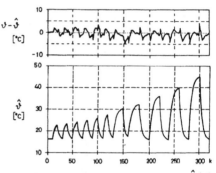

Fig. 7.3.22 Computed temperature $\hat{\vartheta}(k)$ and the modeling error $\big(\vartheta(k) - \hat{\vartheta}(k)\big)$ by fitting a first-order discrete time quasi-linear model with linear $a_1(Q)$ and $b_1(Q)$ functions by least squares parameter estimation from the whole measurement record

Fig. 7.3.23 Computed temperature $\hat{\vartheta}(k)$ and the modeling error $\big(\vartheta(k) - \hat{\vartheta}(k)\big)$ by fitting a first-order discrete time quasi-linear model with linear $a_1(Q)$ and $b_1^{-1}(Q)$ functions by least squares parameter estimation from the whole measurement record

7.3.9 Discrete time identification of the parameters of the theoretically derived model

Equation (7.3.16) is the equation of the theoretically derived model with the normalized signals and the unknown uncertainty factor. In Section 7.3.2 the uncertainty factor was estimated by discrete time identification. Here all the parameters of the model structure given by (7.3.16) are estimated to check whether the estimated model is better than that where only one parameter was tuned.

The parameter estimation led to

$$\Delta\hat{\vartheta}'(k) = (1.01 \pm 0.00773)\Delta\hat{\vartheta}'(k-1) + (0.07705 \pm 0.00252)P'(k-1)$$
$$-(0.5884 \pm 0.03158)Q'(k-1)\Delta\hat{\vartheta}'(k-1)$$

(7.3.43)

The standard deviation of the residuals is $\sigma_{\varepsilon'} = 0.01923$. Using the normalized equations (7.3.9) to (7.3.11) the model equation between the measured signals is

$$\Delta\hat{\vartheta}(k) = 1.01\Delta\hat{\vartheta}(k-1) + 2.56846P(k-1) - 0.002942Q(k-1)\Delta\hat{\vartheta}(k-1)$$

(7.3.44)

Figure 7.3.24 shows the simulated outlet temperature based on the identified model and the modeling error. The model fitting is worse than in Figure 7.3.7 drawn on the basis of the theoretically derived model with only one estimated parameter.

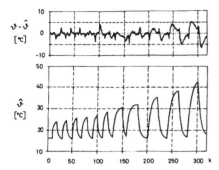

Fig. 7.3.24 Temperature $\hat{\vartheta}(k)$ computed with the estimated parameters based on the theoretically derived model structure and modeling error $\left(\vartheta(k) - \hat{\vartheta}(k)\right)$

7.3.10 Conclusions

1. The heat exchanger was controlled by an on–off regulator, therefore the water load dependence of the power/water temperature difference model was of interest. Since the regulator could set the heating power only at minimum or maximum, only these input levels were applied in the identification experiment.

2. The series of the step responses with constant heating power made the structure and parameter estimation easily possible by the grapho-analytical method.

3. The two-step identification – grapho-analytical identification of the parameters and static regression of the parameter dependence – can be performed only if the signal, on which the parameters depend, is constant during any transient. The one-step method that evaluates the whole record of more transients is suitable also in those cases if the above signal is not constant during the transients.

4. The first order model with water load dependent gain and time constant makes the controller design easy since the water load can be permanently measured and its value can be used for tuning the regulator parameters.

5. Here we sum up the identified nonlinear discrete time model equations and the corresponding mean square errors. The mean square error was chosen instead of any other function of the residuals, because the noise models are completely different in the different situations and the variances of the residuals could not be compared with each other:

   • theoretically derived model with uncertainty factor $\alpha = 0.8$:

   $$\Delta\hat{\vartheta}(k) = \Delta\hat{\vartheta}(k-1) + 3.2625P(k-1) - 0.0037878Q(k-1)\Delta\hat{\vartheta}(k-1)$$

   MSE = 1.704

   • theoretically derived model with uncertainty factor $\alpha = 1.0$:

   $$\Delta\hat{\vartheta}(k) = \Delta\hat{\vartheta}(k-1) + 2.61P(k-1) - 0.0030303Q(k-1)\Delta\hat{\vartheta}(k-1)$$

   MSE = 1.943

   • theoretically derived model with uncertainty factor $\alpha = 1.2$:

   $$\Delta\hat{\vartheta}(k) = \Delta\hat{\vartheta}(k-1) + 2.175P(k-1) - 0.0025252Q(k-1)\Delta\hat{\vartheta}(k-1)$$

   MSE = 2.458

- theoretically derived model with estimated uncertainty factor:

$$\Delta\hat{\vartheta}(k) = \Delta\hat{\vartheta}(k-1) + 2.887P(k-1) - 0.0033521Q(k-1)\Delta\hat{\vartheta}(k-1)$$
$$\text{MSE} = 1.766$$

- theoretically derived model structure with all parameters estimated:

$$\Delta\hat{\vartheta}(k) = 1.01\Delta\hat{\vartheta}(k-1) + 2.56846P(k-1) - 0.002942Q(k-1)\Delta\hat{\vartheta}(k-1)$$
$$\text{MSE} = 1.908$$

- two-step identification of a quasi-linear first-order model with inverse linear water flow dependence of the gain and the time constant:

$$\Delta\hat{\vartheta}(k) = 0.8153\Delta\hat{\vartheta}(k-1) + 1.3132P(k-1)$$
$$+155.08\big[P(k-1)/Q(k-1)\big] - 0.001564Q(k-1)\Delta\hat{\vartheta}(k-1)$$
$$\text{MSE} = 1.682$$

- one-step identification of a quasi-linear first-order model with inverse linear water flow dependence of the gain and the time constant:

$$\Delta\hat{\vartheta}(k) = 0.8858\Delta\hat{\vartheta}(k-1) + 0.9197P(k-1)$$
$$+129.207\big[P(k-1)/Q(k-1)\big] - 0.0013305Q(k-1)\Delta\hat{\vartheta}(k-1)$$
$$\text{MSE} = 1.471$$

- two-step identification of a quasi-linear first-order model with linear water flow dependence of the coefficients of the pulse transfer function

$$\Delta\hat{\vartheta}(k) = 0.8266\Delta\hat{\vartheta}(k-1) + 4.56P(k-1)$$
$$+0.01808P(k-1)Q(k-1) - 0.001133Q(k-1)\Delta\hat{\vartheta}(k-1)$$
$$\text{MSE} = 1.757$$

- one-step identification of a quasi-linear first-order model with linear water flow dependence of the coefficients of the pulse transfer function:

$$\Delta\hat{\vartheta}(k) = 0.9288\Delta\hat{\vartheta}(k-1) + 3.5451P(k-1)$$
$$-0.01131P(k-1)Q(k-1) - 0.001839Q(k-1)\Delta\hat{\vartheta}(k-1)$$
$$\text{MSE} = 1.641$$

- two-step identification of a quasi-linear first-order model with linear and inverse linear water flow dependence of the coefficients of the pulse transfer function:

$$\Delta\hat{\vartheta}(k) = 0.8266\Delta\hat{\vartheta}(k-1) + 266.03\big[P(k-1)/Q(k-1)\big]$$
$$-0.001133Q(k-1)\Delta\hat{\vartheta}(k-1)$$
$$\text{MSE} = 1.848$$

- one-step identification of a quasi-linear first-order model with linear and inverse linear water flow dependence of the coefficients of the pulse transfer function:

$$\Delta\hat{\vartheta}(k) = 0.8159\Delta\hat{\vartheta}(k-1) + 198.477\left[P(k-1)/Q(k-1)\right]$$

$$-0.0003945Q(k-1)\Delta\hat{\vartheta}(k-1)$$

MSE = 1.422

- discrete time identification of a quadratic black box (extended parametric Volterra) model:

$$\Delta\hat{\vartheta}(k) = 0.9149\Delta\hat{\vartheta}(k-1) + 0.5275 + 3.857P(k-1)$$

$$-0.0201Q(k-1) + 0.0001049Q^2(k-1)$$

$$-0.01367P(k-1)Q(k-1) + 0.001503Q(k-1)\Delta\hat{\vartheta}(k-1)$$

MSE = 1.455

- stepwise regression with constant, linear, and quadratic model components:

$$\Delta\hat{\vartheta}(k) = 0.6612\Delta\hat{\vartheta}(k-1) + 4.007P(k-1)$$

$$-0.015734P(k-1)Q(k-1) + 0.006656\Delta\hat{\vartheta}^2(k-1)$$

MSE = 1.893

6. The one- and two-step identification methods gave very similar results – as expected. The comparison of the mean square errors of the one- and two-step identifications of the models with signal dependent parameters shows that the one-step methods have always a bit better output matching. The grapho-analytical method, however, helps to determine the proper model structure.

7. By means of the stepwise regression procedure those model components were automatically selected in the first four steps that were also in the quasi-linear first-order model with linear water flow dependence of the coefficients of the pulse transfer function. This fact shows also the validity of the quasi-linear model with signal dependent parameters.

8. The black box model (the extended parametric Volterra one) gave a good input–output model although the fact that no *a priori* knowledge about the process was used.

9. Two estimated models with the theoretically derived model structure were presented. As is seen, that model was better where only one parameter, the uncertainty factor, was estimated. The model given by (7.3.44) is much worse than that of equation (7.3.15) with the tuned uncertainty factor, because in (7.3.44) three parameters were estimated. It is always practical to estimate only as few parameters as possible and use all *a priori* knowledge given generally by physical laws.

7.4 STEAM HEATED HEAT EXCHANGER

7.4.1 Description of the plant
A steam heated heat exchanger was working in the Institute of Process Engineering and Boilers at the University of Stuttgart (FRG). Its scheme is seen in Figure 7.4.1. The

steam comes from the boiler of the heating power station through a pressure reducer to the top of the heat exchanger. While condensing, the steam transmits its energy to the water in backward flow.

The difference between the water temperatures at the output and the input $\Delta\vartheta(t)$ is proportional to the steam flow $\Phi_S(t)$ and inversely proportional to the water flow. The water flow was kept constant at 8330 [kg/h].

Formerly Baur and Isermann (1976) identified the model $\Phi_S/\Delta\vartheta$ by PRBS test signals. The aim of the present work was to investigate the influence of a bigger change in the steam flow. The task was executed by means of the process computer HP 2100A of the Department of Control and System Dynamics of the Institute. The real time program package OLID-MISO-NOLI (Haber and Bamberger, 1979) was used for the identification of a nonlinear model. The two-step identification method correlation analysis with LS parameter estimation (COR-LS) was used (Haber, 1979, 1988).

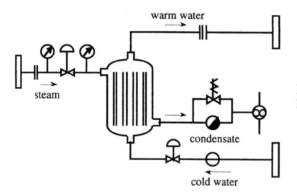

Fig. 7.4.1 Scheme of the steam heated heat exchanger

7.4.2 Process identification at changes in the steam flow

Before having carried out the identification tests, the steady state temperature differences were measured at different steam flows. There was a linear relationship between the steam flow and the current of the digital/analog converter,

$$\Phi_S\left[\text{kg/h}\right] = 111.0 - 15.2\, I_{D/A}\left[\text{mA}\right]$$

and between the temperature difference and the voltage of the analog/digital converter,

$$\Delta\vartheta\left[°C\right] = 1.164 + 0.17\, U_{A/D}\left[V\right]$$

The two calibration curves are seen in Figures 7.4.2 and 7.4.3. The parameter estimation was performed between the electrical values (current and voltages), and since both the current and the voltage are linear functions of the physical values (steam flow and temperature difference) any resulting nonlinear model should show the nonlinear feature of the process itself.

The relationships between the measured current and voltage values are seen in Figure 7.4.4 and between the corresponding physical values in Figure 7.4.5; both are plotted by continuous lines. The static characteristic $\Phi_S/\Delta\vartheta$ is linear only in the close vicinity of the working point $\Phi_S \approx 25$ [kg/h] and it is mildly convex elsewhere. The reason was that only some part of the steam condensed in the heat exchanger. The steam flow was set by a valve before the heat exchanger and at increasing steam flow more steam could condense, while on the other hand the input temperature of the steam increased because of the change in the steam pressure. The above two effects led to the said nonlinearity.

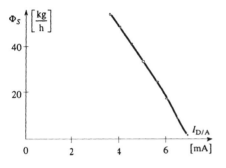

Fig. 7.4.2 Current/steam flow curve calibration

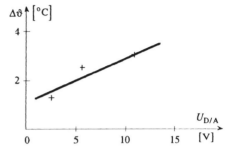

Fig. 7.4.3 Temperature difference/voltage calibration curve

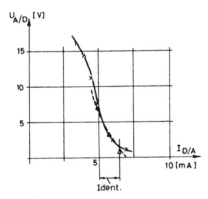

Fig. 7.4.4 Current/voltage characteristic (x: measured, Δ : identified)

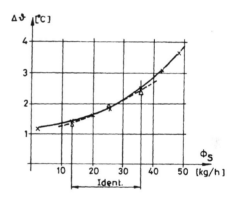
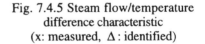

Fig. 7.4.5 Steam flow/temperature difference characteristic (x: measured, Δ : identified)

The domain $I_{D/A} = 5 \div 7.5$ [mA] was selected for the identification where both curves I/U and $\Phi_S/\Delta\vartheta$ were convex. The test signal was a PRTS signal of maximum length and of periodicity 80. The sampling time was 2 [s], which was half of the smallest interval between two possible changes in the test signal. The time functions of the measured current and voltage values are plotted in Figure 7.4.6.

Two multipliers were calculated from the measured current values:

$$x_1(k) = i_{D/A}(k) - E\{i_{D/A}(k)\}$$

and

$$x_2(k) = i^2_{D/A}(k) - E\{i^2_{D/A}(k)\}$$

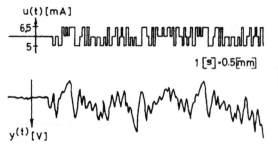

Fig. 7.4.6 Measured current and voltage values

The multipliers were correlated by the measured input signal, by its square, and by the output signal, because the model structure was assumed to be a generalized Hammerstein model. The correlation functions were calculated in the domain $(-8 \le \kappa \le 40)$. The cross-correlation functions between the multipliers and the measured output signal (voltage) are drawn in a line printer plot in Figures 7.4.7 and 7.4.8.

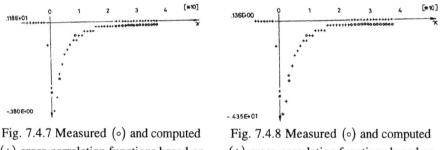

Fig. 7.4.7 Measured (∘) and computed (+) cross-correlation functions based on the estimated model between the multiplier $x_1(k)$ and the output signal (voltage)

Fig. 7.4.8 Measured (∘) and computed (+) cross-correlation functions based on the estimated model between the multiplier $x_2(k)$ and the output signal (voltage)

The constant term of the model was not estimated because the multipliers had zero average value but it was established from the initial steady state values. If equal delay times were taken into account in the linear and the quadratic channels, the identified model was not good, therefore different delay times were also allowed. The loss function of the parameter estimation is plotted in Figure 7.4.9 as a function of the order, considering the best delay time combination at each order. The equation of the best fitting model is (Haber, 1979):

$$\hat{u}'_{A/D}(k) = -\frac{0.01505q^{-1} + 0.425q^{-2}}{1 - 1.396q^{-1} + 0.457q^{-2}} i_{D/A}(k) + \frac{0.004328q^{-1} + 0.01718q^{-2}}{1 - 1.396q^{-1} + 0.457q^{-2}} i_{D/A}(k-2)$$

The output cross-correlation functions calculated from the measured signals and from the output of the estimated model are given in Figures 7.4.7 and 7.4.8. The line printer plots of the input and output signals and the model output signal can be seen in Figure 7.4.10. The constant term of the model was determined from the initial steady state values:

$$U_{A/D} = -1.5 \, [\text{V}] \quad \text{if} \quad I_{D/A} = 5.73 \, [\text{mA}]$$

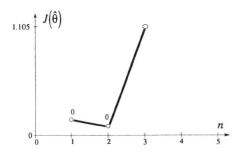

Fig. 7.4.9 Loss functions $J(\hat{\theta})$ of the parameter estimation as a function of the model order n at the best dead time. The numbers in the plot point to the best dead time structure

thus

$$U_{A/D}(k) = U'_{A/D}(k) + 28.3$$

The steady state relation between the current and the voltage is

$$\hat{U}_{A/D}\,[\text{V}] = 28.3 - 7.21\, I_{D/A}\,[\text{mA}] + 0.3524\, I^2_{D/A}\,\left[\text{mA}^2\right]$$

The static characteristics I/U and $\Phi_S/\Delta\vartheta$ are drawn as dotted lines in Figures 7.4.4 and 7.4.5, respectively. The estimated steady state characteristics approximate well the real steady state characteristics (continuous lines). The gain in the working point $K = -2.805$ [V/mA] coincides well with the gain $K = -3.0975$ [V/mA] of an estimated linear model

$$\hat{u}'_{A/D}(k) = \frac{-0.4469q^{-1} + 0.0799q^{-2} + 0.189q^{-3}}{1 - 1.425q^{-1} + 0.4901q^{-2} - 0.007828q^{-3}}\, i_{D/A}(k)$$

The linear model was fitted to the same record as the nonlinear structure.
  The gain of the identified Hammerstein model is

$$K = \frac{\Delta(\Delta\vartheta)}{\Delta\Phi_S} = \frac{2.4 - 1.4}{36.2 - 13.0} = 0.0432\,\left[^\circ\text{C}/(\text{kg/h})\right]$$

in physical units. This value hardly differs from the former results of other researchers obtained by a linear identification.

7.4.3 Conclusions
1. The generalized Hammerstein model is a good input–output model of a complex plant, it approximates at least the nonlinear steady state characteristic well.
2. The parameter estimation of the generalized Hammerstein model is hardly harder than the identification of a linear dynamic model although the information obtained can be much more.
3. The assumption of different delay times in the linear and quadratic channels (instead of equal ones) may lead to better models, but it causes more work when seeking the best fitting structure.
4. The gain of a fitted linear model should be equal to the tangent of the steady state characteristic obtained by fitting a nonlinear dynamic model in the same working point.

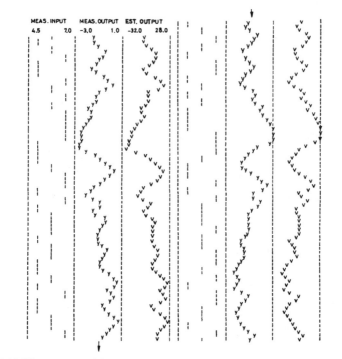

Fig. 7.4.10 The measured input current (I) and output voltage signals (Y) and the
computed output signal (V) based on the estimated model

7.5 DISTILLATION COLUMN SEPARATING METHANOL–WATER MIXTURE

7.5.1 Description of the plant

The investigations were performed at the distillation column located at the Department
of Operations of Chemical Industry, Technical University of Budapest, Hungary. The
scheme of the pilot plant is seen in Figure 7.5.1. The condensed top product flows into
the reflux tank. The condensate pours from the tank under gravitational force. A part of
the flow is fed back to the top as a reflux and the rest is the distillate. The distillate is
fed back to the bottom of the column, i.e., the feed is at the bottom. This means that by
means of the pilot plant the upper half of a usual column can be investigated.

Five plates were placed into the column at a distance of 0.4 [m] from each other. The
type of the trays is sieving with ring valve. The boiler was heated by water steam. The
steam condensed in the horizontal capillaries of the boiler. The pressure of the steam
was kept at a given reference value by means of a PI regulator. The maximum pressure
was 245 [kPa] which value is taken as 100 percent. There is a linear steady state relation
between the pressure and the flow of the steam. The volume of the boiler was 400 [ℓ]
and it was about half full during the experiments. The volume of the liquid in the boiler
was approximately 100 times the hold-up of the plates. The concentration in the boiler
was approximately constant during the experiments, the feedback of the distillate to the
boiler helped to maintain the stationary state.

The temperature of the first (top) plate was measured by a thermoelectric cell. The
corresponding concentrate could be calculated by the calibration curve of Figure 7.5.2.
The reflux and distillate flows were measured by rotameters. They were set manually.

During the experiments methanol–water mixture was distilled at atmospheric pressure.

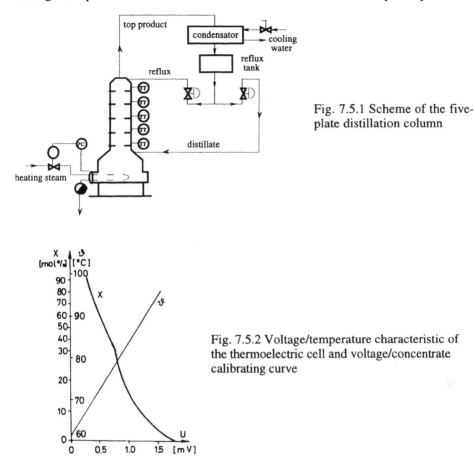

Fig. 7.5.1 Scheme of the five-plate distillation column

Fig. 7.5.2 Voltage/temperature characteristic of the thermoelectric cell and voltage/concentrate calibrating curve

7.5.2 Process identification at changes in the reflux flow

The concentration in the boiler was 3 [mol–percent] and the heating power was $P = 30$ percent of its maximal value. Several steps were made in the reflux flow R in different directions and by various amplitudes.

The new changes were only allowed if the top plate concentration x achieved its steady state value, as is seen in Figure 7.5.3. This strategy made also a grapho-analytical evaluation of the records possible. The measured steady state characteristics, the gains and time constants of the linearized, first-order models are drawn in Figures 7.5.4 to 7.5.6, respectively, as a function of the reflux.

The process is highly nonlinear, which appears in the reflux dependence of the gain and both reflux and direction dependence of the time constant. The physical interpretation of the results is as follows. The concentration of the head product was very low. Here even small increases in the reflux flow lead to high improvement of the head product. (Over a given concentration the curve would saturate because the improvement of the concentration of the head product would be becoming increasingly difficult.) The concentration of the head product increased more slowly than it decreased.

The closed loop behavior of the plant could be observed in the concentration increase; there are two steps in the step response. The first change is a consequence of the increase of the reflux flow and the second one shows the effect when the concentration of the reflux increases, as well.

Because of the very sophisticated structure of the process, we have decided to handle it as a black box. Different nonlinear structures, all being linear in the parameters, were tried out. No delay time was assumed. The possible model components were $1, u(k), u^2(k), u(k)y(k), y^2(k)$. The structure identification resulted in the following second-order nonlinear difference equation:

$$
\begin{aligned}
y(k) = & (0.353 \pm 0.118)y(k-1) + (2.0 \pm 0.31) \\
& + (0.547 \pm 0.54)u(k-1) - (1.656 \pm 0.58)u(k-2) \\
& - (1.68 \pm 0.36)u^2(k-1) + (1.80 \pm 0.35)u^2(k-2) \\
& + (0.713 \pm 0.09)u(k-1)y(k-1) - (0.3 \pm 0.073)u(k-2)y(k-2)
\end{aligned}
$$

(The sampling time was $\Delta T = 0.5$ [min].)

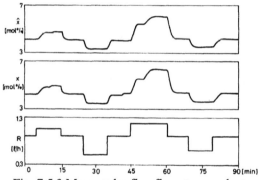

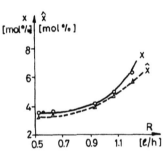

Fig. 7.5.3 Measured reflux flow, top product concentrate, and the computed model output signal

Fig. 7.5.4 Measured ($\circ$) and identified (Δ) steady state relation between the reflux and top product concentrate

The model output coincides well with the measured output signal, both shown in Figure 7.5.3. The steady state characteristic of the model could be calculated by simply replacing the delayed signals by their steady state values

$$
\hat{Y} = \frac{3.091 - 1.714U + 0.188U^2}{1 - 0.635U}
$$

The derivative of the static curve is the gain of the linearized models,

$$
K = \frac{d\hat{Y}}{dU} = \frac{-1.714 + 0.376U + 0.636\hat{Y}}{1 - 0.635U}
$$

The computed steady state characteristic (dashed line) approximates well the measured one (continuous line) in Figure 7.5.4 and the computed gain (dashed line) approximates the grapho-analytically evaluated ones (continuous line) in Figure 7.5.5.

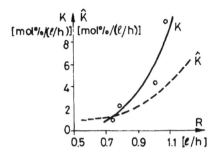

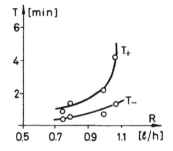

Fig. 7.5.5 Gains of the linearized, first-order reflux/concentrate models estimated by grapho-analytical (∘) and nonlinear (– – –) identification method

Fig. 7.5.6 Grapho-analytically estimated time constants of the linearized first-order reflux/concentrate models

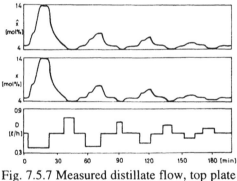

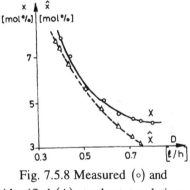

Fig. 7.5.7 Measured distillate flow, top plate concentration and the computed model output signal

Fig. 7.5.8 Measured (∘) and identified (△) steady state relations between the reflux and top product concentrates

7.5.3 Process identification at changes in the distillate flow

The concentration in the boiler was $x = 3.8$ [mol–percent] and the heating power was $P = 30$ percent of its maximum value. Stepwise changes were performed in the distillate flow D. The direction and amplitude of the steps were varied to obtain a good global and valid nonlinear model. To use also linear identification technique, the flow was changed so rarely that the top product concentration could achieve its steady state value. The measured distillate flow and top plate concentration are plotted in Figure 7.5.7. The steady state characteristic of the process is given in Figure 7.5.8 by a continuous line. First-order linear models were assumed and the individual step responses were grapho-analytically evaluated. Both the gain and the time constant decreased by the distillate flow, as seen in Figures 7.5.9 and 7.5.10, respectively. The time constants were less, if the distillate flow was increased, i.e., the top plate concentration was decreased.

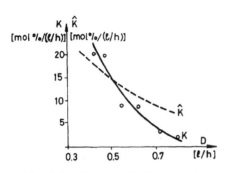

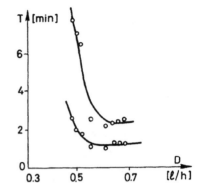

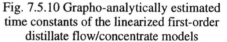

Fig. 7.5.9 Gains of the linearized, first-order distillate flow/concentrate models estimated by grapho-analytical (o) and nonlinear (– – –) identification method

Fig. 7.5.10 Grapho-analytically estimated time constants of the linearized first-order distillate flow/concentrate models

The effects of the change in the distillate flow to the top plate concentration are similar as in the case of changing of the reflux flow but with opposite sign. The liquid streaming from the condensator and the reflux tank is divided into two branches: the reflux and the distillate. Any change in one of them appears in the other with the opposite sign. However, any change in the distillate flow affects the top plate concentration slower than the reflux flow change because the reflux flow cannot change immediately after having set the distillate flow valve. As a consequence of the constant steam generation, the top product condensed into the reflux tank is constant. Therefore a sudden change in the distillate flow appears in the level of the reflux tank with a first-order lag delay. Since the reflux flow pours from the tank as a result of gravitational force, its flow cannot change immediately.

Different nonlinear structures, being linear in the parameters were fitted to the measured data records. The possible model components were: $1, u(k), u^2(k), u(k)y(k), y^2(k)$. No delay time was assumed. The following estimated model fitted to the measured data best

$$\hat{y}(k) = (0.766 \pm 0.027)y(k-1) + (2.02 \pm 0.2) - (2.02 \pm 0.76)u(k)$$
$$-(1.754 \pm 0.87)u(k-1) + (0.871 \pm 0.55)u^2(k)$$
$$+(1.0 \pm 0.58)u^2(k-1) + (0.146 \pm 0.03)u(k)y(k)$$
$$+(0.0277 \pm 0.047)u(k-1)y(k-1) + (0.0432 \pm 0.0012)y^2(k)$$
$$-(0.0358 \pm 0.001)y^2(k-1)$$

(The sampling time was $\Delta T = 0.5$ [min].)

The output signal of the model is drawn in Figure 7.5.7.

The steady state characteristic of the model could be calculated by simply replacing the delayed signals by their steady state values.

$$\hat{Y}^2 + (23.174U - 31.27)Y + 269.85 + U(250.5U - 504.4) = 0$$

The derivative of the static curve is the gain of the linearized models

$$K = \frac{d\hat{Y}}{dU} = \frac{-16.13 + 16.0U + 0.74\hat{Y}}{1 - 0.74U - 0.064\hat{Y}}$$

The computed steady state characteristic (dashed line) approximates well the measured curve (continuous line) in Figure 7.5.8 and the computed gain (dashed line) approximates the grapho-analytically evaluated ones (continuous line) in Figure 7.5.9.

7.5.4 Process identification at changes in the heating power
The concentration in the boiler was $x = 3$ [mol–percent] and the reflux flow was kept constant at $R = 0.925$ [ℓ/min]. Figure 7.5.11 shows some steps in the heating power P and the step responses in the top plate concentration x. The individual step responses were evaluated by a grapho-analytical method. The steady state relation between the heating power and the top plate concentration is presented in Figure 7.5.12. The gains and time constants of the linearized first-order lag models are drawn in Figures 7.5.13 and 7.5.14, respectively. Both gain and time constant decrease by increasing the heating power and the influence of increasing the heating is faster than that of its decreasing.

Nonlinear model identification has not been performed on the data.

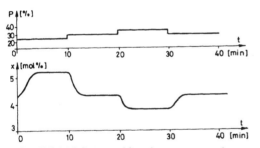

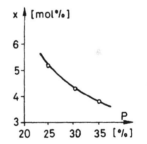

Fig. 7.5.11 Measured heating power and top product concentrate signals

Fig. 7.5.12 Measured steady state relation between the heating power and top product concentrate

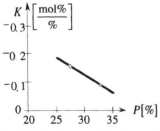

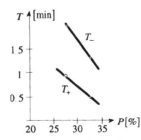

Fig. 7.5.13 Grapho-analytically estimated gains of the linearized first-order pressure/condensate model

Fig. 7.5.14 Grapho-analytically estimated time constants of the linearized first-order pressure/condensate model

7.5.5 Conclusions

1. The distillation pilot plant is a multi-variable process which can be approximated by linearized, first-order models whose parameters (gain, time constant) alter with the working point and the direction of the intervention.
2. Sophisticated models can be handled by black box models. If no *a priori* knowledge is available, a suitable choice of the model components could be, e.g.,: $1, u(k)$, . $u^2(k)$, $u(k)y(k)$, $y^2(k)$ Structure identification is needed to choose the best fitting and acceptable model.
3. In order to obtain models both predictive and linear in the parameters the output signal is allowed to occur in a – at least one discrete time unit – delayed form.
4. The steady state characteristic of a model can be obtained by the best difference equation replacing the actual variables by their steady state values. Experiences showed that not only the best model describes the steady state relation accurately, but any models, which approximate well the measured output signal by its computed model output signal and whose estimated parameters are – approximately – significant to it.

7.6 DISTILLATION COLUMN SEPARATING ETHANOL–WATER MIXTURE

7.6.1 Description of the plant

At the Department of Automation, Technical University of Budapest, Hungary, a pilot plant distillation column was built (Tuschák *et al.,* 1982).

The ten plate bubble cap column was used for separating the ethanol–water mixture. The main technological data are:

feed:	16 [wt-percent],	50 [ℓ/h],
top product:	76 [wt-percent],	10 [ℓ/h],
bottom product:	4 [wt-percent],	40 [ℓ/h],
reflux ratio:	2:1	
pressure:	atmospheric,	
heat (electric):	18 [kW].	

The scheme of the pilot plant is shown in Figure 7.6.1.

The flow rate of the feed is set by the flow rate controller FIC-401. The temperature of the feed can be influenced by the pre-heater and the heater. The pre-heater uses the heat when leaving the condenser. It can be controlled by the valve V2. The regulator TIC-112 controls the feed heater using electric heating.

There are temperature transmitters (TI-101 to TI-110) at each tray of the column. They give the temperature profile of the column and – in the knowledge of the pressure – the composition, too. The pressure of the column is controlled by the regulator PIC-201 located at the top of the column by changing the cooling water rate (V3). The reflux rate is controlled by the regulator FFIC-402. Both the reflux itself and the reflux ratio can be set. There are two pre-coolers for controlling the temperatures of the top- and bottom products. The top distillate is fed immediately to the main tank, but the bottom product only through a bottom tank. The whole process is working in a closed loop. The bottom tank serves for disturbing the concentration of the feed – if necessary – by means of the regulator DIC-501.

The pilot plant is controlled by an SBC 80/20 microcomputer of INTEL. The microcomputer is connected to the process with 16 analog inputs, 4 analog outputs, 48 digital inputs–outputs and 8 interrupt lines.

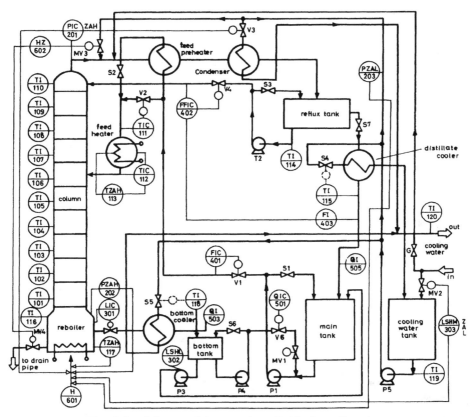

Fig. 7.6.1 Scheme of the ten plate distillation pilot plant

The real time program language used in SBC 80/20 is PCL 80 (Gál *et al.*, 1979). In this language there are several modules, e.g., for data logging and control actions, thus the user has to call them in tasks having priorities and timings.

During the experiments local analog controllers regulated the levels and the feed temperature. The following signals were measured by the microcomputer:

1. pressure p,
2. feed rate F,
3. reflux rate R,
4. distillation rate D,
5. feed temperature ϑ_F,
6. reflux temperature ϑ_R,

7-16. temperatures of the trays 1-10 (ϑ_1 to ϑ_{10}).

The SBC 80/20 controlled the following variables by PI regulators in DDC mode:

1. pressure;
2. feed rate;
3. reflux rate.

The heating power P was set manually.

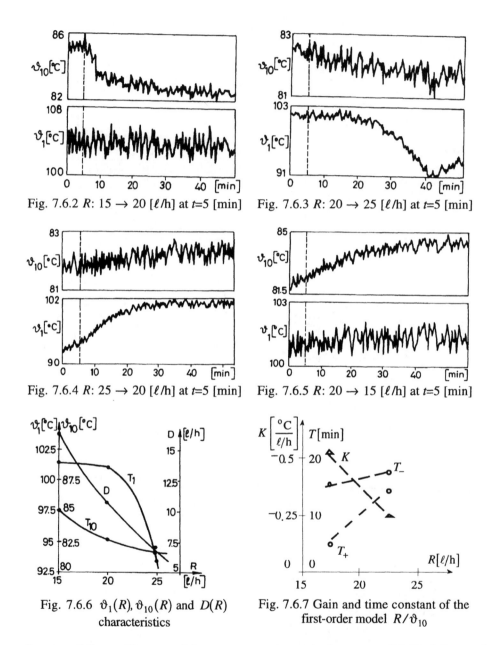

Fig. 7.6.2 R: $15 \rightarrow 20$ [ℓ/h] at t=5 [min]

Fig. 7.6.3 R: $20 \rightarrow 25$ [ℓ/h] at t=5 [min]

Fig. 7.6.4 R: $25 \rightarrow 20$ [ℓ/h] at t=5 [min]

Fig. 7.6.5 R: $20 \rightarrow 15$ [ℓ/h] at t=5 [min]

Fig. 7.6.6 $\vartheta_1(R)$, $\vartheta_{10}(R)$ and $D(R)$ characteristics

Fig. 7.6.7 Gain and time constant of the first-order model R/ϑ_{10}

Because of the small power of the pumps any change in the pressure highly influenced the feed rate. Therefore a feed forward action was used in the feed rate control loop from the pressure.

The purposes of the first experiments were (Tuschák *et al.*, 1982):
1. checking of the measured working point (comparing to the planned values);
2. studying the process dynamics in the small vicinity of the working point;

3. measuring the static nonlinear behavior of the process within the technological limits;
4. identification of the main processes in different working points;
5. pointing out that the approximating quasi-linear models have working point and direction dependent parameters.

The results of the identification should serve at the design of the control loops.

7.6.2 Process identification at step changes in the reflux rate
The following parameters were kept constant: $p = 113$ [kPa], $P = 11.7$ [kW], $F = 75$ [ℓ/h], $\vartheta_F = 65$ [°C], $\vartheta_R = 72$ [°C]. The reflux was changed stepwise in the following domain:

$$R : 15 \rightarrow 20 \rightarrow 25 \rightarrow 20 \rightarrow 15 \; [\ell/h]$$

Figures 7.6.2 to 7.6.5 show the temperature responses of the first (bottom) and 10th (top) plates.
An increase in the reflux causes a decrease in the distillate rate. The reflux ratio is increasing which fact leads to the increase of the temperatures (Figure 7.6.6). The effect of changing the reflux rate can be observed mainly in the top temperature. If identifying the temperature responses of Figures 7.6.2 to 7.6.5 first-order models are obtained whose parameters are given in Figure 7.6.7. The gain is inversely proportional to the reflux rate because the separation becomes more difficult. This is the reason of the increasing time constants, too. In the same reflux change the time constant was less when increasing the reflux rate T_+ than when decreasing it T_-, because the reflux was cooler than the reflux plate and cooled the top temperature faster than the vapor warmed it up when decreasing the reflux rate.

7.6.3 Process identification using pseudo-random binary excitation in the reflux flow
Pseudo-random binary excitation was also applied in the range 15 [ℓ/h] and 20 [ℓ/h]. The PRBS signal was generated with maximal length and periodicity 7. The minimum switching time of the PRBS was 6 [min]. The data were recorded with a sampling time of 15 [s]. The following can be seen in the plot of the measurements in Figure 7.6.8:
- since the temperature is very noisy, filtering is needed;
- the shape of the temperature at the very end of the record (after about 70 [min]) is opposite to the typical shape until then;
- careful observation shows a small linear drift in the temperature;
- the increase of the temperature is faster than its decrease;
- the response of the temperature to changes in the reflux shows a first-order lag character because the response starts without any jump and with a non-zero tangent;
- as a consequence of the last two remarks, a first-order quasi-linear model with direction dependent time constant seems to be a good choice.

The temperature was filtered by a linear digital first-order filter of the form

$$\vartheta_{10}^F(k) = \left[1 - \exp(-\Delta T/T_F)\right]\vartheta_{10}(k-1) + \exp(-\Delta T/T)\vartheta_{10}^F(k-1)$$

Different filter time constants were tried out and $T_F = 50$ [s] was found to be optimal.

To eliminate the nonstationary effects only one period of the excitation was cut out from the whole records. The period chosen for the identification is also seen in Figure 7.6.8. It starts with the 68th sampling and ends with the 236th. Further on we evaluate only the selected part of the measurement record. From physical considerations the temperature should have the same value at the end of the period as at the beginning. The equation of the linear drift was determined by the equation

$$\vartheta_{10}^{DF}(k) = \vartheta_{10}^{F}(1) + \frac{\vartheta_{10}^{F}(N) - \vartheta_{10}^{F}(1)}{N-1}(k-1)$$

where the number of data is $N = 6 \cdot 7/0.25 + 1 = 169$ and the filtered temperature values are at the beginning $\vartheta_F(1) = 82.5$ and at the end of the period $\vartheta_F(169) = 82.34$. The linear drift was subtracted from the filtered temperature signal

$$\vartheta_{10}^{dF}(k) = \vartheta_{10}^{F}(k) - \vartheta_{10}^{DF} \qquad k = 1, 2, \ldots, N$$

In order to obtain numerically better values for the least squares estimation the data were normalized as

$$R'(k) = 4.0 - 0.2R(k) \qquad (\equiv u(k)) \tag{7.6.1}$$

$$\vartheta_{10}^{dF'}(k) = \vartheta_{10}^{dF}(k) - 82.5 \quad (\equiv y(k)) \tag{7.6.2}$$

To fill the memory vector of the least squares estimation the period was extended by 3 initial values. Therefore the number of data pairs became $N = 172$. Figure 7.6.9 shows the reflux flow, the measured and filtered top temperature signals after having eliminated the drift and extended the data by the initial values. Figure 7.6.10 shows the same for the normalized values. As the identification was performed between the normalized values they are denoted furtheron simply by $u(k)$ and $y(k)$.

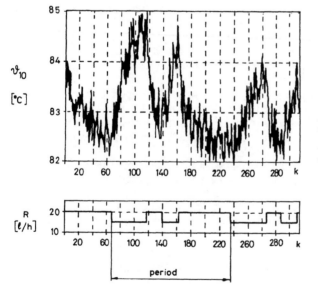

Fig. 7.6.8 Measured reflux R and top temperature ϑ_{10} at PRBS excitation

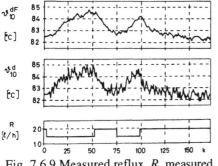
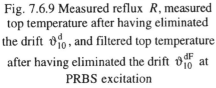

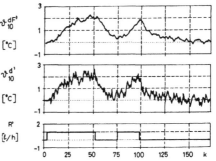

Fig. 7.6.9 Measured reflux R, measured top temperature after having eliminated the drift ϑ_{10}^d, and filtered top temperature after having eliminated the drift ϑ_{10}^{dF} at PRBS excitation

Fig. 7.6.10 Normalized measured reflux R', normalized measured top temperature after having eliminated the drift $\vartheta_{10}^{d'}$, and filtered and normalized top temperature after having eliminated the drift $\vartheta_{10}^{dF'}$ at PRBS excitation

Since the high frequency noise components of the measured top temperature were filtered, the identification was performed only by means of least squares parameter estimation.

Five different models were fitted to the data (Zierfuss, 1987):

- linear model;
- quasi-linear model with constant gain and direction dependent time constant;
- bilinear model;
- different linear models for increasing and decreasing the top temperature;
- quasi-linear model with state dependent gain and direction dependent time constant.

1. Linear model

According to the observation there was no delay time in the record and increasing of the order of the process to greater than one did not improve the fitting. The parameters of the first-order model are as follows

$$\hat{y}(k) = (0.9459 \pm 0.008798)\hat{y}(k-1) + (0.1132 \pm 0.01450)u(k-1)$$

The time constant of the equivalent continuous time model is

$$T = \frac{-0.25}{\ln(0.9459)} = 4.5 \, [\text{min}]$$

The gain of the model is

$$K' = \frac{0.1132}{(1 - 0.9459)} = 2.1$$

This gain is valid between the normalized values, the true gain between the original data is

$$K = -0.2K' = -0.2 \cdot 2.09 = -0.42 \left[{}^\circ\mathrm{C} / (\ell/\mathrm{h}) \right]$$

The normalized reflux flow, measured and filtered top temperature, the computed temperature based on the estimated model and the modeling error are plotted in Figure 7.6.11. The modeling error is defined as the difference of the measured and fitted temperatures $\left[y(k) - \hat{y}(k) \right]$.

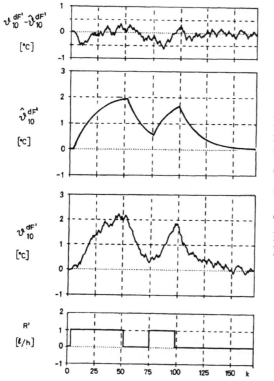

Fig. 7.6.11 Normalized measured reflux R', filtered and normalized top temperature after having eliminated the drift $\vartheta_{10}^{\mathrm{dF}'}$, computed temperature $\hat{\vartheta}_{10}^{\mathrm{dF}'}$ and modeling error $\left(\vartheta_{10}^{\mathrm{dF}'} - \hat{\vartheta}_{10}^{\mathrm{dF}'} \right)$ at fitting a linear first-order model to the PRBS excitation

2. *Quasi-linear model with constant gain and direction dependent time constant*
Based on the plot of the records a trivial assumption was to assume that the gain is constant and the time constant is direction dependent, which could be modeled by

$$K = K_0 \tag{7.6.3}$$
$$T = T_0 + T_1 u(t) \tag{7.6.4}$$

since the reflux has only two values and because of the first-order like feature the sign of the temperature change is proportional to the reflux itself. (It is expedient to use the reflux flow in (7.6.4) to avoid a change being detected in the temperature although only a disturbance occurred.)

The discretization of the first-order quasi-linear differential equation

$$T\dot{y}(t) + y(t) = Ku(t)$$

with (7.6.3) and (7.6.4) leads to the difference equation

$$\left[T_0 + T_1 u(k-1)\right]\frac{y(k) - y(k-1)}{\Delta T} + y(k-1) = K_0 u(k-1) \tag{7.6.5}$$

if the Euler transformation was used. (7.6.5) is equivalent to the difference equation

$$y(k) = a_1 y(k-1) + b_1 u(k-1) + c_1 u(k-1)\left[y(k) - y(k-1)\right] \tag{7.6.6}$$

where a_1, b_1 and c_1 are parameters to be determined. A parameter estimation in the above form did not give satisfactory results that lead to the interpretation that the gain is not constant and other model types have to be tried out.

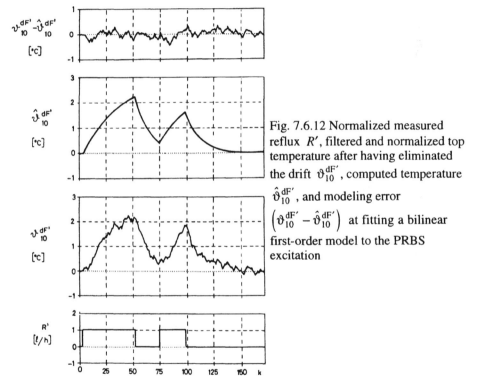

Fig. 7.6.12 Normalized measured reflux R', filtered and normalized top temperature after having eliminated the drift $\vartheta_{10}^{dF'}$, computed temperature $\hat{\vartheta}_{10}^{dF'}$, and modeling error $\left(\vartheta_{10}^{dF'} - \hat{\vartheta}_{10}^{dF'}\right)$ at fitting a bilinear first-order model to the PRBS excitation

3. *Bilinear model*
A small modification of (7.6.6) is

$$y(k) = a_1 y(k-1) + b_1 u(k-1) + c_1 u(k-1)y(k-1) \tag{7.6.7}$$

which is the first-order bilinear model. The result of the least squares parameter estimation

$$\hat{y}(k) = (0.9296 \pm 0.0106)\hat{y}(k-1) + (0.07432 \pm 0.00204)u(k-1)$$
$$+(0.04883 \pm 0.01332)u(k-1)\hat{y}(k-1) \tag{7.6.8}$$

The normalized reflux flow, the filtered measured top temperature, the computed temperature based on the estimated model, and the modeling error are plotted in Figure 7.6.12. The computed temperature increases more slowly than decrease, i.e., the model reproduces the direction dependent time constant character of the process. The modeling error $[y(k) - \hat{y}(k)]$ is also markedly smaller than when a linear model was fitted.

4. *Different linear models for increasing and decreasing the top temperature*
A direction dependent model means that the model is a multi-model, i.e., it is the superposition of two different models, which are either valid for increasing or decreasing top temperature (or reversibly changing the reflux flow), respectively. Both linear models can be identified separately if the regression is allowed only for decreasing or increasing the reflux flow (Haber and Keviczky, 1985). Figure 7.6.13 shows the normalized measured reflux R', the control signal (gate function) for allowing the regression at the increase of the reflux $x_1(k)$ and the control signal (gate function) for allowing the regression at the decrease of the reflux $x_2(k)$. (Both control signals start some samples before the change in the reflux flow occurs to fill the memory vector of the LS parameter estimation.) The estimated first-order linear models are as follows:
- for increasing the reflux flow $[x_1(k) = 1, x_2(k) = 0]$

$$\hat{y}(k) = (0.9762 \pm 0.01581)\hat{y}(k-1) + (0.07634 \pm 0.0213)u(k-1) \qquad (7.6.9)$$

- for decreasing the reflux flow $[x_2(k) = 1, x_1(k) = 0]$

$$\hat{y}(k) = (0.9286 \pm 0.01020)\hat{y}(k-1) + (0.1732 \pm 0.03626)u(k-1) \qquad (7.6.10)$$

Both models are very good, as is seen from the large parameter/(standard deviation) ratios.

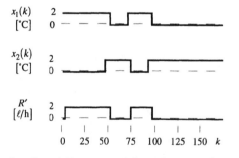

Fig. 7.6.13 Normalized measured reflux R', control signal for allowing the regression at increase of the reflux $x_1(k)$ and control signal for allowing the regression at decrease of the reflux $x_2(k)$

5. *Quasi-linear model with state dependent gain and direction dependent time constant*
If we neglect that the control signals start several samples before the corresponding change in the reflux, then there is a relation between the control signals and the normalized reflux flow,

$$x_1(k) = u(k) \qquad\qquad (7.6.11)$$
$$x_2(k) = 1 - u(k) \qquad\qquad (7.6.12)$$

(This fact is a consequence of the practical normalization of the two-valued input signal and can be checked in Figure 7.6.13.) As the two equations (7.6.9) and (7.6.12) are only valid if the corresponding control signals are equal to one and at a given time only one of the control signals is valid, a general valid model equation can be set up by multiplying the right side of (7.6.9) by $x_1(k)$ and the right side of (7.6.10) by $x_2(k)$ and by adding up the two equations,

$$\hat{y}(k) = \left[0.9762\hat{y}(k-1) + 0.07634u(k-1)\right]x_1(k)$$
$$+\left[0.9286\hat{y}(k-1) + 0.1732u(k-1)\right]x_2(k) \tag{7.6.13}$$

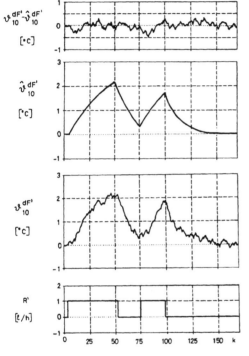

Fig 7.6.14 Normalized measured reflux R', filtered and normalized top temperature after having eliminated the drift $\vartheta_{10}^{dF'}$, computed temperature $\hat{\vartheta}_{10}^{dF'}$, and modeling error $\left(\vartheta_{10}^{dF'} - \hat{\vartheta}_{10}^{dF'}\right)$ at fitting a direction dependent quasi-linear first-order model to the PRBS excitation

If (7.6.11) and (7.6.12) would be replaced into (7.6.13) then the model would not have a *lag* type character. Replace therefore $u(k)$ by $u(k-1)$ in (7.6.11) and (7.6.12) by

$$x_1(k) = u(k-1) \tag{7.6.14}$$
$$x_2(k) = 1 - u(k-1) \tag{7.6.15}$$

and substitute (7.6.14) and (7.6.15) into (7.6.13):

$$\hat{y}(k) = \left[0.9762\hat{y}(k-1) + 0.07634u(k-1)\right]u(k-1)$$
$$+\left[0.9286\hat{y}(k-1) + 0.1732u(k-1)\right]\left[1 - u(k-1)\right] \tag{7.6.16}$$
$$= 0.9286\hat{y}(k-1) + 0.07634u(k-1) + 0.0476u(k-1)\hat{y}(k-1)$$

In (7.6.16) we used that the normalized reflux flow has only two values 0 or 1, and therefore

$$u^2(k) = u(k) \qquad \forall k$$

is valid in every time point. Observe that the model equation (7.6.16) is almost the same as the difference equation of the estimated bilinear model (7.6.8). The normalized reflux flow, the filtered measured top temperature, the computed temperature based on the estimated model, and the modeling error are plotted in Figure 7.6.14. The model fitting is as good as in the case of fitting of the bilinear model.   ∎

From the five model assumptions three models were identified, one of them is linear and therefore valid only for very small changes. The two nonlinear models almost coincide and are valid for larger reflux flow changes (in the investigated domain $15[\ell/h] \leq R \leq 20[\ell/h]$), as well.

The parameter estimation was performed between the normalized data. The difference equations between the measured reflux flow and top temperature can be derived by means of the normalizing equations (7.6.1) and (7.6.2). They are as follows:

- *first-order linear model*

$$\hat{\vartheta}_{10}(k) = 0.9459\hat{\vartheta}_{10}(k-1) + 4.9161 - 0.0226R(k-1)$$

- *first-order bilinear model*

$$\hat{\vartheta}_{10}(k) = 1.125\hat{\vartheta}_{10}(k-1) - 10.603 + 0.7908R(k-1)$$
$$-0.0098R(k-1)\hat{\vartheta}_{10}(k-1)$$

- *first-order direction dependent model*

$$\hat{\vartheta}_{10}(k) = 1.119\hat{\vartheta}_{10}(k-1) - 9.5121 + 0.7701R(k-1)$$
$$-0.0095R(k-1)\hat{\vartheta}_{10}(k-1)$$

The steady state relation can be obtained by equating

$$\vartheta_{10}(k) = \vartheta_{10}(k-1) = \vartheta_{10}$$

and

$$R(k) = R(k-1) = R$$

They are:

- *first-order linear model*

$$\hat{\vartheta}_{10} = 90.87 - 0.42R$$

- *first-order bilinear model*

$$\hat{\vartheta}_{10} = \frac{80.69R - 1021.86}{R - 12.76}$$

- *first-order direction dependent model*

$$\hat{\vartheta}_{10} = \frac{81.06R - 1001.27}{R - 12.53}$$

The estimated (reflux flow)/(top temperature) steady state characteristics are drawn in Figure 7.6.15. The measured values are indicated by circles. They fall in the close vicinity of the estimated curves.

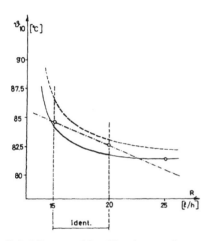

Fig. 7.6.15 Measured and estimated reflux flow/top temperature steady state characteristics: (a) estimated linear characteristics (dashed-dotted line); (b) estimated nonlinear characteristics of the bilinear model (continuous line); (c) estimated nonlinear characteristics of the direction dependent multi-model (dashed line). The measured values are indicated by circles.

7.6.4 Process identification at changes in the heating power
The following parameters were kept constant: $p = 113$ [kPa], $F = 50$ [ℓ/h], $\vartheta_F = 88$ [°C], $R = 20$ [ℓ/h], $\vartheta_R = 70$ [°C]. The heating power was changed stepwise in the following values:

$P:$ $9.0 \to 10.8 \to 12.6 \to 14.4 \to 12.6 \to 10.8 \to 9.0$ [kW]

Figure 7.6.16 shows the step responses for the temperature of plates 1 and 10. The static relations are seen in Figure 7.6.17.

From Figures 7.6.16 and 7.6.17 it can be seen that over 10.8 [kW] the bottom product is practically water, i.e., ϑ_1 cannot be increased furthermore. Between 9 [kW]–10.8 [kW] a first-order lag term describes the ϑ_1 response at the increase of the heating power and a second-order lag term with time delay describes the process in the case of a reverse excitation:

$P: 9 \to 10.8$ [kW]:	$K=1.88$ [°C/kW]	$T_+ = 6.5$ [min]
$P: 10.8 \to 9$ [kW]:	$K=1.88$ [°C/kW]	$T_- = 7.8$ [min]
	$\varsigma_- = 1.3$	$T_{d-} = 7.5$ [min]

The phenomenon can be explained because increasing the heating power the vapor reaches the first plate much faster than the cold reflux when decreasing the heating power.

The reaction between the top temperature ϑ_{10} and the heating power could be described by a first-order lag with no time delay. It is explained by the big vapor speed. The gain and time constants of the models are presented in Figure 7.6.18. The gain is increasing and the time constants are decreasing with the vapor rate, since it is always easier to decrease the ethanol concentrate of the distillate than to increase it. At small heating power the effect of increasing the vapor is slower in ϑ_{10} than the effect of decreasing the heating power because in the latter case the reflux cools the top plate fast. At bigger vapor production the speeds of the effects of the vapor and reflux to the top plate are converging as shown by the time constants at the heating power steps in different directions.

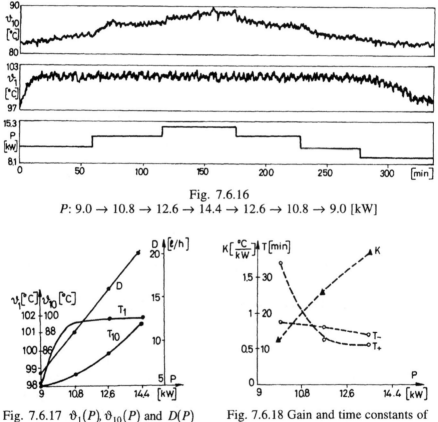

Fig. 7.6.16
P: $9.0 \rightarrow 10.8 \rightarrow 12.6 \rightarrow 14.4 \rightarrow 12.6 \rightarrow 10.8 \rightarrow 9.0$ [kW]

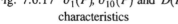
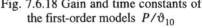

Fig. 7.6.17 $\vartheta_1(P)$, $\vartheta_{10}(P)$ and $D(P)$ characteristics

Fig. 7.6.18 Gain and time constants of the first-order models P/ϑ_{10}

7.6.5 Process identification at changes in the feed rate

The following parameters were kept constant: $p = 113$ [kPa], $P = 11.7$ [kW], $R = 20$ [ℓ/h], $\vartheta_R = 75$ [°C], $T_F = 65$ [°C]. The feed was changed stepwise in the following values:

F: $45 \rightarrow 65 \rightarrow 75 \rightarrow 65 \rightarrow 45$ [ℓ/h]

Figure 7.6.19 presents the temperature responses of the plates 1 and 10 to the feed rate step $45 \rightarrow 65$ [ℓ/h] and Figure 7.6.20 in the case of a feed rate step $65 \rightarrow 45$ [ℓ/h]. The steady state relations are seen in Figure 7.6.21.

The temperature response of the bottom plates can be described by a second-order slow term for feed increase and by a fast first-order lag term for feed decrease:

F: $45 \rightarrow 65$ [ℓ/h]: K=-0.0385 [°C/(ℓ/h)] T_+ = 15 [min] ς_+ = 0.2
F: $65 \rightarrow 45$ [ℓ/h]: K=-0.0385 [°C/(ℓ/h)] T_- = 6 [min]

The temperature response of the top plate can be modeled by a first-order lag term with a direction dependent time constant:

F: $45 \rightarrow 65$ [ℓ/h]: K=-0.053 [°C/(ℓ/h)] T_+ = 15 [min]
F: $65 \rightarrow 45$ [ℓ/h]: K=-0.053 [°C/(ℓ/h)] T_- = 6 [min]

The feed was much colder than the feed plate, thus both top and bottom temperatures decreased when feed rate was increased. However, the effect of the feed changes for the top and bottom products is much less than that of the reflux or the heating power.

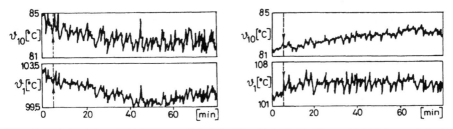

Fig. 7.6.19 F: $45 \rightarrow 65$ [ℓ/h] at t=5 [min] Fig. 7.6.20 F: $65 \rightarrow 45$ [ℓ/h] at t=5 [min]

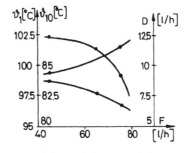

Fig. 7.6.21 $\vartheta_1(F)$, $\vartheta_{10}(F)$ and $D(F)$ characteristics

7.6.6 Conclusions

From the control point of view the column is a multi-variable, nonlinear dynamic plant. The transfer functions between the control actions – reflux rate and heating power – and the controlled variables – top and bottom temperatures – were identified. The models are of first- and second-order quasi-linear lag terms with working point dependent gains and working point and direction dependent time constants.

The (reflux flow)/(top temperature) relation was identified by means of PRBS

excitation. The parameters of the following models were estimated:
- bilinear model;
- different linear models for increasing and decreasing the top temperature.

It was shown that the two direction dependent models can be rewritten into a global valid first-order quasi-linear model with state dependent gain and direction dependent time constant. This nonlinear model has a bilinear structure and its parameters are almost the same as the estimated parameters of the bilinear model.

7.7 FLOOD PROCESS OF RIVER CACHE

7.7.1 Description of the process
The flood process was recorded at the river Cache in southern Illinois (USA) between 1935 and 1951. Eight separate measured rainfall excesses $u(k)$ and surface run offs $y(k)$ are drawn in Figure 7.7.1 (Diskin and Boneh, 1973). Both quantities are measured in inch/day and the sampling period was one day.

Based on physical considerations, the model of the flood process has the following features: (Diskin and Boneh, 1973; Diskin et al., 1975):
1. The model outputs (surface run offs) are always positive;
2. The total rainfall excesses are equal to the total surface run offs for each flood, i.e., the static gain of the model is equal to one.

7.7.2 Process identification by Volterra series
Diskin and Boneh (1972, 1973) modeled the flood process by a quadratic Volterra series:

$$y(k) = h_0 + \sum_{\kappa_1=0}^{m} h_1(\kappa_1)u(k-\kappa_1) + \sum_{\kappa_1=0}^{m} \sum_{\kappa_2=0}^{m} h_2(\kappa_1,\kappa_2)u(k-\kappa_1)u(k-\kappa_2)$$

The following considerations can be deduced from the *a priori* physical information:
1. Zero rainfall excess causes no surface run off, i.e., there is no constant term in the model

$$h_0 = 0$$

2. The number of first-order kernels cannot be more than the maximum number of non-zero output samples in the records. Denote it by m_{max}, and in our case $m_{max} = 17$.
   Then

$$h_1(\kappa_1) = 0 \qquad \kappa_1 \geq m_{max}$$

3. All first-order kernels are positive

$$h_1(\kappa_1) > 0 \qquad \kappa_1 = 0, 1, ..., m_{max} - 1$$

4. The sum of the first-order kernels is equal to one

$$\sum_{\kappa_1=0}^{m_{max}-1} h_1(\kappa_1) = 1$$

5. The memory of the second-degree kernels cannot be greater than the maximum number of non-zero output samples in the records

$$h_2(\kappa_1, \kappa_2) = 0 \quad \text{if} \quad x_{\kappa_1} \geq m_{max} \quad \text{or} \quad x_{\kappa_2} \geq m_{max}$$

6. The distance between the indices of the second-degree kernels has to be less than the maximum number of non-zero input samples in the records. Denote it by m_{min}, and in our case $m_{min} = 5$. Then

$$h_2(\kappa_1, \kappa_2) = 0 \quad \text{if} \quad \kappa_1 - \kappa_2 \geq m_{min} \quad \text{or} \quad \kappa_2 - \kappa_1 \geq m_{min}$$

7. The sum of quadratic kernels at any off-diagonal has to be zero,

$$\sum_{\kappa_1=0}^{m_{max}-1-\kappa} h_2(\kappa_1, \kappa_1 + \kappa) = 0 \quad \kappa = 0, 1, \ldots, m_{max} - 1$$

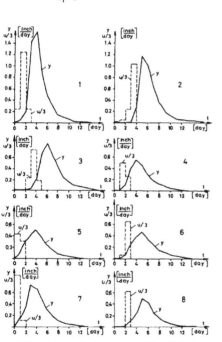

Fig. 7.7.1 Eight measured rainfall excesses and surface run offs at the river Cache

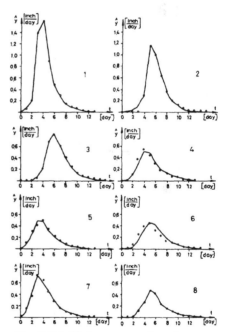

Fig. 7.7.2 Eight computed surface run offs based on identified Volterra series and physical assumptions (Diskin and Boneh, 1973). The circles show the measured run offs

Diskin and Boneh (1972, 1973) identified the process by the nonparametric Volterra series having taken the *a priori* physical limitations into account. The computation time was very big, in consequence of the large number (74) of unknown parameters. Therefore, the optimization procedure of Fletcher and Powell was applied instead of the LS method. The computed surface run offs are listed in Diskin and Boneh (1973) and are drawn in Figure 7.7.2.

7.7.3 Process identification by parametric models
The aim of the investigation was to check whether simple nonlinear parametric models
can be used for the flood model. The following models were fitted (Haber, 1982):
   1. linear (L);
   2. generalized Hammerstein (H);
   3. parametric Volterra (V κ);
   4. Volterra weighting function series (VS κ).

κ denotes the allowed maximum distance between two cross-product terms
$u(k - \kappa_1)u(k - \kappa_2)$, i.e., $\kappa = \max\{|\kappa_1 - \kappa_2|\}$. The Volterra weighting function series
model is a nonparametric model and was formed from the parametric Volterra model by
$A(q^{-1}) = 1$. In both models κ denotes the number of off-diagonals of the quadratic
kernels and all off-diagonals were taken into account till the same memory, the so-called
model order. The eight individual rainfall excesses and surface run offs were chained to
one input $\{u(k)\}$ and one output $\{y(k)\}$ record as seen in Figure 7.7.3. No delay time
was assumed and, what is more, the transfer functions were assumed to have derivative
terms, i.e., the first term of the numerator of the pulse transfer function was involved in
the model $(b_0 \neq 0)$. Furthermore, no constraints – neither of equality or unequality
type – were taken into account to ensure a simple LS parameter estimation procedure.
     The normalized loss functions σ_ε of the parametric models are seen in Figure 7.7.4
and those of the nonparametric ones in Figure 7.7.5, as a function of the model order.
There is a clear break in the curves of Figure 7.7.4 that indicates a second-order model.
Each fitted model is characterized by two numbers: the first shows how well the model
output coincides with the measured output and the second shows the significance of the
estimated model parameters. The indices are qualitative measures, 0 denotes bad and 1
good. The parameters of the second-order model, as well as of a Volterra weighting
function series with memory of 5, are listed in Table 7.7.1.
     The computed outputs of the best models are seen in Figure 7.7.3. The static
relationships of the best fitting models are seen in Figure 7.7.6. All curves approximate
the straight lines at 45°, i.e., the static gain of the process is about one, as expected.
The second-order parametric model is a periodic one, as can be seen from the poles of the
pulse transfer functions of the different channels,

$$1 + a_1 q^{-1} + a_2 q^{-2} = 0$$

being real if

$$a_1^2 - 4a_2 > 0$$

which is fulfilled by the best, second-order models (Table 7.7.1). Finally, Figure 7.7.7
presents the eight computed surface run offs based on the identified parametric Volterra
models separately. They approximate well both the measured surface run offs
(Figure 7.7.1) and the model outputs of the nonparametric Volterra series of Diskin and
Boneh (1973) (Figure 7.7.2).

TABLE 7.7.1 Estimated parameters of the different flood models of
the river Cache

	L	H	V1	VS1
$y(k-1)$	0.59±0.05	0.63±0.04	0.85±0.045	
$y(k-2)$	0.0051±0.039	-0.03±0.03	-0.16±0.03	
$u(k)$	0.034±0.008	0.085±0.02	0.092±0.014	0.096±0.016
$u(k-1)$	0.21±0.008	-0.012±0.02	-0.047±0.016	0.048±0.018
$u(k-2)$	0.17±0.014	0.177±0.02	0.16±0.015	0.173±0.018
$u(k-3)$				0.23±0.018
$u(k-4)$				0.156±0.018
$u(k-5)$				0.129±0.018
$u^2(k)$		-0.0097±0.0066	-0.015±0.005	-0.016±0.006
$u^2(k-1)$		0.078±0.007	0.059±0.006	0.045±0.007
$u^2(k-2)$		-0.0071±0.007	-0.022±0.005	0.019±0.007
$u^2(k-3)$				-0.014±0.007
$u^2(k-4)$				-0.01±0.007
$u^2(k-5)$				-0.021±0.007
$u(k)u(k-1)$			0.029±0.01	0.0066±0.014
$u(k-1)u(k-2)$			0.13±0.012	0.14±0.014
$u(k-2)u(k-3)$			-0.088±0.014	0.025±0.014
$u(k-3)u(k-4)$				-0.036±0.014
$u(k-4)u(k-5)$				-0.037±0.014
$u(k-5)u(k-6)$				0.029±0.01
σ_ε	0.054	0.0425	0.0318	0.0376
$\hat{Y}$	0.9758U	0.63113U + 0.153U²	0.6779U + 0.288U²	0.83475U + 0.1286U²
$a_1^2 - 4a_2$	0.3277	0.5169	1.3625	

7.7.4 Conclusions

1. The flood process was originally described by the Volterra series that is a nonparametric description. The number of parameters was 74. It was shown that the second-order quadratic generalized Hammerstein and parametric Volterra models give a sufficient model of the flood process with only 8 or 11 parameters, respectively.

2. The nonlinear behavior of the flood process appears not in a nonlinear steady state characteristic but in a nonlinear dynamic behavior, as shown by the significant cross-product terms in the parametric Volterra model.

3. The elimination of the physical restrictions from the estimation procedure allowed the usage of the LS technique. The identified model, however, describes the process well. The good results are owed to the low noise/signal ratio, as well.

7.8 FLOOD PROCESS OF RIVER SCHWARZA

7.8.1 Description of the process

The flood process was recorded at the river Schwarza in Thuringia (GDR) in October 1960. The input signal is the rainfall (in [mm/(6 h)]) and the output signal is the run off flow (in $\left[m^3/s \right]$). The measured records are seen in Figure 7.8.1. The data were sampled 53 days by 6 hour sampling time. Based on physical considerations, the flood model has the following features:

1. The process is aperiodic;
2. At zero rainfall the flow of the river does not vanish, i.e., the model has a constant term;
3. Small rainfalls will be swallowed by the earth and the run off flow increases only for large rainfall. This fact points to a convex steady state character ahead.

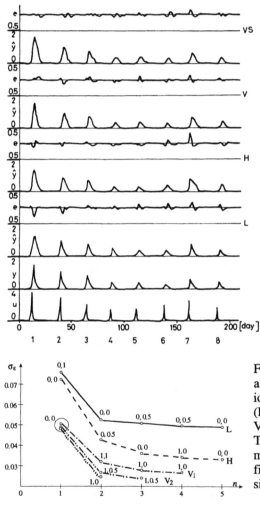

Fig. 7.7.3 Measured rainfall excesses, measured and computed (collected) surface run offs based on the identified models at the flood of the river Cache. The numbers 1 to 8 show the serial numbers of the individual rainfall excesses and surface run offs

Fig. 7.7.4 Normalized loss functions as a function of model order at the identification of the river Cache (L: linear, H: generalized Hammerstein, V κ : parametric Volterra).
The data pairs show the quality of the model fitting (first number the output fit, second number the parameter significance).

7.8.2 Previous process identification results

The flood process was described by a physical model that consisted of several partial models. The identification results are in Becker (1975). Hoffmeyer–Zlotnik *et al.*, (1978b and 1979) identified the flood process by a Hammerstein model and a Volterra series.

Hoffmeyer–Zlotnik *et al.*, (1978a) present how the physical constraints can be taken into account at the identification by parametric models. The denominator of the pulse-transfer function has to satisfy the inequality

$$a_1^2 - 4a_2 > 0$$

in order to have real poles in the second-order denominator

$$1 + a_1 q^{-1} + a_2 q^{-2} = 0$$

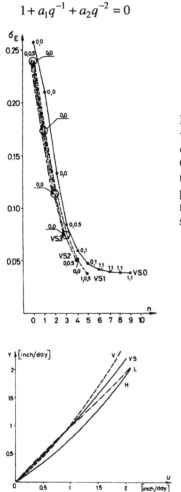

Fig. 7.7.5 Normalized loss functions of the Volterra weighting function series model VS κ as a function of the model order at the identification of the river Cache. (κ denotes the maximum distance between the delayed input terms in the cross-products) The data pairs show the quality of the model fitting (first number the output fit, second number the parameter significance).

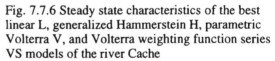

Fig. 7.7.6 Steady state characteristics of the best linear L, generalized Hammerstein H, parametric Volterra V, and Volterra weighting function series VS models of the river Cache

7.8.3 Process identification by different nonlinear parametric models

The following models were fitted to the measured records (Haber, 1982):

1. linear (L);
2. generalized Hammerstein (H);
3. parametric Volterra (V κ).

The nonlinear models were of quadratic degree. κ denotes the allowed maximum distance between two cross-product terms $u(k - \kappa_1)u(k - \kappa_2)$, i.e., $\kappa = \max\{|\kappa_1 - \kappa_2|\}$. The structure identification resulted in zero delay time. The normalized loss functions σ_ε are plotted – as a function of model order – in Figure 7.8.2. The parameters of the most significant models that turned to be of second-order are listed in Table 7.8.1. All models in it are aperiodic, as is seen from value $(a_1^2 - 4a_2)$. The computed model outputs are drawn in Figure 7.8.1. The parametric Volterra model V3 coincides best with the measured run off flow, except the first data points that are owed to the initial transients caused by the constant term. The steady state characteristics of the second-order models are seen in Figure 7.8.3. The parametric model V3 has the convex curve with positive intersection at the axis Y as expected.

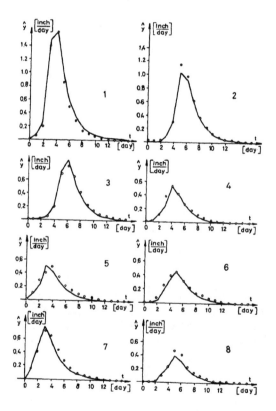

Fig. 7.7.7 Eight computed surface run offs based on the identified parametric Volterra model. The circles show the measured run offs

7.8.4 Conclusions

1. The parametric models give a good approximation of the flood process with many fewer parameters than the Volterra series.
2. The parametric Volterra model is prior to other – linear, generalized Hammerstein – models.
3. The identified parametric Volterra model shows both the convex steady state characteristic and the oscillation-free dynamic feature of the process, as expected from physical considerations.

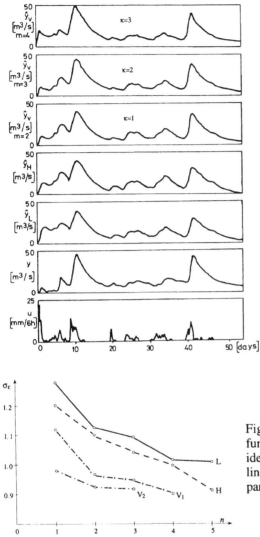

Fig. 7.8.1 Measured and computed flows based on the identified models at the flood of the river Schwarza

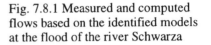

Fig. 7.8.2 Normalized loss functions as a function of model order at the identification of the river Schwarza (L: linear, H: generalized Hammerstein, V κ: parametric Volterra)

7.9 OPEN-CIRCUIT CEMENT GRINDING MILL PILOT PLANT

7.9.1 Description of the plant

In cement factories the grinding process consumes much energy. The energy consumption hardly depends on the flow of raw material, therefore it is expedient to maximize the material flow. To design a good control algorithm, one needs the raw material flow model of the cement mill. Previous experiments and theoretical considerations have shown that the mill can be modeled by a first-order lag term with no dead time, unit gain, and a time constant decreasing with the material flow.

The unit static gain can be explained by the fact that the input and output material flows should be equal in the steady state. There is a relation between the residence time

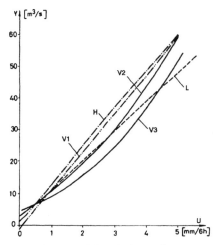

Fig. 7.8.3 Steady state characteristics of different second-order models of the river Schwarza (L: linear, H: generalized Hammerstein, Vκ: parametric Volterra)

TABLE 7.8.1 Estimated parameters of the different flood models of the river Schwarza

	L	H	V1	V2	V3
$y(k-1)$	-1.4218±0.058	-1.378±0.06	-1.2623±0.059	-1.2019±0.066	-1.0799±0.068
$y(k-2)$	0.45618±0.057	0.41591±0.058	0.30916±0.057	0.25033±0.063	0.13974±0.064
c_0	0.032431±0.15	-0.058734±0.15	0.058447±0.14	0.14349±0.136	0.27796±0.134
$u(k-1)$	0.33623±0.035	0.49494±0.074	0.29332±0.077	0.19706±0.078	-0.027±0.092
$u(k-2)$	-0.009951 ±0.0043	0.044883±0.08	0.19158±0.08	0.30105±0.089	0.36597±0.090
$u(k-3)$				-0.18967±0.076	-0.16984±0.1
$u(k-4)$					-0.011405 ±0.076
$u^2(k-1)$		-0.010886 ±0.0042	-0.011245 ±0.0039	-0.0029357 ±0.0044	0.024944 ±0.0076
$u^2(k-2)$		-0.0053037 ±0.0042	-0.040656 ±0.0074	-0.024373 ±0.0087	-0.025504 ±0.009
$u^2(k-3)$				0.013895±0.008	0.033323 ±0.009
$u^2(k-4)$					0.000639 ±0.008
$u(k-1)u(k-2)$			0.065929 ±0.012	0.025866±0.016	0.027226 ±0.015
$u(k-2)u(k-3)$				-0.028587 ±0.014	-0.073007 ±0.019
$u(k-3)u(k-4)$					-0.020793 ±0.015
$u(k-1)u(k-3)$				0.063937 ±0.014	0.079595 ±0.016
$u(k-2)u(k-4)$					0.066061 ±0.021
$u(k-1)u(k-4)$					-0.029433 ±0.013
σ_ε	1.139	1.111	1.038	0.988	0.922
$\hat{Y}$	0.943+9.493U	-1.549+14.24U -0.427U^2	1.247+10.348U +0.299U^2	2.963+6.369U +0.987U^2	4.569+2.592U+ +1.365U^2
$a_1^2-4a_2$	0.197	0.235	0.357	0.443	0.605

and material flow their product is proportional to the mass of the material in the mill (Lippek and Espig, 1983). Since the time constant is proportional to the residence time, the time constant is inversely proportional to the material flow.

In the frame of a cooperation between the Research Institute for Processing (FIA, Freiberg, GDR) and the Central Research and Design Institute for Silicate Industry (SZIKKTI, Budapest, Hungary), a laboratory scale cement mill was identified. The basic parameters of the mill are as follows: diameter 0.76 [m], length 1.5 [m]. The mill worked in open loop during the investigations. Two experiments were carried out: first the input material flow was increased stepwise from 100 [t/h] to 500 [t/h] steps and then the material flow was decreased in the same rate to its initial value. Both the input $u(t)$ and output $y(t)$ material flows were recorded and sampled in every 5 [min] (Figure 7.9.1.).

TABLE 7.9.1 The grapho-analytically estimated time constants and poles of the linearized pulse transfer functions in different working points and at different input feed step directions

Direction	100 → 200 [kg/h]		200 → 300 [kg/h]		300 → 400 [kg/h]		400 → 500 [kg/h]	
of change	T[min]	a_1	T[min]	a_1	T[min]	a_1	T[min]	a_1
↑	17.4	-0.7502	7.6	-0.5179	8.0	-0.5353	6.0	-0.4346
↓	16.0	-0.7316	9.0	-0.5738	7.6	-0.5179	4.6	-0.3372

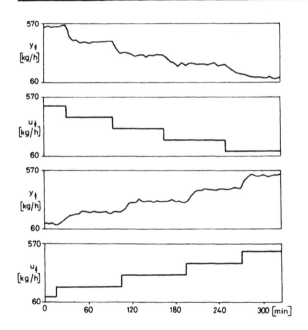

Fig. 7.9.1 The measured input u and output y material flows

7.9.2 Process identification by the grapho-analytical method
The stepwise changes in the output signal allow the use of grapho-analytical methods. Figures 7.9.2 and 7.9.3 show the step responses at increasing and decreasing input signals, respectively. Although the measurements are highly disturbed, it can be assumed that the input–output model is of first-order,

$$T\dot{y}(t) + y(t) = Ku(t)$$

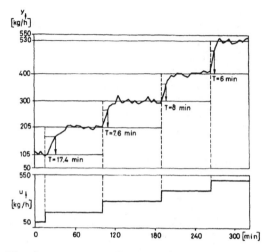

Fig. 7.9.2 Input and output material flows and the estimated time constants at input feed increase

The time constants T were calculated from the time intervals when 63 percent of the changes were achieved. They are also drawn in Figures 7.9.2 and 7.9.3. Furthermore, the parameters of the equivalent difference equation,

$$y(k) + a_1 y(k-1) = b_1 u(k-1) \qquad\qquad (7.9.1)$$

were calculated, where

$$a_1 = \exp(-\Delta T / T) \qquad\qquad (7.9.2)$$
$$b_1 = K\left[1 - \exp(-\Delta T / T)\right] \qquad\qquad (7.9.3)$$

In (7.9.1) and (7.9.2) ΔT is the sampling time and $-a_1$ is the pole of the pulse transfer function

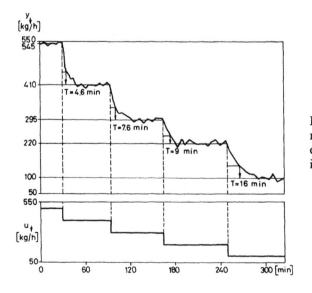

Fig. 7.9.3 Input and output material flows and the estimated time constants at input feed decrease

$$\frac{b_1 q^{-1}}{1 + a_1 q^{-1}}$$

The parameters T and a_1 are summarized in Tables 7.9.1 and are plotted in Figures 7.9.4 and 7.9.5 as a function of the material flow. The mean values of the input and output material flows were considered as the working points.

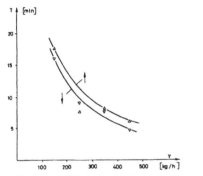

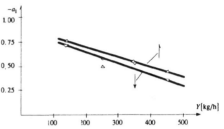

Fig. 7.9.4 Working point and direction dependence of the time constants estimated by grapho-analytical method

Fig. 7.9.5 Working point and direction dependence of the pole of the pulse transfer function of the quasi-linear model estimated by grapho-analytical method

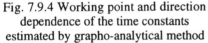

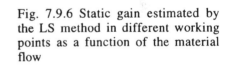

Fig. 7.9.6 Static gain estimated by the LS method in different working points as a function of the material flow

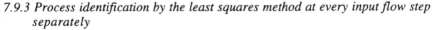

7.9.3 Process identification by the least squares method at every input flow step separately

Both measurement records contain four input feed steps and the corresponding step responses of the output material flow. Every step response was identified separately,. always assuming a linear structure and constant term. Table 7.9.2 summarizes the estimated parameters of the pulse transfer functions and the equivalent continuous first-order models. As before, the step response equivalent discrete/continuous transformation was used. The gain, time constant, and the pole of the pulse transfer functions are plotted in Figures 7.9.6 to 7.9.8 as a function of the working point and direction of the input feed change, respectively.

7.9.4 Conclusions drawn from the identification of the linear models in different working points

The results of the identification of the linearized models in the different working points and at different input feed step directions proved the correctness of the *a priori* theoretical assumptions:

1. The gain fluctuates by 20 percent around one, and its average value is one.
2. The time constant is approximately inversely proportional to the material flow.
3. The pole of the linear pulse transfer functions at different working points decreases linearly with the material flow.
4. The time constants (and the poles of the pulse transfer functions) differ with the direction of the change in the material flow; both parameters are bigger with decreasing and smaller with increasing material flow.

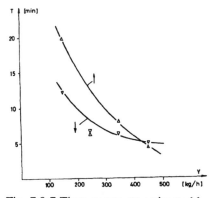

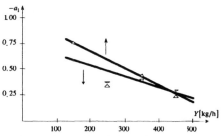

Fig. 7.9.7 Time constants estimated by the LS method in different working points as a function of the material flow

Fig. 7.9.8 Pole of the pulse transfer function of the quasi-linear model estimated by the LS method in different working points as a function of the material flow

To eliminate the measurement errors that caused a bias in the estimation of the static gain (see Figure 7.9.6) the levels of the input material flows were corrected according to Table 7.9.3. Furthermore, to prove the numerical conditions of the LS estimation, both input and output material flows were normalized by dividing them by 300. Therefore, the range of the material flows between (100–500) changed to (0.333–1.667).

TABLE 7.9.2 Parameters of the linearized first-order models estimated by the LS method in different working points and at different input feed step directions

Direction	$100 \to 200$ [kg/h]				$200 \to 300$ [kg/h]				$300 \to 400$ [kg/h]				$400 \to 500$ [kg/h]			
of change	a_1 $\pm\sigma(a_1)$	b_1 $\pm\sigma(b_1)$	K	T [min]	a_1 $\pm\sigma(a_1)$	b_1 $\pm\sigma(b_1)$	K	T [min]	a_1 $\pm\sigma(a_1)$	b_1 $\pm\sigma(b_1)$	K	T [min]	a_1 $\pm\sigma(a_1)$	b_1 $\pm\sigma(b_1)$	K	T [min]
↑	-0.7756 ±0.0345	0.2901 ±0.0666	1.29	19.68	-0.4474 ±0.0652	0.4896 ±0.1209	0.89	6.22	-0.5373 ±0.0592	0.5167 ±0.0782	1.12	8.05	-0.3226 ±0.0923	0.7730 ±0.0903	1.14	4.42
↓	-0.6656 ±0.0600	0.3511 ±0.0760	1.05	12.28	-0.4628 ±0.1085	0.4421 ±0.0987	0.82	6.49	-0.4521 ±0.0568	0.6224 ±0.0703	1.14	6.30	-0.3538 ±0.3883	0.5744 ±0.5200	0.89	4.82

Direction	Corrected input flow values (u_c) [kg/h]				
of change	100	200	300	400	500
↑	105	205	300	400	530
↓	100	220	295	410	545

TABLE 7.9.3 Corrections of the input material flow

7.9.5 Discrete time process identification of the global valid nonlinear model
The cement mill can be described by a quasi-linear first-order model with constant (unit) gain and signal dependent time constant. From Figure 7.9.4 it can be seen that a linear dependence is not enough, the hyperbolic function should be approximated by a quadratic polynomial. The difference equation of the first-order lag term with

$$K = K_0$$
$$T = T_0 + T_1 x(t) + T_2 x^2(t)$$

is

$$y(k) = -\left(\frac{\Delta T}{T_0} - 1\right) y(k-1) + K_0 \frac{\Delta T}{T_0} u(k-1)$$
$$-\frac{T_1}{T_0} x(k)\left[y(k) - y(k-1)\right] - \frac{T_2}{T_0} x^2(k)\left[y(k) - y(k-1)\right]$$

if Euler discretization is assumed. Another way could be the assumption of an inverse linear signal dependence,

$$K(t) = K_0$$
$$T^{-1}(t) = T_0^{-1} + \bar{T}_1^{-1} x(t)$$

the same Euler discretization technique results in

$$y(k) = -\left(\frac{\Delta T}{T_0} - 1\right) y(k-1) + K_0 \frac{\Delta T}{T_0} u(k-1)$$
$$+ K_0 \frac{\Delta T}{\bar{T}_1} x(k) u(k-1) - \frac{\Delta T}{\bar{T}_1} x(k) y(k-1)$$

In the first case four, in the second case three, coefficients have to be estimated in a form that is linear in the parameters. The number of unknown coefficients can be further reduced if the knowledge $K_0 = 1$ is used.

Figures 7.9.5 and 7.9.8 have shown that the pole of the pulse transfer function is a linear function of the material flow. Using this knowledge about the structure of the nonlinear model, a difference equation of the global valid process can be set up as follows (Haber *et al.*, 1986a, 1986b).

The unit static gain leads to a relation between the coefficients b_1 and a_1

$$b_1 = 1 + a_1$$

The pole $-a_1$ depends linearly on the sequence $\{x(k)\}$,

$$a_1 = a_{10} + a_{11} x(k) \tag{7.9.4}$$

Thus the resulting difference equation of the model is:

$$y(k) - u(k-1) = a_{10}\left[u(k-1) - y(k-1)\right] + a_{11} x(k)\left[u(k-1) - y(k-1)\right]$$

It is worth mentioning that the number of coefficients is now equal to the number of unknown parameters. The discrete time model corresponds to a continuous time first-order lag term with

$$K = 1$$

and

$$T(t) = -\frac{\Delta T}{\ln\left(a_{10} + a_{11}x(t)\right)} \qquad (7.9.5)$$

The signal $x(t)$ the parameters depend on can be either the input or the output material flow. Both assumptions have their advantages:
1. the input material flow does not include noise;
2. the output material flow corresponds physically better to the working point of the mill.

TABLE 7.9.4 Estimated signal dependence of the poles and the calculated time constants at the nonlinear model identification of the whole record

Direction of change	Pole depends on	σ_ε	a_{10} ±σ(a_{10})	a_{11} ±σ(a_{11})	100 → 200 [kg/h] a_1	T [min]	200 → 300 [kg/h] a_1	T [min]	300 → 400 [kg/h] a_1	T [min]	400 → 500 [kg/h] a_1	T [min]
↑	u	0.0363	-0.9581 ±0.1088	0.3608 ±0.0927	-0.7174	15.05	-0.5973	9.70	-0.4772	6.76	-0.3566	4.85
↑	y	0.0341	-0.9190 ±0.0796	0.3910 ±0.0776	-0.7235	15.45	-0.5933	9.58	-0.4628	6.49	-0.3325	4.54
↓	u	0.0497	-0.8772 ±0.1272	0.4982 ±0.1734	-0.7113	14.68	-0.5450	8.24	-0.3790	5.15	-0.2131	3.23
↓	y	0.0505	-0.9462 ±0.1719	0.4757 ±0.1949	-0.7084	14.50	-0.5500	8.36	-0.3912	5.33	-0.2327	3.47

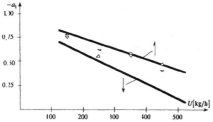

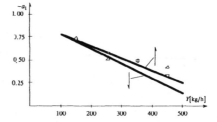

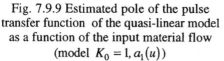

Fig. 7.9.9 Estimated pole of the pulse transfer function of the quasi-linear model as a function of the input material flow (model $K_0 = 1, a_1(u)$)

Fig. 7.9.11 Estimated pole of the pulse transfer function of the quasi-linear model as a function of the output material flow (model $K_0 = 1, a_1(y)$)

The working point can be approximated by the input feed only because the amplitude of the step is – relatively – not too big. We have used both approximations; the results of the parameter estimation are seen in Table 7.9.4. This contains the estimated coefficients a_{10} and a_{11}, their standard deviations and the normalized loss functions σ_ε. The pole of the pulse transfer function of the linearized models in the different working points and the time constants can be calculated according to (7.9.4) and (7.9.5).

The best first-order model is

$$T\dot{y}(t) + y(t) = Ku(t)$$

where the gain is

$$K = 1$$

and for increasing the feed the time constant is

$$T_+(t) = \frac{5}{\ln(0.09190 - 0.3910y(t))}$$

and for decreasing the feed it is

$$T_-(t) = \frac{5}{\ln(0.9162 - 0.4547y(t))}$$

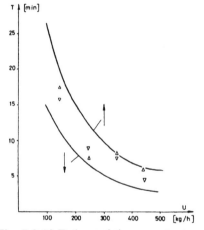

Fig. 7.9.10 Estimated time constant as a
function of the input material flow
(model $K_0 = 1, a_1(u)$)

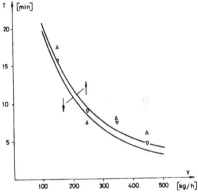

Fig. 7.9.12 Estimated time constant as a
function of the output material flow
(model $K_0 = 1, a_1(y)$)

In this case we have assumed that the parameters depend on the input feed, then the input material flow after the step is replaced $x(t)$ in (7.9.4) and (7.9.5). In the other case when we assumed that the parameters depend on the output material flow, the working point was characterized by the mean value between the steady state value of the input flow before and after the step, i.e., this value was replaced $x(t)$ in (7.9.4 and 7.9.5) by filling in Table 7.9.4.

Figures 7.9.9 and 7.9.10 show the signal dependence of the pole of the linear pulse transfer functions and the time constants on the input material flow, respectively, and Figures 7.9.11 and 7.9.12 show the same for the case of output material flow signal dependence.

In Figures 7.9.9 to 7.9.12 the continuous lines are the estimated signal dependencies that fit well to the marked individual time constants and poles of grapho-analytical evaluation. The coincidence is better in the case when the working point was put to the material flow

$$x(t) = y(t)$$

This fact can be well understood since the output material flow characterizes the state of the mill and not the input feed.

Finally, Figures 7.9.13 and 7.9.14 present the measured and corrected material flows and the computed outputs of the identified models. The coincidence between the measured $\{y(k)\}$ and fitted $\{\hat{y}(k)\}$ output material flow sequences are good.

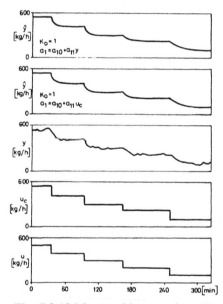

Fig. 7.9.13 Measured input and output records and calculated output flow based on the identified nonlinear model at decreasing input feed

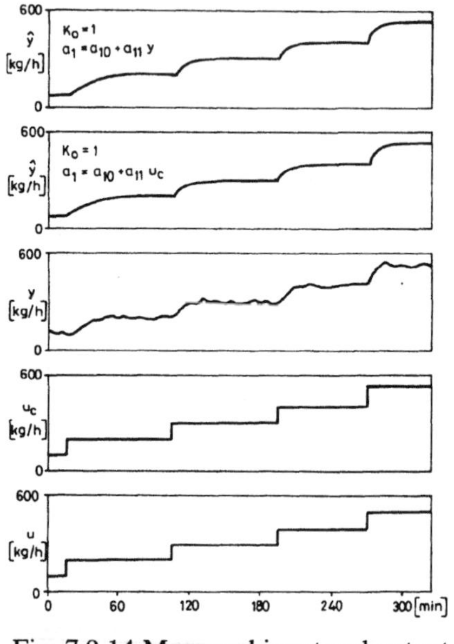

Fig. 7.9.14 Measured input and output records and calculated output flow based on the identified nonlinear model at increasing input feed

7.9.6 Conclusions

1. It should be mentioned that the nonlinear identification algorithm can be performed for any – persistently exciting – feed time function, however, the grapho-analytical method requires the steps in the feed.
2. The incorporation of a priori physical knowledge into the identification may lead to the reduction of the number of unknown parameters.
3. Grapho-analytical methods help in determining the proper nonlinear structure.

7.10 REFERENCES

Antomonov, Y.G. and A.B. Kotova (1968). Mathematical description of the dynamics of ion conductances and action potential. *Mathematical Biosciences*, Vol. 2, pp. 451–463.
Baur, U. and R. Isermann (1976). On-line identification of a steam heated heat exchanger with a process

computer – Case study. *Prepr. 4th IFAC Symposium on Identification and System Parameter Estimation*, (Tbilisi: USSR), pp. 252–293.

Boros, A. (1982). Investigations on the adaptive control of fermentation processes. Discrete time identification (in Hungarian). *Report*, Department of Agricultural Chemical Technology, Technical University of Budapest, (Budapest: Hungary).

Diskin, M.H. and A. Boneh (1972). Estimation of kernels for second order Volterra series in hydrology. *Prepr. Symposium on Nonlinear Estimation Theory*, (San Diego: USA), pp. 13.

Diskin, M.H. and A. Boneh (1973). Determination of optimal kernels for second order stationary surface run off systems. *Water Resources Research*, Vol. 9, pp. 311–325.

Diskin, M.H., A. Boneh and A. Golan (1975). On the impossibility of a partial mass violation in surface run off systems. *Water Resources Research*, Vol. 11, pp. 236–244.

Fedina, L. (1974). Expansion of the stimulus in the membrane (in Hungarian). *Publication of the Department of Biology of the Hungarian Academy of Sciences*, Vol. 19, pp. 423–460.

Gál, T., J. Hetthéssy, L. Keviczky and L. Varga (1979). A Simple real-time language for 8080 microcomputers. *Prepr. MIMI'79*, (Zurich: Switzerland).

Haber, R. (1979). An identification procedure for nonlinear dynamic models by means of a process computer – Modelbuilding, theory, program package, testing (in German). *Report-Kfk-PDV 175*, Department of Control and Processdynamics (IVD), University of Stuttgart, (Stuttgart:FRG).

Haber, R. (1982). Comparison of different nonlinear dynamic models at the identification of a flood process. *Int. Journal of Systems Science*, Vol. 8, pp. 95–102.

Haber, R. (1988). Parametric identification of nonlinear dynamic systems based on nonlinear crosscorrelation functions. *IEE Proceedings*, Vol. 135, Pt. D, 6, pp. 405–420.

Haber, R. and W. Bamberger (1979). OLID-MISO-NOLI-Interactive software package for identification and extremum control of nonlinear dynamic processes by means of a process computer. *Prepr. 2nd IFAC Symposium on Software for Computer Control*, (Prague: Czechoslovakia), C-VIII-1 - C-VIII-4.

Haber, R. and L. Keviczky (1979). Parametric description of dynamic systems having signal dependent parameters. *Prepr. JACC*, (Denver: USA), pp. 681–686.

Haber, R. and I. Wernstedt (1979). New nonlinear dynamic models for the simulation of electrically excited biological membrane. *Int. Journal of Systems Science*, Vol. 5, pp. 74–76.

Haber, R. and L. Keviczky (1985). Identification of *linear* systems having signal-dependent parameters. *Int. Journal of Systems Science*, Vol. 16, 7, pp. 869–884.

Haber, R., L. Keviczky and M. Hilger (1986a). Modeling and identification of a cement mill by a linear model with signal dependent parameters. *Prepr. IMACS/IFAC Symposium on Modeling and Simulation for Control of Lumped and Distributed Parameter Systems*, (Lille: France), pp. 273–276.

Haber, R., L. Keviczky and M. Hilger (1986/b). Identification of dynamic processes from sampled data. Nonlinear systems (in Hungarian). *Research Monograph Series on Process Identification and Control in the Silicate Industry*, No. 85. Publication of SZIKKTI, (Budapest: Hungary).

Haber, R. and R. Zierfuss (1987). Identification of an electrically heated heat exchanger by several nonlinear models using different structure. *Report TUV-IMPA-87/4*, Institute of Machine- and Processautomation, Technical University of Vienna, (Vienna: Austria).

Haber, R. and R. Zierfuss (1991). Identification of nonlinear models between the reflux and the top temperature of a distillation column. *Prepr. 9th IFAC Symposium on Identification and System Parameter Estimation*, (Budapest: Hungary), pp. 486–491.

Hodgkin, A.L. and A.F. Huxley (1952). Current carried by sodium and potassium ions through the membrane of the giant axon of Loligo. *Journal of Physiology*, Vol. 116, pp. 449–472.

Hodgkin, A.L., A.F. Huxley and B. Katz (1957). Measurement of current–voltage relations in the membrane of the giant axon of Loligo. *Journal of Physiology*, Vol. 116, pp. 424–448.

Hoffmeyer–Zlotnik, H.J., J. Wernstedt and A. Kurz (1978a). Applications of methods of statistic identification for developing models of the catchment area of a river in a high and low level range. *Prepr. 7th IFAC World Congress*, (Helsinki: Finland), pp. 1453–1459.

Hoffmeyer–Zlotnik, H.J., J. Wernstedt, A. Becker and P. Braun (1978b). Modeling of hydrologic systems by means of statistical methods (in German). *Messen, Steuern, Regeln*, Vol. 7, pp. 385–389.

Hoffmeyer–Zlotnik, H.J., P. Otto, A. Seifert, H. Heym and H. Luckers (1979). Application of experimental models for automatic control and prediction of water flows in the mountain of medium height (in German). *24. Internat. Wissenschaft. Kolloquium*, TH (Ilmenau: GDR), pp. 173–177.

Hoyt, R.C. (1963). The squid giant axon: Mathematical models. *Biophysical Journal*, Vol. 3, pp. 399.

Lippek, E. and D. Espig (1983). Relation between residence time and time behavior of tube mills (in German), *20. Diskussionstagung Zerkleinern und Klassieren*, (Berlin: GDR), Poster.

Nyeste, L. and L. Szigeti (1982). Dynamic optimization of fermentation processes. pp. 663–673.

Tischmeyer, M. (1974). Experimental modelbuilding and technical simulation of nonlinear biological systems and its realization on the process of electrically excited biological membrane (in German). *Dissertation*, Technical University of Ilmenau, (Ilmenau: GDR).

Tuschák, R., T. Bézi, G. Tevesz, J. Hetthéssy and R. Haber (1982). Practical experiences on the setup and identification of a distillation pilot plant. *Prepr. 6th IFAC Symposium on Identification and System Parameter Estimation*, (Washington: USA), pp. 663 – 668.

Zierfuss, R. (1987). Identification of nonlinear dynamic processes. *Thesis*, Institute of Machine- and Processautomation, Technical University of Vienna, (Vienna: Austria).

AUTHOR INDEX

SUBJECT INDEX

MATHEMATICAL MODELLING:
Theory and Applications

---

1. M. Křížek and P. Neittaanmäki: *Mathematical and Numerical Modelling in Electrical Engineering*. Theory and Applications. 1996
   ISBN 0-7923-4249-6

2. M.A. van Wyk and W.-H. Steeb: *Chaos in Electronics*. 1997
   ISBN 0-7923-4576-2

3. A. Halanay and J. Samuel: *Differential Equations, Discrete Systems and Control*. Economic Models. 1997   ISBN 0-7923-4675-0

4. N. Meskens and M. Roubens (eds.): *Advances in Decision Analysis*. 1999
   ISBN 0-7923-5563-6

5. R.J.M.M. Does, K.C.B. Roes and A. Trip: *Statistical Process Control in Industry*. Implementation and Assurance of SPC. 1999
   ISBN 0-7923-5570-9

6. J. Caldwell and Y.M. Ram: *Mathematical Modelling*. Concepts and Case Studies. 1999   ISBN 0-7923-5820-1

7. 1. R. Haber and L. Keviczky: *Nonlinear System Identification - Input-Output Modeling Approach*. Volume 1: Nonlinear System Parameter Identification. 1999   ISBN 0-7923-5856-2; ISBN 0-7923-5858-9 Set

   2. R. Haber and L. Keviczky: *Nonlinear System Identification - Input-Output Modeling Approach*. Volume 2: Nonlinear System Structure Identification. 1999   ISBN 0-7923-5857-0; ISBN 0-7923-5858-9 Set

---

KLUWER ACADEMIC PUBLISHERS – DORDRECHT / BOSTON / LONDON